A C S   S Y M P O S I U M   S E R I E S   **468**

Emerging Technologies in Hazardous Waste Management II

D. William Tedder, EDITOR
Georgia Institute of Technology

Frederick G. Pohland, EDITOR
University of Pittsburgh

Developed from a symposium sponsored
by the Division of Industrial and Engineering Chemistry, Inc.,
of the American Chemical Society
at the Industrial and Engineering Chemistry Special Symposium,
Atlantic City, New Jersey,
June 4–7, 1990

American Chemical Society, Washington, DC 1991

Library of Congress Cataloging-in-Publication Data

Emerging technologies in hazardous waste management II / D. William
  Tedder, editor, Frederick G. Pohland, editor.

  p.   cm.—(ACS   symposium   series,   ISSN   0097–6156;   468)
  "Developed from a symposium sponsored by the Division of Industrial
and Engineering Chemistry, Inc., of the American Chemical Society at
the Industrial and Engineering Chemistry Special Symposium, Atlantic
City, New Jersey, June 4–7, 1990."

  Includes bibliographical references and indexes.

  ISBN 0–8412–2102–2

  1. Hazardous   wastes—Management—Congresses.   2. Sewage—
Purification—Congresses. 3. Refuse and refuse disposal—Congresses.

  I. Tedder, D. W. (Daniel William), 1946–   . II. Pohland, F. G.
(Frederick George), 1931–   . III. American Chemical Society. Division
of Industrial and Engineering Chemistry. IV. Series.

TD1020.E443   1991
628.4′2—dc20   91–20107
                                                        CIP

Foreword

THE ACS SYMPOSIUM SERIES was founded in 1974 to provide a medium for publishing symposia quickly in book form. The format of the Series parallels that of the continuing ADVANCES IN CHEMISTRY SERIES except that, in order to save time, the papers are not typeset, but are reproduced as they are submitted by the authors in camera-ready form. Papers are reviewed under the supervision of the editors with the assistance of the Advisory Board and are selected to maintain the integrity of the symposia. Both reviews and reports of research are acceptable, because symposia may embrace both types of presentation. However, verbatim reproductions of previously published papers are not accepted.

Contents

BIOLOGICAL TREATMENT

SOLID WASTE MANAGEMENT

Preface

TOXIC MATERIALS, ONCE CREATED, must either be detoxified or effectively isolated from the environment. The complexities lie primarily with issues of technology selection, process integration, and practical constraints, such as costs. In many instances, the best strategy may be to avoid waste production or reduce generation rates by modifying the manufacturing processes.

The symposium on which this book is based featured approximately 50 papers, covering a range of topics. Manuscripts submitted during the symposium, and several topical reviews solicited by the editors, were reviewed and revised before acceptance. The final selection of the 21 chapters included here is based on peer review, scientific merit, the editors' perceptions of lasting value or innovative features, and the general applicability of either the technology itself or the scientific methods and scholarly details provided by the authors.

This volume is a continuation of a theme initiated in 1989. Its predecessor, *Emerging Technologies for Hazardous Waste Management*, ACS Symposium Series No. 422 (1990), is a related collection of 22 manuscripts on waste management, but these two volumes are remarkably different. No single volume can do justice to this broad subject, and no topic is afforded comprehensive treatment in either volume, but there is virtually something for everyone. The authors include physicists, chemists, biologists, and soil scientists with an assortment of mechanical, civil, environmental, nuclear, and chemical engineers.

The contributions in this volume are divided into four sections: thermal treatment and abiotic emissions control, water management, biological treatment, and solid waste management. A novel incinerator modification, fundamental studies of single particle adsorption, and SO_2 emissions control are discussed in the first section. The second section begins with a review of photochemistry and continues with contributions on oxidation, followed by several describing water purification by ion exchange, and ending with a review of essential cost factors for treatment comparison. The third section includes experimental contributions describing microbiological processes for treating munitions wastes, a biofilter design, and pesticide detoxification with an immobilized enzyme system. A modeling study of batch reactor sequencing for denitrification concludes the section. The final section of the book includes reviews and discussion of species migration, a soil venting model, fixation technology, and the design of earth-mounded concrete vaults.

Acknowledgments

The symposium on which this book is based was supported by several organizations that are committed to excellence, solving waste problems, and reducing environmental pollution. Their generosity was essential to the overall success of the symposium, and we gratefully recognize it here.

Our symposium supporters were the Nalco Chemical Company of Naperville, Illinois; the Petroleum Research Fund of Washington, D.C.; and Waste-Tech Services, Inc., of Golden, Colorado.

We also gratefully acknowledge the generous support of several organizations that sponsored events. These are the Ciba–Geigy Corporation of Toms River, New Jersey; Eli Lilly and Company of Indianapolis, Indiana; the Hoechst Celanese Corporation of Chatham, New Jersey; and Merck and Company, Merck Chemical Manufacturing Division, Rahway, New Jersey.

D. WILLIAM TEDDER
Georgia Institute of Technology
Atlanta, GA 30332–0100

FREDERICK G. POHLAND
University of Pittsburgh
Pittsburgh, PA 15261–2294

March 26, 1991

Chapter 1

Emerging Technologies for Hazardous Waste Management

Overview

D. William Tedder[1] and Frederick G. Pohland[2]

[1]School of Chemical Engineering, Georgia Institute of Technology, Atlanta, GA 30332-0100
[2]Department of Civil Engineering, University of Pittsburgh, Pittsburgh, PA 15261-2294

International developments in recent years have resulted in rapid changes in the political climate, especially in eastern Europe. A new openness in the Warsaw Pact countries and the Soviet Union (U.S.S.R.) has revealed numerous smokestack industries still operating with 1940s technologies, and pollution problems that are no longer tolerated in the West (1–4). This new openness is also closing low-cost doors for Western waste disposal (5,6). Perhaps for the first time, the realities of industrialization without modern pollution control and waste management are recognized nearly worldwide; the clamor for improved waste management and pollution abatement now echoes in regions that have long been silent. More than ever before, hazardous waste management is an international concern (7).

With the apparent end of the cold war in sight, thoughts are turning toward the legacy of the arms race (8). Eyes are now focusing on the stark realities of dismantling large stockpiles of weapons, including significant inventories of chemical and biological agents of mass destruction. Agreements are already in place between the U.S. and the U.S.S.R. to reduce nuclear capabilities, and those of chemical and biological warfare, but the most appropriate management technologies are still emerging.

The costs of the cold war have been enormous, although the total costs may not yet be calculable. In a recent interview, for example, a Russian official made the casual observation that "the cost of producing chemical weapons is much less than the cost of safely destroying them" (anon., recent U.S. televised report). Clearly, these aging inventories of chemical and biological weapons cannot be ignored, and must be safely detoxified. Hazardous waste management has significant economic ramifications.

This volume addresses some of these problems, but necessarily focuses on a few approaches that are under development. Since these are emerging technologies and management techniques, neither process safety nor economic considerations are discussed in detail, although several contributions also include these factors. While both factors are important, additional research and

0097–6156/91/0468–0001$06.00/0

development are needed before they can be assessed with acceptable levels of confidence.

Thermal Treatment and Abiotic Emissions Control

In the U.S., hazardous waste incinerators and nuclear power reactors share the same dubious distinction. Much of the public perceives them as being risky, and both are subject to the "not in my backyard" or "NIMBY" syndrome. This perception is at least partly due to their respective potentials for generating airborne pollutants which may travel some distance from the generating site before deposition. Although technologies for the treatment of airborne emissions are highly developed (9), there is public concern that harmful emissions will be generated and released to the environment. In certain instances, however, the most effective strategies for managing hazardous wastes will utilize thermal technologies; incineration will be unavoidable in some cases. Consequently, the development of improved waste incineration methods and off-gas treatment technologies is essential.

In Chapter 2, investigators from the Institute of Gas Technology in Chicago present a novel concept for utilizing internal radiant energy within a gas-fired combustion chamber to maintain higher combustion temperatures. This technology has been patented and appears important for the thermal destruction of many chlorinated species where removal efficiencies are highly dependent upon the combustion chamber temperatures and residence times. Short residence times are desirable to reduce the production of nitrogen oxides, but shorter residence times also require higher temperatures. Therefore, these promising results should prove significant in waste incineration.

A second contribution describes fundamental studies using an electrodynamic balance (EDB) to measure single-particle desorption rates, a key phenomenon for the design of thermal treatment of soils and solid residues. In Chapter 3, Stephanopoulos et al. present results for three types of solid particles—montmorillonite, and two synthetic chars with different pore structures. Significant differences were identified among the various solid-organic compound pairs examined in adsorption-desorption sequences in the EDB. Not surprisingly, these differences are related to the solid pore structures. Continuing studies should define minimal temperature requirements more precisely and enable the design of more efficient and controllable thermal treatment processes.

Jozewicz and Kirchgessner discuss lime treatment methods in Chapter 4 that can be used to maintain high reactivities for the removal of sulfur dioxide. This technology is important for emissions control from coal-fired boilers, but also has implications for the removal of acidic species from hazardous waste incinerator effluents. In particular, a small amount of SO_2 in the calcination gas yields a CaO that is more reactive with SO_2 upon furnace injection than a CaO calcined in the absence of SO_2.

Water Management

Potential improvements in water management may result from better detoxification techniques through chemical reactions and purification. Groundwater pollution (e.g., from landfills and land-disposed leachates) remains a concern. It is increasingly difficult for generators to abandon wastes, and short-term benefits must be weighed more carefully against long-term liabilities. The resulting polluted waters usually contain only trace levels of restricted species, but appropriate management is still required. Moreover, traditional methods are often ineffective and new treatment methods are needed to economically manage these wastes.

Photochemical and ultraviolet (UV) treatment technologies are still being actively investigated and three contributions describe aspects of this important topic. In Chapter 5, Larson, Marley and Schlauch review the status and several important applications. Light can promote the decomposition of dissolved pollutants by a variety of mechanisms. Although some classes of organic chemicals absorb sunlight strongly and are rapidly photolyzed in natural waters or wastewaters, for others, some means of indirect photodegradation is necessary. Classical sensitized photolysis with its singlet oxygen mechanism is probably of limited importance in water, but sensitizers such as riboflavin may be active toward compounds with which they can form light-absorbing complexes.

Solar radiation can be concentrated and, in certain instances, effectively combined with catalysts to enhance destruction rates. Graham et al. describe laboratory and small-scale field studies in which thermal heating and UV radiation combine to provide enhanced destruction rates for 1,2,3,4-tetrachlorodibenzo-*p*-dioxin at temperatures below 800 °C. Products of incomplete combustion are formed with lower yields, and are destroyed at lower temperatures than in conventional incinerators. It appears that a simultaneous exposure to concentrated sunlight and high temperature can significantly increase pollutant destruction rates when compared to heating at similar temperatures alone.

The relative merits of chemical treatment (e.g., H_2O_2 or O_3) vs photochemical-induced oxidation and reaction is of continuing interest. A number of papers relating to this topic were presented during the 1989 Symposium (*10*) and additional contributions were received this year. Heeks et al. (Chapter 7) present a particularly interesting comparison of oxidation technologies conducted at the Xerox facility in Webster, NY. The Ultrox, Perox-Pure, and Rayox systems are all commercially available and were compared for their efficacy in destroying several common chlorinated solvents and toluene.

It was concluded that UV oxidation technology is applicable on a commercial scale for groundwater purification. It was particularly effective in the destruction of toluene and unsaturated chlorinated solvents, but less efficient for the destruction of saturated chlorinated species such as dichloro- and trichloroethanes.

Water purification by solute separation and retention is another important option. The main problems with such technologies relate to the extremely low pollutant concentrations, the relatively high decontamination factors that are required, and the need to either regenerate or stabilize spent separating agents. Modeling capabilities are also needed to guide research and the selection of alternatives.

Ion exchange technology represents one alternative that can be important, even for the removal of low pollutant concentrations. In multicomponent systems, competition may occur for active sites between the various solutes. This situation complicates the analysis and design. In Chapter 8, Robinson, Arnold and Byers present data and analysis techniques for such cases, using the system Ionsiv IE-96 chabazite zeolite to remove ^{90}Sr and ^{137}Cs from wastewaters containing parts per billion concentrations. Binary isotherms obtained by batch measurements were fitted using a modification of the Dubinin-Polyani model (11).

Chapter 9 includes a related ion exchange study by Hall, Watson and Robinson. Traditionally a batch unit operation, ion exchange can be operated continuously. Higgins (12,13) developed one of the earlier designs. The present unit operates similarly, primarily for strontium removal. It offers substantially reduced secondary waste production rates compared to the available alternatives. These authors present their multicomponent equilibrium model, the use of the Thomas model for predicting breakthrough and their scaleup techniques.

The selection of wastewater treatment technology requires the evaluation of many parameters and costs. Such comparisons are usually lengthy, but can be shortened somewhat by the strategic application of spreadsheet-based computer simulation programs. In Chapter 10, Counce et al. describe an economic model that can assist in the evaluation of volatile organic chemical (VOC) air-stripping options for treating groundwater. They clearly review the necessary steps in such evaluations and explain its implementation into spreadsheets. Their simulator consists of three general parts: (1) process design, (2) estimation of fixed capital and annual operating costs, and (3) operating lifetime analysis. This chapter is a useful reference and overview of the evaluation procedures. Their examples, jet fuel and trichloroethylene spills, illustrate the value of parametric methods.

Biological Treatment

The biological treatment of wastewaters and solids continues as an emerging technology (*14,15*), and four contributions are included in the present volume. Solids treatment, often referred to as "bioremediation," is finding many applications. One of the more effective methods for remediating oil spills, developed from the Exxon Valdez experience (*16*), is simply to add phosphate and nitrate fertilizers to the environment, and allowing naturally occurring microorganisms to metabolize the hydrocarbons (Lessard, R. R., I&EC Special Symposium Dinner Speaker, Atlantic City, NJ, 1990).

Although microbiological processes involve incredibly complex chemistry, they are often surprisingly simple to implement and are cost effective. Biotechnology may frequently be the best choice, especially if the waste is not particularly labile and time is not a limiting remediation consideration. Other factors, such as the average annual temperature and active growth period, may also determine overall effectiveness.

In Chapter 11, Fernando and Aust describe the biodegradation of munitions wastes using *Phanerochaete chrysosporium,* a white rot fungus. After 30 days incubation, about half of the ^{14}C in their 2,4,6-trinitrotoluene (TNT) samples degraded to $^{14}CO_2$ and only 3% could be recovered. Similarly, about two thirds of the initial hexahydro-1,3,5-trinitro-1,3,5-triazine (RDX) concentration was converted to $^{14}CO_2$ after the same incubation period. Ground corn cobs served as the nutrient for this fungus; soil cultures apparently degraded TNT and RDX at higher rates than liquid cultures. *Pseudomonas* has also been used to degrade TNT (*17*). *Pseudomonas* and *Aceinomycetes* are reported to degrade aromatics (*18–21*). Polycyclic aromatics and biphenyls are degraded by *Phanerochaete chrysosporium* (*22,23*).

In Chapter 12, Utgikar and Govind discuss the use of a biofilter to control volatile organic chemicals. They present a mathematical model and summary calculations for 90% VOC removal. Their approach utilizes a countercurrent packed column (or trickle bed filter) to transfer VOCs from the air to a liquid phase. A microbial culture is attached to the packed column (e.g., activated carbon) and nutrients are provided through recirculation of the liquid phase which is introduced into the top of the packed column. Experimental data are presented for the removal of toluene and methylene chloride. They used biomass (a mixed culture) from an activated sludge plant that had been acclimated to toluene and methylene chloride.

A third experimental study of biodegradation is presented by Havens and Rase in Chapter 13. They examined the use of an enzyme, parathion hydrolase, derived from an overproducing strain of *Pseudomonas diminuta*. This enzyme was partially purified and immobilized on several rigid supports, several controlled pore glasses and a beaded polymer. These supports were compared with respect to parathion degradation; the hydrolase attacks the phosphoryl oxygen.

The immobilized enzyme was capable of reducing the concentrations of several organophosphorous pesticides to levels below 1 ppm.

In Chapter 14, Baltzis, Lewandowski, and Sanyal present a mathematical model for describing biological denitrification in a sequencing batch reactor. Their model assumes noninhibitory kinetics for nitrate reduction, inhibitory kinetics for nitrite reduction, and incorporates possible toxicity effects of nitrite on the biomass. It has been successful in qualitatively describing experimental data from a 1200-gallon pilot unit. Not surprisingly, sensitivity analyses indicate that kinetic parameters have the greatest impact on reactor performance. They demonstrate how their model can be used to select operating parameters; results are currently being applied to the design of a larger unit for treating munitions wastes.

Solid Waste Management

The disposal of solid wastes is often considered the alternative of last resort. Appropriately, more emphasis is being placed on waste minimization and process modifications that avoid waste production altogether. Moreover, efficient processes that minimize waste production are essential to maintain a competitive posture.

In the meantime, substantial inventories of hazardous solids must be stabilized and eliminated. Superfund sites are perhaps the most widely discussed and represent the most significant potential risks, but landfills are much more pervasive and many contain low concentrations of hazardous materials from illegal or unregulated practices. Thus, stabilization is a common problem at hazardous waste sites and landfills, and with mine tailings.

It is appropriate to overview the state-of-the-art with respect to solidification and waste stabilization. Chapter 15 provides a starting point for this last section in the book. Much of the impetus results from the Resource Conservation and Recovery Act, the Comprehensive Environmental Response, Compensation and Liability Act and the Superfund Amendments and Reauthorization Act. Ultimately, these legislative actions translate into economic realities that must be faced by the waste generators, primarily those who are intimately involved with manufacturing. While stabilization is often effective, Bishop points out that not every waste can be stabilized. A number of factors (e.g., pH and redox potential) may affect the leachabilities of the resulting waste forms. At one extreme, borosilicate glasses are contemplated for high-level fission product wastes. At the other extreme, minimal treatment, such as immobilization in a pozzolan-based cement, is proposed.

In Chapter 16, Chawla et al. overview a related issue, the interactions between soils, contaminants, and surfactants. This topic is particularly important whenever soil washing is contemplated (20). Soil matter can bind with organic pollutants through various physical and chemical interactions. Adsorption, hydrogen bonding, and ion exchange can all be important. The interactions between a soil, the contaminant, and any surfactant that is considered for soil washing are of paramount importance.

The chemistry of species migration is complex and site-dependent. This consideration has long been a dominant consideration in radioactive waste management and in the selection of a federal repository for high-level waste disposal. In the long-term, it is necessary to avoid water and seek highly adsorbent geologic formations. In the short-term, much useful information can be obtained from transport studies in real time. In Chapter 17, for example, Sandhu and Mills examine the mechanisms of mobilization and attenuation of inorganic species in coal ash basins. The site is the Savannah River Plant, a well-known plutonium production facility, but also a generator of ash from coal-fired power generation. Sediment cores, collected from ash basins of various ages, were sectioned and analyzed to elucidate the mobilization mechanisms. There were significant releases and mobilization of most elements to the lower horizons of impounded ashes; this effect was most pronounced in the old basin. Low pH, generated by decomposing organic matter, appears partly responsible, but complexation by organic ligands may also play a role.

In Chapter 18, Doepker discusses a similar finding at mine sites where soils close to metal smelters have elevated concentrations of lead, cadmium, copper, zinc and other heavy metals. Acetate ion (e.g., ammonium acetate at a pH of 4.5) effectively mobilizes cadmium and, to a lesser extent, zinc and lead. Although the western soil conditions are quite different from those in South Carolina, acidity and pH are key factors in metals leachabilities.

Soil venting is another important method for site remediation that is finding many applications. It is broadly accepted for soils contaminated with volatile or semivolatile species. It is complicated, however, by a need to understand the site hydrology and air flow in some detail. In Chapter 19, Kuo et al. describe a three-dimensional soil venting model that can be used to understand a site and develop venting strategies. Their predictions are compared with experimental measurements at several remediation sites.

While much has been said about fixation, there are many species requiring immobilization and innumerable matrices and constraints affecting the problem. Chu et al., for example, compare fixation techniques for soil containing arsenic in Chapter 20. The waste soils contained from 1200 to 2100 mg/kg. They compared Portland cement, fly ash, silicates, and ferric and aluminum hydroxides, and found that silicates were most effective in reducing arsenic leaching rates.

Low-level radioactive waste disposal is discussed by Darnell et al. in Chapter 21. In particular, they analyze the long-term structural and radiological performance of abovegrade earth-mounded concrete vaults. Vault and grout waste degradation are modeled over extended time periods using computer codes. The resultant radiological doses are calculated. They conclude that this disposal strategy will satisfy the performance objectives set by the U.S. Department of Energy.

Summary

The emerging themes for rational waste management strategies are relatively straightforward. Toxic materials, once created, must either be detoxified or effectively isolated from the environment. The complexities lie primarily with technology selection issues, process integration, and practical constraints (e.g., costs). In many instances, the best strategy may be to avoid waste production or reduce generation rates by modifying the manufacturing processes.

While combustion technologies have the potential for dispersing hazards and increasing effective population doses, rational long-term management precludes the disposal of thermally or chemically unstable waste forms. Intuitively, the preferred waste forms are those which are thermodynamically stable. The complexities begin with selecting conversion technologies (e.g., chemical vs. thermal or combustion).

Many wastewater problems result from the presence of exceedingly low pollutant concentrations. The common challenge is to find effective technologies for removal and concentration before detoxification. In this regard, low-temperature oxidation technologies [either chemical, photochemical, or other water irradiation techniques (24)] continue to attract considerable attention from numerous investigators. They are a continuing theme in this volume as well as its predecessor (10).

The use of biological systems for waste treatment is another strategy that is clearly emerging, and the possibilities seem as varied as the number of microorganisms one may employ. As demonstrated by contributions in this volume, gaseous, liquid, and solid wastes may be treated using bioengineering techniques. While many of these methods can be complex, they can also be surprisingly simple. Increasingly, they are perceived as affordable, but bioremediation may also be slow. Thus, they are emerging as mid- to long-term treatment options.

Species migration and waste solidification must often be considered concomitantly. Thermodynamic stability is a primary concern in both instances, and this subject has not been adequately discussed, especially with respect to complex matrices such as contaminated soils. Of course, pH plays a key role and, as noted previously (10), species are more labile in acidic, rather than in basic media. Redox potential is also important for species undergoing valence changes; reducing conditions generally seem to make immobilization more favorable.

Literature Cited

1. Siuta, J. State, conditions and indispensable activities in the sphere of environmental protection. *Z Nauk Polit L Bud*, 40:5–23, 1988.
2. Ziegler, C. E. Bear's view: Soviet environmentalism. *Technol Rev*, 90(3):44–51, April 1987.
3. Bazell, R. Science and society: red forest. *New Republic*, 198(2):11, May 3, 1988.
4. Colchester, N. Industrial waste. *The Econ*, 315(2):PS12, June 30 1990.
5. Simons, M. In Leninallee, cans, bottles and papers: it's the West's waste! *New York Times*, 139, July 5 1990.
6. East European loss will boost EC waste market. *Euro Chem News*, 55(1):P20, July 23 1990.
7. West assists East with pollution programme. *Euro Chem News*, 53(1):P14, Nov 13 1989.
8. Marshall, E. Radiation exposure: hot legacy of the cold war. *Science*, 249(1):474, Aug 3 1990.
9. Goossens, W. R. A., Eichholz, G. G., and Tedder, D. W., editors. *Treatment of Gaseous Effluents at Nuclear Facilities*. Volume 2 of *Radioactive Waste Management Handbook*, Harwood Academic Publishers, New York, 1991.
10. Tedder, D. W. and Pohland, F. G., editors. *Emerging Technologies in Hazardous Waste Management*. Volume 422 of *ACS Symposium Series*, American Chemical Society, Washington, DC, 1990.
11. Ruthven, D. M. *Principles of Adsorption and Adsorption Processes*. John Wiley & Sons, New York, 1984.
12. Higgins, I. R. Countercurrent liquid-solid mass transfer method and apparatus. *U.S. Patent No. 2,815,322*, Dec 3, 1957.
13. Long, J. T. *Engineering for Nuclear Fuel Reprocessing*. American Nuclear Society, La Grange, IL, 2nd edition, 1978.
14. Forgie, D. J. L. Selection of the most appropriate leachate treatment methods. Part 1: A review of potential biological leachate treatment methods. *Water Pollut Res J Can*, 23(2):308–328, 1988.
15. Lee, M. D., Thomas, J. M., Borden, R. C., Bedient, P. B., and Ward, C. H. Biorestoration of aquifers contaminated with organic compounds. *CRC Crit Rev Environ Con*, 18(1):29–89, 1988.
16. Lee, D. B. Tragedy in alaska waters. *National Geographic*, 176(2):260–263, August 1989.
17. Selivanovskaya, S. Y., Gorkunova, T. A., and Naumova, R. P. Effect of clay on aerobic bacterial decomposition of trinitrotoluene. *Sov J Water Chem Tech (Eng Trans Khim Tekn Vo)*, 9(1):31–34, 1987.
18. Inoue, A. and Horikoshi, K. A pseudomonas thrives in high concentrations of toluene. *Nature*, 388:264–266, 1989.

19. Zimmermann, W. Degradation of lignin by bacteria. *J Biotech*, 13(2–3):119–130, Feb 1990.
20. McDermott, J. B., Unterman, R., Brennan, M., Brooks, R. E., Mobley, D. P., Schwartz, C. C., and Dietrich, D. K. Two strategies for PCB soil remediation: biodegradation and surfactant extraction. In *Proc AIChE Nat Meeting, New York, 1988*, American Institute of Chemical Engineers, New York, 1988.
21. Johnston, J. B. and Renganathan, V. Production of substituted catechols from substituted by a *Pseudomonas*. *Enzym Microb Tech*, 9(12):706–708, Dec 1987.
22. Eaton, D. C. Mineralization of polychlorinated biphenyls by *Phanerochaete chrysosporium*: a ligninolytic fungus. *Enzym Microb Tech*, 7:194–196, 1985.
23. Bumpus, J. A. Biodegradation of polycyclic aromatic hydrocarbons by *Phanerochaete chrysosporium*. *Appl Envir Microb*, 55:154–158, 1989.
24. Fleming, R. W. and Tedder, D. W. Water purification by radiation induced oxidation (thesis excerpts). *J Env Sci Health*, A25(4):425–446, 1990.

RECEIVED April 5, 1991

THERMAL TREATMENT AND ABIOTIC EMISSIONS CONTROL

Chapter 2

Hazardous Material Destruction in a Self-Regenerating Combustor–Incinerator

Tian-yu Xiong, Donald K. Fleming[1], and Sanford A. Weil[2]

Institute of Gas Technology, 3424 South State Street, Chicago, IL 60616

A novel concept of internal radiant regeneration of heat within a gas-fired combustion chamber, developed at the Institute of Gas Technology (IGT), was both analytically and experimentally proven. A portion of the combustion heat is recovered from combustion products and transferred to the inlet combustion air through radiant regeneration, resulting in super-adiabatic combustion temperatures. In a natural gas-fired test combustor/incinerator, combustion temperatures as high as 2066°C were achieved without using air preheat or oxygen enrichment. Incineration tests performed in the test incinerator, using carbon tetrachloride (CCl_4) as a surrogate for polychlorinated biphenyls (PCBs), have demonstrated that the destruction and removal efficiency (DRE) is strongly dependent on the combustion temperature. Results showed that the CCl_4 remaining was as low as 2×10^{-7}, at temperatures of about 2000°C with residence times as short as 40 milliseconds. This patented technique has potential application in compact, high-efficiency destruction of hazardous material.

As public environmental awareness increases, dissatisfaction with the conventional practices of hazardous and industrial waste disposal intensifies. Consequently, waste disposal and the public's acceptance of it present serious public policy issues. The preferred alternative is a more efficient, immediate, and economic conversion of waste materials to innocuous byproducts. One method of waste destruction is incineration, particularly for toxic dioxin and PCB-containing wastes. In the Toxic Substances Control Act (TSCA) regulations, a DRE of six-9s (99.9999%) is required for the incineration of some PCB-containing materials (1). In general, higher temperatures ensure more

[1]Current address: M-C Power Corp., 8040 S. Madison, Burr Ridge, IL 60521
[2]Current address: P.O. Box 221, Edwards, NY 13635

0097–6156/91/0468–0012$06.00/0
© 1991 American Chemical Society

complete incineration of wastes, thereby increasing the DRE and reducing stack gas emissions. To achieve the required DRE, the TSCA also requires either a 2-second dwell time at 1200°C (±100°C) and 3% oxygen or a 1.5-second dwell time at 1600°C (±100°C) and 2% oxygen in the stack (*1*). Incineration at extreme temperature, therefore, offers an attractive advanced approach for achieving a higher DRE. It also offers better economics because the residence time required for sufficient destruction can be reduced.

Different types of high-temperature incineration techniques, including oxygen-enriched firing (*2,3*), recuperative air preheat, thermochemical fuel modification, and electric pyrolysis (*3*), are currently being developed. Some of them are supported for demonstration by the U.S. Environmental Protection Agency (EPA) (*3*). All of these approaches, however, need substantial external equipment and as a result will increase capital and operating costs.

The approach developed at IGT is novel because the recuperation for combustion air preheat is achieved within the combustor (*4*). Thus, it is relatively simple to construct, requires minimal external equipment, and is cost-effective. This unique approach will also allow the development of an efficient and compact transportable incineration system for the destruction of toxic wastes on a site-by-site basis, thus eliminating the problem of transporting toxic wastes. This feature is also vigorously supported by the EPA (*3*).

This paper describes the concept of internal radiant regeneration of heat within a gas-fired combustion chamber and presents experimental results obtained in a bench-scale, extreme-temperature combustor/incinerator using CCl_4 as a surrogate for PCBs.

Concept and Model

Internal Radiant Regeneration Concept. Heat radiation from the walls of any type of furnace is important in transferring heat from the flame to the load inside the furnace. A conventional method to increase energy conversion from the combustion products to the load through the walls is to enhance convective heat transfer from the flame to the walls. Based on this concept, the flat-flame burners, the high-velocity impingement burners, and the high-momentum burners have been developed. However, these approaches use high-velocity combustion gases to achieve high convective heat transfer to the furnace walls. This results in high combustion noise levels and substantial static pressure drop through the furnace.

The use of intensive heat radiation from a porous plate placed in a gas flow duct to enhance energy conversion to a load has been investigated by Echigo (*5*). A similar concept of a porous wall burner that would improve the thermal efficiency of a gas-fired furnace was proposed by Hemsath (*6*). Obviously, the radiant porous walls heated by the passing combustion gases can achieve higher energy conversion to the load. However, combustion temperature in all the combustion chambers is not influenced by the wall radiation because the inlet combustion air is not preheated. The concept of internal radiant regeneration of heat within a gas-fired combustion chamber is aimed at increasing the combustion temperature.

In the novel concept developed at IGT (Figure 1), the combustor consists of two porous plates placed opposite each other. The combustion air enters the combustion chamber through the inlet porous plate. It mixes with fuel and burns in the chamber, and the combustion products exit through the exit porous plate. If there are no heat losses from the system, the final combustion products leaving the exit porous plate are at the adiabatic combustion temperature. Because the inner surface of the inlet porous plate is colder than the outlet porous plate

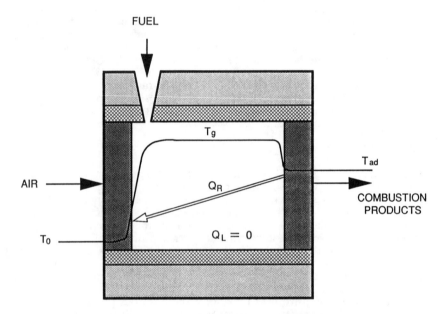

Figure 1. Internal radiant regeneration of heat concept.

heated by the combustion products, heat will radiate within the chamber, from the inner surface of the outlet plate to the surface of the inlet plate. This radiation will cause heating of the inlet plate which, in turn, will preheat the combustion air. A steady-state system exists in which the inlet air has been indirectly preheated by this radiation (through convective heat transfer from the porous solids to the incoming air), combustion occurs at temperatures greater than the normal adiabatic combustion temperature, and the gases are cooled from this ultra-hot combustion as they pass through the exit porous plate. Thus, a portion of combustion heat is recovered from the combustion products and transferred to the inlet combustion air through the convective-radiative heat transfer mode within the combustion chamber. This results in a super-adiabatic combustion temperature ("ultra-hot").

The Theoretical Model. A theoretical model has been developed to prove the concept and explore the parametric effects on combustor performance. The combustion chamber, as illustrated in Figure 1, consists of two parallel porous plates as opposite walls of the combustor. One porous plate serves as the inlet for combustion air; the other is the outlet for combustion products. The basic heat transfer mechanism within the porous medium is conduction and convection. Inside the combustion chamber, convective heat transfer between the gas flow and the sidewalls and radiation heat transfer between the gas phase and the solid surfaces as well as wall-to-wall heat transfer were taken into account. All solid surfaces were assumed to be gray bodies, and the gas phase within the combustor was considered gray gas. Only steady-state flow and heat transfer were considered in the modeling. The temperature distribution on each wall and within the gas phase was assumed to be uniform; this does not compromise the main purpose of the modeling, which is concerned only with the bulk heat transfer in the system.

The major variables characterizing the performance of the system that were involved in the modeling are —

- Configuration of the combustor: cylindrical or rectangular chamber, the plate spacing, the length-to-diameter (L/D) ratio of the combustor

- Fuel type (hydrogen, hydrocarbon, alcohol), oxidant components, and the stoichiometry and pressure of the system

- Effect of chemical dissociation in the combustion process on combustion temperature

- Emissivities (absorptivities) of the porous plates and the walls of the combustor

- Thickness, porosities, and thermal conductivities of the porous plates

- Heat flux in the system

- Emissivities of radiating species in the gas phase and luminosity of the combustion products

- Heat loss through the walls of the combustor.

The concept of super-adiabatic temperature combustion by internal radiant regeneration of heat has been proven based on the numerical solutions of the analytical modeling. To achieve a higher combustion temperature, the following conditions are desired:

- High emissivities of both the inlet and the outlet plates

- Small L/D ratio of the combustion chamber

- Near-stoichiometric and intermediate firing intensity (about 1.4 MW/m^2)

- Transparent flame

- Minimum heat loss from the system.

The theoretical analysis did not reveal any inherent errors in the original concept; no "fatal flaws" were found. However, the analytical results indicated two significant technical challenges in achieving super-adiabatic temperature combustion: First, the need for materials with satisfactory extreme-temperature operation and desired emissivities, and second, the assurance of a transparent flame in the combustor for effective internal radiant regeneration of heat. In addition, NO_x emissions from the extreme-temperature combustion could be a barrier for application. Finally, extreme-temperature measurement was also a serious challenge in the investigation.

Testing of the Ultra-Hot Combustor/Incinerator

An experimental study was conducted on a bench-scale combustor/incinerator to prove the concept of the internal radiant regeneration of heat and to evaluate its capability for the destruction of hazardous material at extreme temperatures. The experimental system, including both combustion and measurement subsystems, was developed in several versions for the proof-of-concept studies.

Combustor. The test combustor was designed for a 15-kW firing rate. Figure 2 is a schematic drawing of the down-firing test combustor. The combustion chamber was 102 mm in diameter by 76 mm high and was adjustable. Zirconia spinel honeycomb plates (102 mm diameter by 25.4 mm thick) were used to construct the inlet and outlet porous plates. The honeycomb has a design of 16 x 16 squares per inch with 70% open area. The emissivity of the spinel is assumed equal to zirconia at 0.35. High-density, vacuum-formed zirconia cylinders were used to insulate the extreme-temperature system. Combustion air was introduced in the combustor through the top plate. Natural gas was injected into the chamber through six nozzles positioned at different angles to ensure excellent mixing with the combustion air preheated by the inlet plate. The resulting flame was uniform and transparent.

Incineration System. CCl_4 was selected as the surrogate material for testing the incinerator capability. CCl_4 is generally believed to be more stable than PCBs and relatively nontoxic, and will not result in products of incomplete combustion (PICs).

The CCl_4 incineration system was integrated with the combustor, as shown in Figure 3. Liquid carbon tetrachloride was supplied and metered by the Masterflex adjustable-speed pump and then injected into the combustion air stream. The temperature in the chamber where the injected CCl_4 mixes with the combustion air flow was kept slightly higher than the boiling temperature of CCl_4, which is 80°C, to achieve complete evaporation of the CCl_4. Sodium

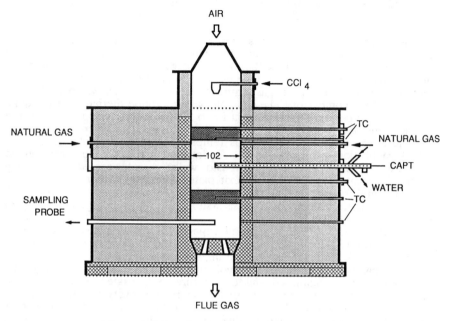

Figure 2. The ultra-hot combustor/incinerator.

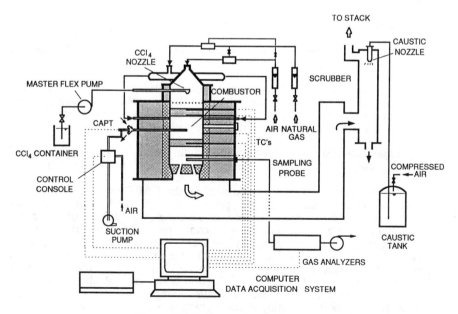

Figure 3. The experimental system.

hydroxide (NaOH) solutions were used to neutralize the hydrogen chloride (HCl) produced in the CCl_4 incineration in the methane flame.

Measurement System. In the measurement system, also shown in Figure 3, the flow rates of the combustion air and natural gas were measured by two rotameters, respectively. Two thermocouples were installed at the outer and inner surfaces of the inlet plate: One thermocouple was located within the outlet plate to measure the exhaust gas temperature, and the other was placed at the side wall of the combustor. The computer-assisted pulse thermometry (CAPT), capable of temperature measurement up to 2500°C, was developed to measure the extreme gas temperature in the combustor. A conventional suction pyrometer was used for the flue gas temperature measurement and also for CAPT calibration of up to 1760°C. An infrared pyrometer was also used to measure the temperature at the combustor walls. A water-cooled gas sampling probe was located underneath the exit plate. Gas composition, including O_2, CO, CO_2, and NO_x in the flue gas, was analyzed for combustion diagnostics. A computer data acquisition and analysis system, also shown in Figure 3, was used in the experimental study.

To evaluate CCl_4 incineration capacity in the combustor/incinerator, high-temperature gas samples were drawn from the combustion system through a modified EPA Method 5 analytical train, and the residual CCl_4 was trapped in double Tenax beds in series. Sample volumes of 20 liters were drawn, measured by a meter that was calibrated, and temperature-compensated. Double tubes of absorbent were used in the traps to assure complete recovery; a system of blanks and sample spikes was used to verify the analytical technique. The sample traps were analyzed by gas chromatography/mass spectrometry (GC/MS) techniques. This approach was shown to have a sensitivity of about 0.7 micrograms of CCl_4 in the sample. The sensitivity was sufficient to measure seven-9s (99.99999%) destruction.

Experimental Results

Super-Adiabatic Combustion Temperature. To prove the concept, the unit was operated in two modes:

1. Operation of the combustor with porous plates at both the inlet and outlet of the chamber. This is the basic design mode of the concept, with hot gases from the chamber heating the exit porous plate, which, in turn, radiate to the inlet porous plate for air preheat.

2. Operation without the outlet porous plate. This comparison mode should provide a nearly adiabatic combustion temperature in the chamber.

Combustion temperatures measured by the CAPT for the two operating regimes at different excess air are presented in Figure 4. The calculated adiabatic combustion temperatures, considering the effect of chemical dissociation, are also shown in Figure 4 for comparison. The temperatures in the combustor without the exit porous plate were 100° to 150°C lower than the calculated adiabatic combustion temperature due to heat loss from the system. However, the combustion temperatures in the chamber with the exit porous plate installed were 50° to 130°C higher than the calculated adiabatic combustion temperatures. The maximum temperature measured in the combustor was 2066°C, resulting in the melting of a zirconia spinel chip and an alumina tube placed in the chamber. Therefore, an increase in combustion

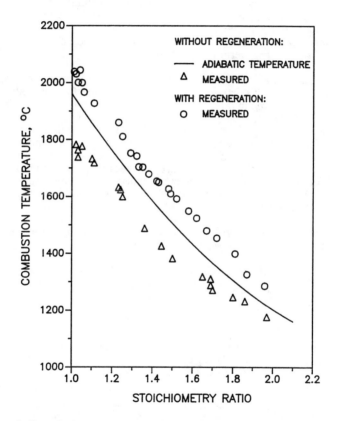

Figure 4. Proof of the internal radiant regeneration of heat in the bench-scale combustor.

temperature as high as 280°C was achieved through the internal radiant regeneration of heat. This is approximately equal to 600°C of combustion air preheating at low excess air operation.

Parametric effects of the firing rate, air/fuel stoichiometric ratio, as well as the L/D ratio of the combustor on the combustion temperatures were investigated. Figure 5 shows increments of the combustion temperatures due to the internal radiant regeneration of heat versus the air/fuel ratio and the L/D ratio of the combustion chamber. The combustion temperature is higher in the combustor with a smaller L/D ratio because of higher radiation heat transfer from the exit plate to the inlet plate. When the air/fuel ratio is lower, the increment of the combustion temperature is higher because less of the combustion air can be preheated to a higher degree. The results of this parametric study provide solid data for technical evaluation and scale-up design of the internal radiant regenerating combustor.

Combustion Emissions. It is anticipated that achieving acceptable combustion emissions, particularly NO_x, could be a serious challenge for application of the ultra-hot combustion technique. NO_x concentrations in the flue gas were measured at varying operating conditions, such as combustion temperature and firing rate. The results, presented in Figure 6, demonstrate that NO_x emissions dramatically increased with increased combustion temperature and significantly reduced at lower residence times because the formation of NO_x is primarily a kinetically controlled process. Therefore, by reducing the residence time — which is adequate for high-efficiency incineration at extreme temperatures — NO_x emissions can be controlled below 300 ppm (corrected to 0% O_2) when the combustion temperature is as high as 2040°C.

Incineration of CCl_4. In general, the overall reaction of CCl_4 in the methane/air flame can be expressed as follows:

$$CH_4 + \beta CCl_4 + 2\alpha O_2 + 7.569\alpha N_2 \rightarrow$$

$$4\beta HCl + 2(1 - \beta) H_2O + (1 + \beta) CO_2$$

$$+ 2(\alpha - 1) O_2 + 7.569\alpha N_2$$

where β is the stoichiometric ratio of the CCl_4/CH_4 reaction and α is the stoichiometric ratio of the CH_4/O_2 combustion reaction. β can be calculated from the measured CCl_4 flow rate, $G(CCl_4$-kg/min), and CH_4 flow rate, $V(CH_4$-Nm^3/min). That is,

$$\beta = \frac{G/153.81}{V/22.41} = 0.1457\frac{G}{V}$$

An overall evaluation of the incineration capability of CCl_4/CH_4 is the residual of destructed material defined by the ratio of destructed material recovered to the input rate. If ΔG (mg/Nm^3) is the amount of destructed material recovered in the products of combustion that is detected from the sampling gas at temperature, $T_s(K)$, the total remainder of CCl_4 in the products of combustion is determined by —

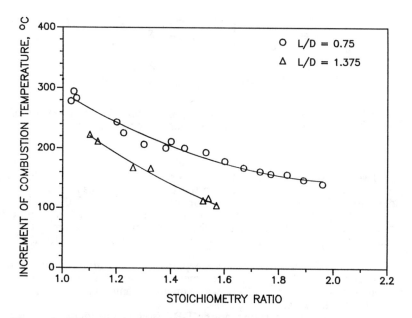

Figure 5. Effect of the L/D ratio of the combustor and the air/fuel ratio on internal radiant regeneration of heat.

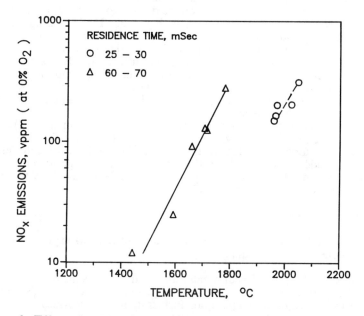

Figure 6. Effect of combustion temperature and residence time on NO_x emissions.

$$\eta = \Delta G \left[\frac{A_{g,o} V}{G} \right] \cdot \left[\frac{273}{T_s} \right] \times 10^{-6}$$

$$= 0.1457 \ \Delta G \left[\frac{A_{g,o}}{\beta} \right] \cdot \left[\frac{273}{T_s} \right] \times 10^{-6}$$

where $A_{g,o}$ = $9.569\alpha + 5\beta - 1$ is the molar amount of dry products of the CH_4/CCl_4/air reaction. The DRE can be determined as follows:

$$DRE = 1 - \eta$$

Two important parameters, temperature and residence time, are expected to have an essential effect on the DRE. An increase in incineration temperature will allow reduction of the residence time of the incinerated material in the unit, which means higher incineration intensity. On the other hand, higher temperature will increase the destruction efficiency during the same residence time. The main goal of the tests is to explore the effect of both temperature and residence time on CCl_4 destruction.

The CCl_4 incineration tests were performed at different temperatures (from 1413° to 2048°C) and residence times (from 16 to 170 milliseconds). The stoichiometric ratio of CCl_4/CH_4 was generally maintained at 0.27. The incineration intensity was 2340 kg/h-m³.

The effect of temperature on the DRE is shown in Figure 7. It is clear that temperature has a strong impact on the destruction efficiency. A CCl_4 residual as low as 2 X 10⁻⁷, or nearly seven-9s (99.99998%) conversion, was achieved at 2010°C with residence times as short as 41 milliseconds. An increase in temperature of about 440°C (for example, from 1540° to 1980°C) can improve the destruction efficiency by one order of magnitude.

The effect of residence time on the destruction efficiency was investigated at constant temperatures, as shown in Figure 8. By increasing the reaction temperature by 390°C, the residence time required for a certain DRE could be reduced to 1/4. In addition, at a higher incinerating temperature (1926°C), the effect of residence time is greater than that at a relatively lower temperature (1537°C) because the chemical kinetics at the high temperature are more sensitive to the reaction duration. Incineration at a higher temperature, therefore, has a wider operating range than lower temperature incineration for the same destruction efficiency.

In summary, incineration at extreme temperature can either 1) promote a destruction efficiency that will be an essential advantage for some types of hazardous materials or 2) reduce the residence times for sufficient destruction. Shorter residence times allow the use of smaller equipment than lower temperature incinerators, and they help to hold NO_x emissions to acceptable levels at higher temperatures.

Application Potential

Technical and economic evaluations were performed to explore the potential applications of this technique.

Technical Evaluation. Experimental and analytical results have demonstrated that the developed internal radiant regeneration technique is significantly beneficial for extreme-temperature incineration and provides a considerable energy savings.

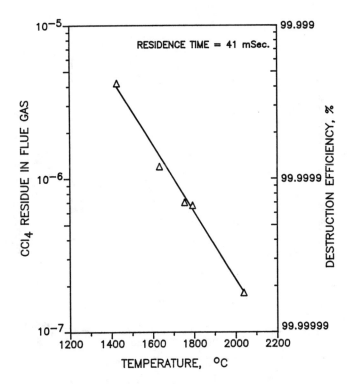

Figure 7. Effect of temperature on CCl_4 destruction efficiency (residence time: 41 ms).

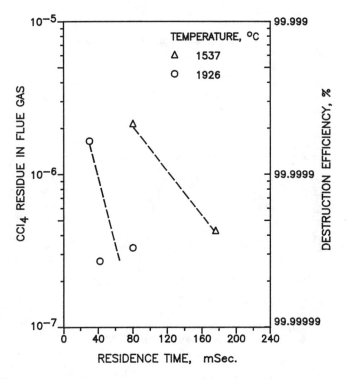

Figure 8. Effect of residence time on CCl$_4$ destruction efficiency.

The tests have shown that the combustion temperature measured in the bench-scale test combustor was increased by up to 280°C by the internal radiant regeneration of heat; the combustion air entering in the combustion reaction was heated up to 600°C without external air preheating. For a scale-up 300-kW unit, the combustion temperature is predicted to be increased by 160°C, which is equivalent to 330°C of air preheating. This increase will be even higher if emissivities of the inner surfaces of the inlet and exit porous plates are higher than 0.35 for the current zirconia spinel honeycomb plates.

For incineration applications, extreme temperatures can expand the range of materials that need to be destroyed at high temperatures as well as increase incineration intensity because the residence time for sufficient destruction can be minimized, as demonstrated in the testing. Compared with the conventional techniques such as oxygen enrichment and external air preheating to increase combustion temperature, the internal radiant regeneration of heat allows not only capital cost savings, but also refractory wall temperature reduction because all walls except the inlet plate in the combustor emit radiation to the inlet plate, where the temperature is lowest. For example, in the 300-kW combustor, the temperature at the side walls is 360°C below the combustion temperature.

The internal radiant regeneration of heat can also be applied to enhance flame stabilization of extremely lean (low-calorie) gas combustion because the combustion air or mixture can be preheated over 600°C. Therefore, a potential for incinerating low-heating-value off-gas in the radiant regenerating combustor is expected.

Because the efficiency of the internal radiant regeneration of heat is dependent on the spacing between the two porous plates and the luminosity of the flame, the capacity of the unit, the type of fuel, and the material to be incinerated in the unit are restricted. Based on analytical results, the capacity limit of the ultra-hot combustor applying this technique could be a 1.5-MW firing rate in practice. Correspondingly, the capacity limit of incineration will be equivalent to 500 kg/h CCl_4. By using higher emissivity (up to 0.8) porous materials for the inlet and the exit plates, the capacity can be doubled. In addition, only gaseous and pre-evaporated liquid materials can be incinerated in the ultra-hot combustor to ensure a transparent flame.

In summary, the technical evaluation suggested potential applications in compact or transportable high-efficiency incinerator systems with up to 1.5 to 3-MW firing capacity.

Economic Evaluation. The preliminary economic evaluation was conducted for a compact, 300-kW commercial combustor/incinerator system using IGT's ECONANALYSIS financial analysis software and using sensitivity analysis to identify critical cost components.

The conceptual design of the ultra-hot combustor/incinerator was developed based on the analytical and experimental results obtained from the bench-scale unit. The core of the incinerator is a 12-inch-long horizontally firing, cylindrical combustion chamber with an inner diameter of 16 inches. The combustion chamber is bonded at both ends by two 3-inch-thick porous grids. The materials of construction were selected to withstand the extreme operating temperatures. The capacity of incineration is equivalent to 90 kg/h CCl_4. The combustion temperature is predicted to be 2010°C, and the residence time of the combustion products is 60 to 70 milliseconds.

Figure 9 is a schematic diagram of this compact 300-kW firing-capacity incinerator system. Liquid waste is evaporated in the recuperator and then injected into the combustion air flow. The incinerator flue gas is vented to the

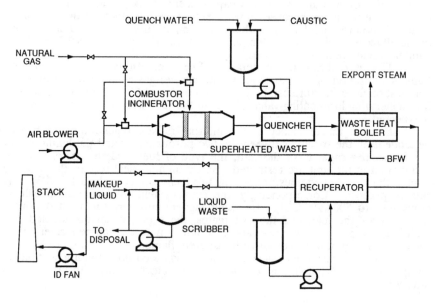

Figure 9. The ultra-hot incinerator system.

atmosphere after recovery of heat in the waste heat boiler and the recuperator and cleanup through neutralization and scrubbing. The plant is assumed to be located where waste streams are generated and utilities and general facilities are available.

Costing of each individual piece of equipment was done using available cost estimate correlations from literature or vendors' quotations. The direct installation costs of equipment were then calculated using a factored estimate method. Capacities and costs of the equipment are summarized in Table I.

Table I. Summary of Equipment Capacities and Costs of a 300-kW Ultra-Hot Incinerator System

Item	Capacity	Equipment Cost	Direct Installed Equipment Cost
Ultra-Hot Combustor/Incinerator	450 kW	$ 42,610	$ 55,390
Waste Heat Boiler	390 kW	19,180	28,770
Recuperator	35 kW	7,850	14,920
Combustion Air Blower	12.7 m^3/min	940	1,320
Induced Draft Fan	34 m^3/min	2,350	3,290
Quench Water Pump	3.8 L/min	1,890	3,020
Liquid Waste Pump	3.8 L/min	1,890	3,020
Scrubber Recirculation Pump	38 L/min	2,020	3,230
Quencher	190 kW	19,440	29,160
Stack	150 mm dia.	2,480	2,980
Quench Water Tank	3.5 m^3	3,350	5,030
Liquid Waste Tank	2.7 m^3	2,950	4,430
Scrubber	22 m^3/min	10,740	17,180
Total Costs		$117,690	$171,740

Notes:
1. 1988 fourth-quarter dollars.
2. Equipment costs include instrument/control, taxes, and freight charges.
3. Installed costs include foundations and supports, electrical, piping, insulation, handling, and erection.

The capital cost for the entire system is estimated at $171,740. Constant annual operating costs consist of operating and maintenance labor, which is $220,000 per year, based on two operators per shift. The levelized cost of service is therefore estimated to be $300/ton of CCl_4 waste destroyed. However, the levelized cost of service could be greatly reduced for the larger size units because a major part of the operating costs for labor will remain the same. The sensitivity study has suggested that if the firing capacity can be increased from 300 to 900 kW, the levelized cost will be reduced from $300 to $120/ton CCl_4. According to *Process Engineering*, the cost for PCB incineration is $140 to $1,500/ton (7). Therefore, the present technology is economically competitive. The technical and economic evaluations demonstrate a strong potential application of the internal radiant regenerating combustor to compact or transportable extreme-temperature incineration markets.

Conclusions

The novel concept of internal radiant regeneration of heat within a gas-fired combustor was both analytically and experimentally proven. Super-adiabatic combustion temperatures were achieved in the test combustor/incinerator without using air preheating or oxygen enrichment. It was demonstrated that this technique can be applied in extreme-temperature destruction of hazardous materials to achieve very high DREs (up to seven 9s) and incineration intensity. Because the residence time needed for a high DRE is very short, NO_x emissions can be controlled within an acceptable range.

IGT has designed a 300-kW commercial incineration system based on its successful bench-scale studies. Technical and economical evaluations have shown that the potential for developing an advanced, high-intensity, high-efficiency, compact, and transportable incinerator system is excellent.

Acknowledgments

The authors would like to thank Mr. Robert Sheng for his contribution to the preliminary scale-up design of the commercial incineration system, Mr. Richard Biederman for the economic analysis, and Mr. Lloyd McHie for his technical assistance.

Literature Cited

1. ASME; *Hazardous Waste Incineration, A Resource Document*; January 1988.
2. Achanya, P.; *Incineration of Hazardous Waste in a Mobile System*; International Symposium on Alternative Fuels and Hazardous Wastes; AFRC; October 9-11, 1984.
3. *The Superfund Innovative Technology Evaluation Program, Progress and Accomplishments*; Report to Congress; U.S. EPA, EPA/540/5-88/001; February 1988.
4 Weil, S. A.; Xiong, T-Y.; Fleming, D. K.; *Process and Apparatus for High Temperature Combustion*; U.S. Patent No. 4,828,481; May 9, 1989.
5. Echigo, R.; *Effective Energy Conversion Method Between Gas Enthalpy and Thermal Radiation and Application to Industrial Furnaces*; The 7th International Heat Transfer Conference; 1982, Vol. VI.
6. Hemsath, K. M.; *Application and Development of a Porous Wall Burner*; Final Report by Indugas, Inc.: Toledo, OH, June 1986.
7. Editorial Staff; "Toxic Waste: The Burning Question"; *Process Engineering*. Spring 1989. *Vol. 70*, No. 4a.

RECEIVED April 5, 1991

Chapter 3

Incineration of Contaminated Soils in an Electrodynamic Balance

M. Flytzani-Stephanopoulos[1], A. F. Sarofim[1], L. Tognotti[1,3],
H. Kopsinis[2], and M. Stoukides[2]

[1]Department of Chemical Engineering, Massachusetts Institute
of Technology, Cambridge, MA 02139
[2]Department of Chemical Engineering, Tufts University,
Medford, MA 02155

Understanding contaminant evolution from landfill soils is important in several in-situ remediation processes as well as in thermal treatment and incineration of contaminated top soils. To delineate the rate-limiting processes in the absence of interparticle effects, single surrogate soil particles are examined in this work. The adsorption-desorption characteristics of toluene and carbon tetrachloride on single, surrogate soil particles have been studied using an electrodynamic balance (EDB) under ambient conditions (P=latm; T=298K). The EDB offers high mass sensitivity ($\Delta m \sim 10^{-9}$g) in the absence of external mass transfer limitations and interparticle effects. In this work, three types of solid particles, 100-170μm in diameter, were examined, namely montmorillonite, a clay, and two synthetic chars, Spherocarb and Carbopack, of very different pore structures. Three different values of relative pressures, P/P_o, were tested for each liquid by changing the saturator bath temperatures. Significant differences were identified among the various solid-organic compound pairs examined in adsorption-desorption sequences in the EDB. These are strongly correlated with differences in the solid pore structures.

At the present time, alternatives to conventional hazardous waste treatment methods are actively being sought in response to stringent landfill regulations and associated high landfill costs. Incineration is a proven solution for treating

[3]Current address: Department of Chemical Engineering, University of Pisa, Pisa, Italy

top soils containing organic contaminants and well-designed incineration systems provide the highest overall degree of destruction of hazardous waste streams (1). Among other soil treatment methods, currently under development, are the in-situ decontamination by vacuum extraction and air stripping, and thermal desorption of contaminants by heating the soil to temperatures well below those typical in incinerators. The latter solution is more easily accepted by the public than incineration. Overall, significant growth in the use of incineration and other thermal soil treatment methods is anticipated in the near future (1, 2).

A typical incinerator system consists of a primary combustor, where contaminants are primarily volatilized and partially burned, and the secondary combustion chamber where the thermal destruction is completed (3). More than 20% of the total hazardous wastes generated in the U.S. can be considered incinerable (1, 4). According to the experience gained to date, rotary kiln incinerators seem to show superior applicability and versatility compared to the other existing types (1, 2). A rotary kiln incinerator consists of a cylindrical refractory-lined shell mounted on a slight incline. The shell rotates to provide for transfer of waste and to enhance mixing. Wastes undergo partial volatilization followed by destructive distillation and partial combustion. Following the primary combustor, an afterburner is used where the gas phase oxidation reactions are completed. Transient phenomena involving rapid release of waste vapor into the kiln environment may occur during the batch-mode operation of a rotary kiln and may cause failure of the incinerator system. Such phenomena, called "puffs", are frequently encountered and have been studied by a number of investigators (5-7).

In the rotary kiln the solids can be considered as a bed of many layers of particles that are being slowly stirred (3). In kilns operated at lower temperatures, the soil is picked up and dropped by baffles known as flights in order to augment the contacting between the soils and the gas stream. Contaminants may exist either adsorbed onto the internal pore structure of the particles or adsorbed on the external surface of the particles or as a liquid phase within the bed (3). Hence, both intraparticle and interparticle effects contribute to the high complexity of the rotary kiln system (3, 6). Numerous particles with variable properties are involved and the isolation of individual effects is not always easy.

It is clear that a fundamental study of the evolution of contaminants from soil particles can contribute significantly to understanding the transport of hazardous chemicals in soils and the processes limiting the operation of rotary kilns. In particular, it would be very helpful if the characteristics of a single particle reactor were first examined because in this configuration interparticle effects are eliminated. To this end, an electrodynamic balance (EDB) appears to be a very useful tool. The EDB consists of electrodes that can suspend a single charged particle in space, here modified to permit particle heating by a

CO_2 laser, temperature measurement by a two- or three-color infrared-pyrometer and continuous particle weighing by a position-control system (8, 9).

The above apparatus has been used successfully in a number of applications including measurement of temperature and weight of charcoal particles undergoing oxidation (10, 11), measurement of adsorptivities and heat capacities of single particles (12) and measurement of buoyancy forces at low Grashof numbers. Recently, the same device has been used as a basic tool in developing a novel droplet imaging system that offers unique capabilities for characterizing size, mass, density and composition of individual droplets (17). The EDB has also been used to obtain water activities in single electrolyte solutions (18).

In the present communication we report data on adsorption and desorption characteristics of two organic compounds; namely, toluene and carbon tetrachloride on single particles of surrogate soils in the EDB.

Experimental

A schematic diagram of the EDB system is shown in Figure 1. The electrodynamic balance is a chamber that can hold a single charged particle suspended in space by means of electric fields (19, 20). The balance consists of two endcap electrodes and a ring electrode. A DC potential is applied across the endcap electrodes and a AC voltage is applied to the ring electrode. The EDB can measure the diameter, mass, density, excess charge and surface area of a single levitated particle (21). A 5mW He-Ne laser illuminates the particle for position sensing and a 5-20W CO_2 laser is used for particle heating. An optical microscope is used for viewing the particle and for manual particle position control (Figure 1). Additional details on the experimental apparatus and procedure can be found elsewhere (8-10, 21).

The EDB was used to measure the relative weight change of a suspended particle during adsorption and desorption of organic vapors at room temperature. A single, dry particle was suspended in the chamber, through which a finite dry nitrogen flowrate was maintained. The particle was degassed before the adsorption runs by heating it with the CO_2 laser. The voltage was then adjusted in order to balance the particle at its correct position. Under no flow conditions, the voltage required to levitate the particle is recorded as V_{nf}. Prior to an adsorption experiment, the particle is balanced in dry nitrogen flow at a selected flowrate. The required voltage is recorded as V_o, i.e. the initial voltage (t=0) at zero adsorption. For the adsorption experiments, a dry nitrogen stream saturated with the organic vapor (toluene or carbon tetrachloride) was introduced into the EDB chamber. Saturation of the nitrogen stream took place by passing it through a bubbler holding the organic liquid at a constant temperature.

(a)

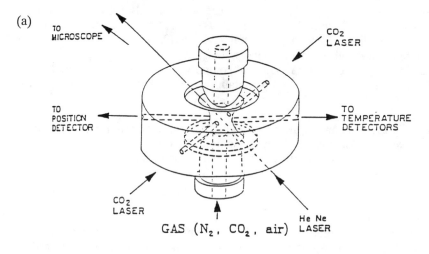

(b)

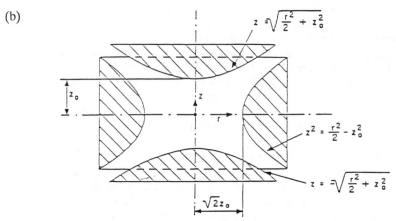

Figure 1. (a) Expanded view of the electrodynamic balance. (b) Cross-sectional view of the electrodes in the electrodynamic balance.

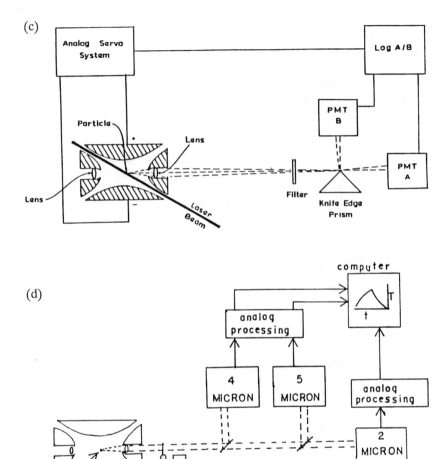

Figure 1 (continued). (c) Position control system for the electrodynamic balance. (d) Temperature measurement system.

The balancing voltage V gradually increases during adsorption to reach a final constant value recorded as V_a. Using the above definitions one can calculate (9) particle weight changes:

$$\frac{m_a - m_o}{m_o} = \frac{V_a - V_o}{V_{nf}} = X_a$$

where m_o is the mass of the dry solid particle, m_a is the sum of m_o plus the mass of organic adsorbed, and X_a is the maximum amount (in grams) of contaminant adsorbed per gram of solid particle.

In order to determine potentially different solid-hydrocarbon affinities and the effect of pore structure, three types of solids have been examined in this work with two types of organic adsorbates, namely toluene and carbon tetrachloride. A porous clay, montmorillonite, and two synthetic char particles, namely "Spherocarb" and "Carbopack" were used. The physical properties of these materials are given below (9):

Material	Diameter (μm)	Surface Area (m^2/g)	Bulk Density (g/cm^3)	Porosity
(S)	125-150	860 (1050*)	.63	.525
(C)	150-180	10.4	.95	.615
(M)	90-125	192	.65	.802

S: Spherocarb; C: Carbopack; M: Montmorillonite.
* determined by high-pressure CO_2 adsorption-desorption measurements in the EDB (22)

Three different saturator temperatures were used, -- 25, 0 and -21° C -- and thus the ratio P/P_o of the partial pressure of the organic vapor over its vapor pressure P_o at 25°C (which was always the temperature in the electrodynamic balance chamber) was varied. For toluene the P/P_o values were 1.0, 0.23 and 0.05 for saturator temperatures 25, 0 and -21°C, respectively, while for carbon tetrachloride the corresponding P/P_o values at the same saturator temperatures were 1.0, 0.29 and 0.08, respectively.

A series of adsorption and desorption experiments with different nitrogen flowrates were run to establish the importance of external mass transfer. Using flowrates in the range of 14-28 sccm no significant differences in the rates of adsorption-desorption were measured (8, 9). Thus, the recorded weight changes in the EDB were free of gas-film diffusion limitations.

Results

Figure 2 shows results for toluene adsorption on a Spherocarb particle of 170μm diameter for $P/P_o = 1$ and for a total volumetric flowrate maintained at 21.4 sccm. Figure 3 depicts the desorption profile of toluene at room temperature in dry nitrogen. An initial slow decrease in V/V_o is observed which probably corresponds to the slowly diminishing vapor pressure of toluene in the chamber during the initial displacement of the saturated gas from the chamber. This is followed first by a fast and then a slow decrease in V/V_o.

In order to compare the rates of adsorption and desorption in various experiments the data were normalized using as a variable the fractional attainment Y of the maximum adsorption, where $Y = (V - V_o)/(V_a - V_o)$.

Figure 4a compares the X_a values for Spherocarb, montmorillonite and Carbopack particles during adsorption of toluene at $P/P_o = 1$. Figure 4b shows a similar comparison during desorption of toluene from each type solid. In this case normalized data are shown. Each experimental point in these figures corresponds to an average value from at least three particles examined. Figures 5a and 5b show the data from carbon tetrachloride adsorption and desorption on the three types of particles again for $P/P_o = 1$. Each curve shown in Figures 4 and 5 can provide information about the characteristic time of either the adsorption or the desorption of the organic compound from the particle. The times required for the particles to adsorb 50% and 90% of the maximum amount that can be adsorbed under these conditions were defined as $\tau_{0.5a}$ and $\tau_{0.9a}$, respectively. Similarly, the times required to desorb 50% and 90% of the adsorbed compound were defined as $\tau_{0.5d}$ and $\tau_{0.9d}$, respectively. Table I contains X_a values for several experiments as well as average values of the above denoted characteristic times.

Experiments were also conducted with the saturator temperature kept at 0°C while the temperature in the electrodynamic balance chamber (and therefore the solid particle temperature as well) was kept at 25°C. In this case, the relative pressure for toluene was $P/P_o = 0.23$, while for CCl_4 the relative pressure was higher, $P/P_o = 0.29$. Results for adsorption and desorption of either compound on montmorillonite and Spherocarb particles are shown in Table II. Results with Carbopack are not shown because the adsorbed amount was very low. Table III contains similar results for montmorillonite and Spherocarb particles with the saturator temperature kept at -21°C. At this temperature the ratio P/P_o for C_7H_8 was 0.05 and for CCl_4 it was 0.08. Typical experimental results for all three saturator temperatures employed are shown in Figures 6a and 6b and Figures 7a and 7b, respectively, for the C_7H_8-Spherocarb and C_7H_8-montmorillonite pairs. Similar results are shown in Figures 8a and 8b and 9a and 9b, respectively, for the adsorption and desorption of CCl_4 on Spherocarb and montmorillonite.

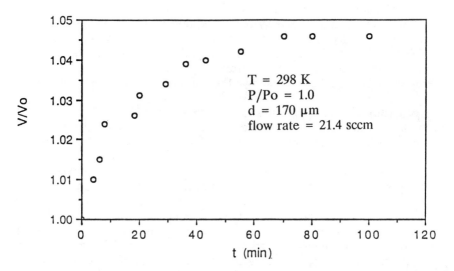

Figure 2.   Adsorption of C_7H_8 on Spherocarb particle.

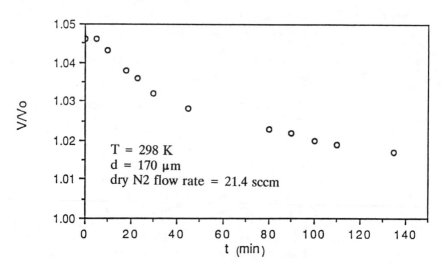

Figure 3.   Desorption of C_7H_8 from Spherocarb particle.

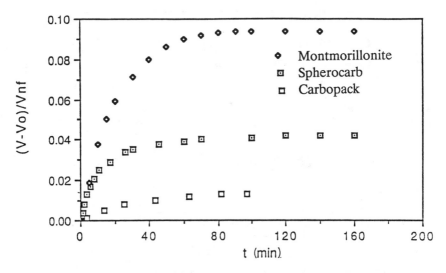

Figure 4a. Average adsorption curves of C_7H_8 for different materials; T = 298 K, P/P_o = 1.0.

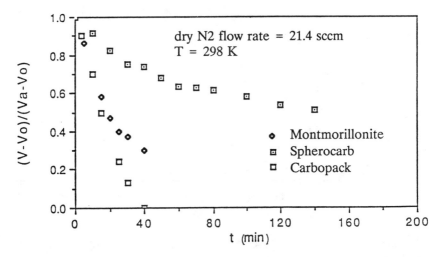

Figure 4b. Average desorption curves of C_7H_8 for different materials.

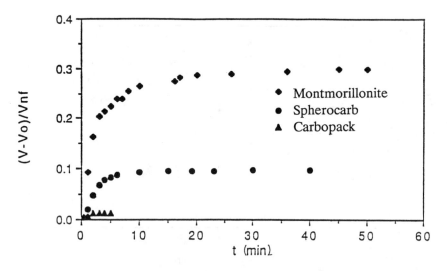

Figure 5a. Average adsorption curves of CCl$_4$ for different materials; T = 298 K, P/P$_o$ = 1.0.

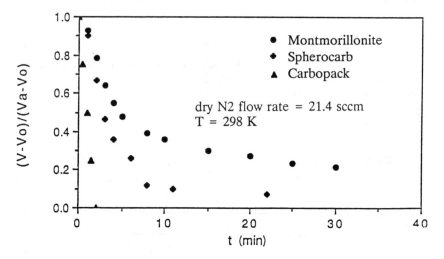

Figure 5b. Average desorption curves of CCl$_4$ for different materials.

Table I. EDB Data of Ambient Adsorption (at $P/P_o = 1$) And Desorption of Toluene and Carbon Tetrachloride on Various Solid Paricles

Solid-Adsorbate	Xa (g Liq/g Solid)	$\tau_{0.5a}$ (min)	$\tau_{0.9a}$ (min)	$\tau_{0.5d}$ (min)	$\tau_{0.9d}$ (min)
S-C$_7$H$_8$	0.029 0.061 0.048 0.045 0.025 0.045	8	45	140	>200
M-C$_7$H$_8$	0.130 0.070 0.100 0.075	13	47	17	>40
C-C$_7$H$_8$	0.009 0.017	16	62	15	32
S-CCl$_4$	0.109 0.092 0.107 0.087	2	7	5	>30
M-CCl$_4$	0.292 0.290 0.272 0.347 0.260 0.280 0.360	2	17	3	11
C-CCl$_4$	0.012 0.012 0.012	1	2	0.5	1.5-2.0

S: Spherocarb; M: Montmorillonite; C: Carbopack

TABLE II.   EDB Data of Ambient Adsorption (at intermediate P/Po values) and Desorption on Single Solid Particles

P/Po	Solid-Adsorbate	Xa (g liq/g solid)	$\tau_{0.5a}$ (min)	$\tau_{0.9a}$ (min)	$\tau_{0.5d}$ (min)	$\tau_{0.9d}$ (min)
0.23	S-C$_7$H$_8$	0.021 0.018 0.013 0.009 0.017	10	35	25	50
	M-C$_7$H$_8$	0.060 0.050 0.052 0.050	16	45	14	25
0.29	S-CCl$_4$	0.045 0.054 0.054 0.053 0.066	1	3	1	4
	M-CCl$_4$	0.188 0.198 0.178 0.172 0.164	1	5	1	5

S: Spherocarb; M: Montmorillonite; C: Carbopack

TABLE III. EDB Data of Adsorption (at low P/Po values) and Desorption of Toluene and Carbon Tetrachloride on Single Solid Particles

P/Po	Solid-Adsorbate	Xa (g liq/g solid)	$\tau_{0.5a}$ (min)	$\tau_{0.9a}$ (min)	$\tau_{0.5d}$ (min)	$\tau_{0.9d}$ (min)
0.05	S-C_7H_8	0.008	4	10	3	8
		0.011				
		0.012				
		0.004				
		0.016				
		0.014				
	M-C_7H_8	0.015	5	20	4	10
		0.021				
		0.025				
		0.026				
0.08	S-CCl_4	0.041	2	3	1	4
		0.026				
		0.048				
		0.061				
	M-CCl_4	0.086	2	4	1	5
		0.085				
		0.086				

S: Spherocarb; M: Montmorillonite; C: Carbopack

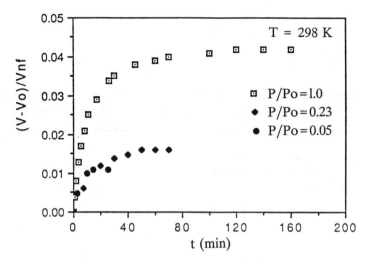

Figure 6a.   Spherocarb-C_7H_8 adsorption curves for different relative pressures.

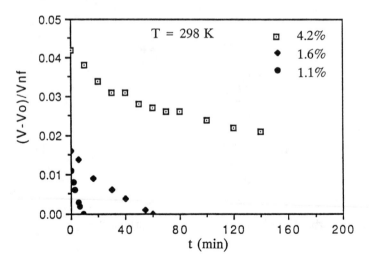

Figure 6b. Spherocarb-C_7H_8 desorption curves for the different adsorbed amounts of C_7H_8 shown in Figure 6a.

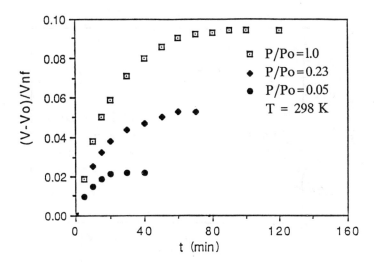

Figure 7a. Montmorillonite-C_7H_8 adsorption curves for different relative pressures.

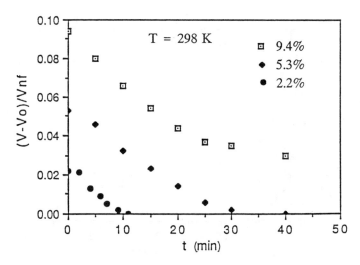

Figure 7b. Montmorillonite-C_7H_8 desorption curves for the different adsorbed amounts of C_7H_8 shown in Figure 7a.

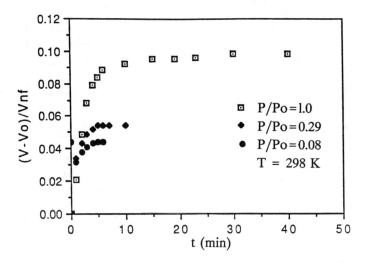

Figure 8a. Spherocarb-CCl₄ adsorption curves for different relative pressures.

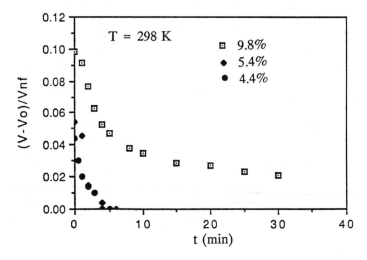

Figure 8b. Spherocarb-CCl₄ desorption curves for the different adsorbed amounts of CCl₄ shown in Figure 8a.

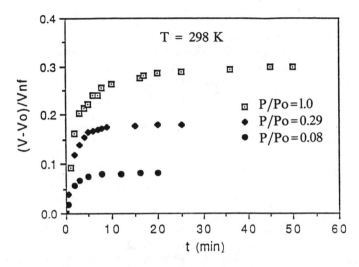

Figure 9a. Montmorillonite-CCl₄ adsorption curves for different relative pressures.

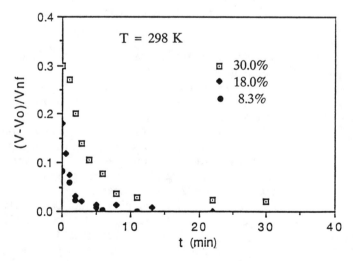

Figure 9b.   Montmorillonite-CCl₄ desorption curves for the different adsorbed amounts of CCl₄ shown in Figure 9a.

Discussion

There are several observations to be made in discussing the experimental results obtained in this work: (a) time scales were long for both adsorption and desorption, indicating hindered diffusion of the contaminant in the solid (Table I); (b) time scales were longer for toluene than for carbon tetrachloride; (c) the amount of contaminant adsorbed increased with the relative pressure, P/P_o (Tables I-III); (d) the particle-to-particle variability was greatest for Spherocarb (as indicated by variation in the values of X_a, Tables I-III); and (e) the desorption times for Spherocarb were significantly longer than the corresponding adsorption times at high P/P_o, but approached symmetry at low P/P_o (Tables I-III).

The amounts of toluene and carbon tetrachloride adsorbed on each type of solid particle at $P/P_o = 1.0$, corresponded to approximately monolayer coverage for Spherocarb, and slightly above monolayer coverage for the montmorillonite and Carbopack particles (9). Carbopack adsorbed the least amount of either organic compound, in agreement with its low surface area and large pores. The higher (approximately twofold) X_a values for CCl_4 adsorption on all solids are commensurate with the density difference between CCl_4 (1.595 g/cm^3) and C_7H_8 (0.866 g/cm^3). Montmorillonite adsorbed the highest amount of both C_7H_8 and CCl_4, even though it has a much lower surface area than Spherocarb (192 vs~1050 m^2/g). For either solid, the characteristic times of adsorption and desorption were much higher (min) than what would correspond to external film diffusion (10^{-3}-10^{-2} sec) or pore diffusion even of the Knudsen type (0.1-1 sec).

These results point to hindered diffusion in the smallest pores of the materials and capillary condensation in pores with radius in the range > 10 Å and < 250 Å as is well established in the literature (23). The pore size distribution of Spherocarb particularly supports this hypothesis. In earlier work carried out in this laboratory (24), Spherocarb particles were found to contain large vesicular pores (of radius r > 250 Å) in their interior. Large variability in macroporosity (indicated by a twofold density variation) was found, which can explain the particle-to-particle variability in adsorption reported here. Pore volume and pore surface distributions for Spherocarb (25) show the volume of micropores (r < 10 Å) and mesopores (10 $< r < 250$ Å) together not to exceed 33% of the total, while 90% of the surface area lies in micropores and 10% in mesopores. The low amounts and large time scales associated with contaminant adsorption on Spherocarb (Table I, Figures 2, 4a, 5a) can then be explained by condensation in the micropores interconnecting and leading to the vesicular macropores.

Similar agreements hold for the observed diffusion transients involving slow migration of liquid molecules in the small pores. Montmorillonite is expected to have a much larger voidage fraction in the mesopore regime than

Spherocarb and a smaller fraction in the macropore, which can explain its higher 'adsorption' due to capillary condensation.

In summary, the EDB was shown to be a useful tool for studying the fundamental processes of contamination and decontamination of porous soils. The ability of the EDB to study small weight changes on a single particle in the absence of external diffusion limitation and interparticle effects has been proven. Significant differences among various solids and contaminant compounds have been identified in this "first-generation" study. It is clear that a systematic study of the adsorption-desorption characteristics and their dependence on the physical and chemical properties of the soil particles is necessary. Work toward this goal is currently underway.

Acknowledgements

We gratefully acknowledge the Center for Environmental Management of Tufts University for support of this research under grant #CR-813481-02-0. The apparatus used at MIT has been developed with support from the Exxon Corporation.

Literature Cited

1. Oppelt, E.T. "Incineration of Hazardous Waste" *J. Air Pollut. Control Assoc.* 1987, *37*, p.558.

2. Oppelt, E.T. "Hazardous Waste Destruction" *Environ. Sci. Technol.* 1986, *20*, p.312.

3. Lighty, J.S., Britt, R.M., Pershing, D.W., Owens, W.D. and Cundy, V.A. "Rotary Kiln Incineration II. Laboratory-Scale Desorption and Kiln-Simulator Studies-Solids" *J. Air Pollut. Control Assoc.* 1989, *39*, p.187.

4. "New Jersey Hazardous Waste Facilities Plan" New Jersey Waste Facilities Siting Commission, published by the Environmental Resources Management, Inc., p.170, Trenton, N.J., 1985.

5. Linak, W.P., McSorley, J.A., Wendt, J.O. L. and Dunn, J.E. "On the Occurence of Transient Puffs in a Rotary-Kiln Incinerator Simulator: II. Contained Liquid Wastes on Sorbent" *J. Air Pollut. Control Assoc.* 1987, *37*, p.934.

6. Linak, W.P., McSorley, J.A., Wendt, J.O.L. and Dunn, J.E. "Hazardous Waste Management - On the Occurence of Transient Puffs in a Rotary Kiln Incinerator I. Prototype Solid Plastic Wastes" *J. Air Pollut. Control Assoc.* 1987, *37*, p.54.

7.   Wendt, J.D.L., Linak, W.P. and McSorley, J.A. Paper presented at the AFRC Int'l Symposium on Incineration of Hazardous Municipal and Other Wastes, Palm Springs, CA, Nov. 2-4, 1987.

8.   Tognotti, L., Flytzani-Stephanopoulos, M., Sarofim, A.F., Kopsinis, H. and Stoukides, M. "Study of Adsorption-Desorption of Contaminants on Soil Particles Using The Electrodynamic Thermogravimetric Analyzer" *Environ. Sci. Technol.* 1990, in print.

9.   Kopsinis, H. "Study of Adsorption-Desorption of Contaminants on Soil Particles Using the Electrodynamic Thermogravimetric Analyzer" MS Thesis, Tufts University, Medford, MA, 1990.

10. Dudek, D.R. "Single Particle, High Temperature, Gas-Solid Reactions in an Electrodynamic Balance" PhD Thesis, Massachusetts Institute of Technology, Cambridge, MA, 1988.

11. Bar-Ziv, E., Jones, D.B., Spjut, R.E., Dudek, D.R., Sarofim, A.F. and Longwell, J.P. "Measurement of Combustion Kinetics of a Single Char Particle in an Electrodynamic Thermogravimetric Analyzer" *Comb. and Flame* 1989, *75*, p.81.

12. Monazam, E.R., Maloney, D.J. and Lawson, L.O. "Measurements of Heat Capacities, Temperatures and Absorptivities of Single Particles in an Electrodynamic Balance" *Rev. Sci. Instrum.* 1989, *60*, p.3460.

13. Greene, W.M., Spjut, E.R., Bar-Ziv, E., Sarofim, A.F. and Longwell, J.P. "Photophoresis of Irradiated Spheres: Absorption Centers" *J. Opt. Soc. Amer.* 1985, *82*, p.998.

14. Greene, W.M., Spjut, E.R., Bar-Ziv, E., Longwell, J.P., and Sarofim, A.F. "Photophoresis of Irradiated Spheres: Evaluation of the Complex Index of Refraction" *Langmuir* 1985, *1*, p.361.

15. Spjut, R.E., Sarofim, A.F. and Longwell, J.P. "Laser Heating and Particle Temperature Measurement in an Electrodynamic Balance" *Langmuir* 1985, *1*, p.355.

16. Spjut, R.E., Bar-Ziv, E., Sarofim, A.F. and Longwell, J.P. "Electrodynamic Thermogravimetric Analyzer" *Rev. Sci. Instrum.* 1986, *57*, p.1604.

17. Maloney, D.J., Flashing, G.E., Lawson, L.O. and Spann, J.F. "An Automated Imaging and Control System for the Continuous

Determination of Size and Relative Mass of Single Compositionally Dynamic Droplets" *Rev. Sci. Instrum.* 1989, *60*, p.450.

18. Cohen, M.D., Flagan, R.C. and Seinfeld, J.H. "Studies of Concentrated Electrolyte Solutions Using the Electrodynamic Balance I. Water Activities for Single-Electrolyte Solutions" *J. Phys. Chem.* 1987, *91*, p.4568.

19. Wuerker, R.F., Shelton, H. and Langmuir., R.V. "Electrodynamic Containment of Charged Particles" *J. Appl. Phys.* 1959, *30*, p.342.

20. Davis, E.J. and Ray, A.K. "Single Aerosol Particle Size and Mass Measurements Using an Electrodynamic Balance" *J. Coll. Interf. Sci.* 1980, *75*, p.566.

21. Arnold, S., Amani, Y. and Orenstein, A. "Photophoretic Spectrometer" *Rev. Sci. Instrum.* 1980, *51*, p.1202.

22. Bar-Ziv, E., Longwell, J.P. and Sarofim, A.F. "Determination of the Surface Area of Single Particles from High Pressure CO_2 Adsorption-Desorption Measurements in an Electrodynamic Chamber" accepted for publication.

23. Gregg, S.J. and Sing, K.S.W. "Adsorption, Surface Area and Porosity;" Academic Press: London, 1982; 2nd Edition, p.113.

24. D'Amore, M., Dudek, R.D., Sarofim, A.F. and Longwell, J.P. "Apparent Particle Density of a Fine Particle" *Powder Technology* 1988, *56*, p.129.

25. D'Amore, M. personal communication.

RECEIVED April 5, 1991

Chapter 4

Calcination of Calcium Hydroxide Sorbent in the Presence of SO$_2$ and Its Effect on Reactivity

Wojciech Jozewicz[1] and David A. Kirchgessner[2]

[1]Environmental Systems Division, Acurex Corporation, P.O. Box 13109, Research Triangle Park, NC 27709
[2]Air and Energy Engineering Research Laboratory, U.S. Environmental Protection Agency, Research Triangle Park, NC 27711

When calcium hydroxide [Ca(OH)$_2$] is calcined in an isothermal flow reactor with 300 ppm or less sulfur dioxide (SO$_2$) present, the structure of the sorbent is characterized by retention of higher pore volumes and surface areas than when calcined in the absence of SO$_2$. Upon reinjection into the isothermal flow reactor (1,000 °C, 3,000 ppm SO$_2$, 0.8 s) a higher level of reactivity with SO$_2$ is observed for the SO$_2$-modified calcine than for the unmodified sorbent. Structural evidence suggests that a slower initial rate of sintering in the modified calcine accounts for its enhanced reactivity. It is postulated that deposition of small amounts of calcium sulfate (CaSO$_4$) on pore faces and along crystal boundaries retards the solid state diffusion process responsible for sintering. The effects on sintering and reactivity of various combinations of residence time (0.6-2.0 s) and SO$_2$ concentration (10-50 ppm) during precalcination have been tested.

The limestone injection multistage burner (LIMB) process developed by EPA is designed to control sulfur dioxide (SO$_2$) emissions from coal-fired boilers through the injection of dry calcium-based sorbents into the high temperature zone of the furnace (about 1,230 °C). Research has shown that calcium hydroxide [Ca(OH)$_2$] is more reactive with SO$_2$ in this process than are other commercially available sorbents (1,2,3,4). Once this fact had been established, considerable experimental effort was devoted to understanding the fundamental kinetic behavior of Ca(OH)$_2$ in high temperature injection processes, so that SO$_2$ capture could be predicted.

Calcination of both calcium carbonate (CaCO$_3$) and Ca(OH)$_2$ has been studied extensively at the laboratory scale and modeled (5,6,7). A rate constant appropriate for the calcination of Ca(OH)$_2$ has been developed (6,8). Sintering of the resulting calcium oxide (CaO) is not as well understood. Sintering is important because the loss of specific surface area and porosity through this solid state diffusion process

0097–6156/91/0468–0050$06.00/0

competes directly with the process of sulfation. An empirical sintering model has been proposed (5), however, and has been shown to effectively correlate experimental data (9).

Of particular interest in the sintering process is the effect of combustion gas composition, since this is the environment of practical concern. Wide agreement has been reached on the catalytic effect of CO_2 on sintering of CaO. This effect has been documented (8,10,11,12,13,14) and is incorporated in the rate equation (5). The catalytic effect of water on sintering has been shown (8,12) and can also be included in the equation (5). Presently, the equation (5) does not account for the effect of SO_2 on sintering.

There appears to be a discrepancy in the available data on the effect of SO_2 on sintering. The ability of calcium sulfate ($CaSO_4$) product to promote sintering has been noted (15). SO_2 has been suggested to promote sintering as well (16). It has also been observed, however, that a small amount of SO_2 in the calcination gas yields a CaO that is more reactive with SO_2 upon furnace injection than a CaO calcined in the absence of SO_2 (17,18). Before the effect of SO_2 on sintering can be incorporated into the rate equation the effect must first be verified, and then quantified. The purpose of this paper is to verify the inhibition of sintering by SO_2 as noted earlier (18). A secondary purpose is to determine the practical application of precalcining sorbents in the presence of small amounts of SO_2 prior to furnace injection as suggested before (17).

Experimental System

Hydration. The parent material for the $Ca(OH)_2$ used in this study is a commercial CaO produced by the Longview Lime Division of Dravo Lime Company, Saginaw, Alabama. Hydroxides are produced in a steam-jacketed Ross ribbon blender loaded with 6.8 kg pulverized CaO and allowed to reach a temperature of 30-50 °C. A total of 3.7 kg water is than added at a constant rate for 40 min. The blender is run for an additional 20 min while the heat of reaction drives off excess water, leaving a dry $Ca(OH)_2$ product.

Reactor. The electrically heated isothermal flow reactor into which the hydrates are injected was used for structural and sulfation studies (Figure 1). The reaction chamber consists of two concentric, 3.35 m long quartz tubes heated by three Lindberg tube furnaces. The inner tube has a 15 mm ID, and the outer tube has a 50.8 mm OD. Premixed gases enter the reactor at the bottom of the annulus between the inner and outer tubes and are heated as they move upward. At the top of the reactor they enter the inner tube with the sorbent and move downward through the reaction zone. Gases and solids exit the reactor through an air-cooled heat exchanger that quenches the reactant stream to approximately 280 °C. Solids are captured in a cyclone followed by a glass-fiber filter. Sorbent is fed from a fluidized bed feed tube through 1.9 mm ID syringe tubing. For this study the reactor is operated at nominal residence times of 0.6-2.0 s. Residence time is adjusted by varying gas flow, at standard temperature and pressure [25 °C, 760 mm Hg (101 kPa)] from 25.54 to 5.86 L/min. Gas composition is 95 vol percent N_2, 5 vol percent O_2, and from 10 to 3,000 ppm SO_2. Although they would be of interest, SO_2 concentrations

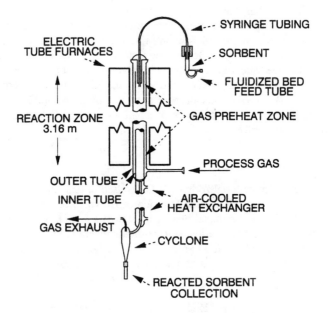

Figure 1. Isothermal flow reactor.

below 10 ppm cannot be well controlled. Due to variability in the reactor system, selected samples are replicated until standard deviation falls to less than 10 percent of the mean.

Analysis. Calcium utilization or conversion of reactor products is calculated after analyses for calcium by atomic absorption spectrophotometry, and for sulfate by ion chromatography. Calcium utilization analyses have a precision of ±5 percent. No sulfur species other than sulfate have been detected. Specific surface area and porosity are measured by N_2 adsorption/desorption in a Micromeritics Digisorb 2600 autoanalyzer. Machine error is less than ±1 percent.

Results

In order to isolate the effect of SO_2 on sintering from the effects of time and temperature on sintering, $Ca(OH)_2$ was initially calcined in the presence of SO_2 concentrations varying from 10-5,000 ppm, while holding time and temperature constant at 0.8 s and 1,000 °C, respectively (Figure 2). In the absence of SO_2, $Ca(OH)_2$ will calcine to CaO having a maximum surface area of about 76 m^2/g (*19,20*), and then sinter to successively lower surface areas. At 1,000 °C and 0.8 s, the CaO has sintered to a surface area of 24 m^2/g. Since $CaSO_4$ has a volume approximately three times that of the CaO reactant, one would expect any degree of sulfation to immediately reduce surface area below 24 m^2/g through the processes of pore filling and pore occlusion. It can be seen (Figure 2), however, that calcination of $Ca(OH)_2$ in the presence of SO_2 concentrations of 10-300 ppm results in higher surface areas than calcination in the absence of SO_2. The calcium conversions accompanying these SO_2 concentrations are 3.4-11.0 percent and must cause fine pore filling and blocking. For surface areas greater than 24 m^2/g to occur, therefore, the resistance to sintering which accompanies these conversions must be significant. An SO_2 concentration of 10 ppm causes the greatest amount of surface area to be retained, with successively more surface area being lost at higher SO_2 concentrations and calcium conversions. Above 300 ppm SO_2, retained surface area falls below 24 m^2/g due to buildup of the $CaSO_4$ product layer. The behavior of porosity (Figure 3) under the same conditions as above mirrors the trend in surface area reduction throughout the range of SO_2 concentrations and calcium utilizations as expected.

   In the next step, the effect on structure of calcination or contact time in the presence of SO_2, were determined. In the absence of SO_2 (Figure 4), thermal sintering reduces surface area over time as one would expect. In the presence of 10 and 50 ppm SO_2, thermal sintering reduces surface area at approximately the same rate as in the absence of SO_2 over the 0.6-2.0 s time frame investigated. In both cases, however, the surface areas retained in the presence of SO_2 are higher than in the absence of SO_2. This implies that, during the 0-0.6 s time period, the rate of sintering in the absence of SO_2 is greater. At all points, the surface area retained in the presence of 10 ppm SO_2 is greater than that retained in the presence of 50 ppm SO_2. The porosity behavior (Figure 5) reflects the trends in the surface area data.

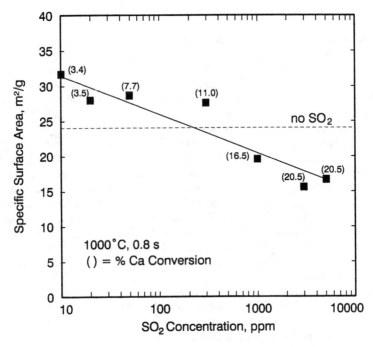

Figure 2. Surface area retained after calcination with and without SO_2.

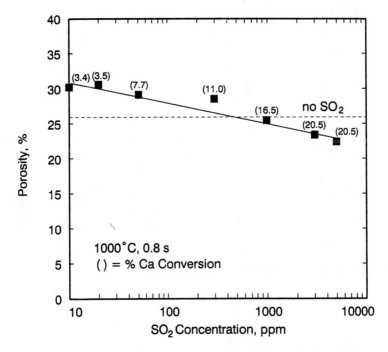

Figure 3. Porosity retained after calcination with and without SO₂.

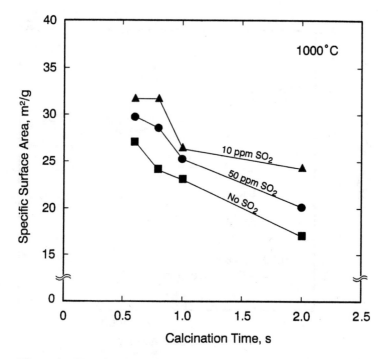

Figure 4. Variation in surface area with SO_2 levels and time during precalcination.

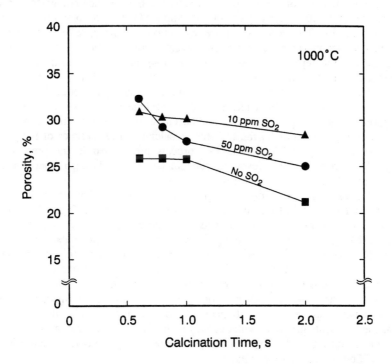

Figure 5. Variation in porosity with SO_2 levels and time during precalcination.

The loss of surface area through sintering is known to be a significant factor in reducing the reactivity of $Ca(OH)_2$ with SO_2 during furnace injection. Since the presence of small amounts of SO_2 during calcination appears to retard the initial rate of sintering, it has been suggested that this behavior may have a practical application in sorbent injection processes. It was suggested that precalcination of sorbents in the presence of small amounts of SO_2 will produce precalcines that are highly reactive with SO_2 during reinjection into a boiler (17). This would be of interest because high surface area precalcines have been shown to be less reactive with SO_2 during furnace injection than their uncalcined counterparts (21).

To test this hypothesis, sorbents were precalcined with and without 10 ppm SO_2 for times of 0.6 to 2.0 s. These sorbents were then reinjected into the flow reactor at 1,000 °C for 0.8 s with 3,000 ppm SO_2 to determine differences in calcium utilization (Figure 6). The lowermost curve (Figure 6) reflects the level of calcium conversion achieved at various precalcination times in the presence of 10 ppm SO_2. The uppermost curve (Figure 6) shows the levels of additional calcium conversions achieved when these same sorbents were reinjected at 1,000 °C for 0.8 s with 3,000 ppm SO_2. These values are reported separately because they would reflect the only SO_2 capture of interest if the precalcined sorbents were to be used in a furnace injection process. The middle curve shows the calcium conversion obtained during reinjection by the sorbents precalcined in the absence of SO_2. It is clear that, for all precalcination times, sorbents precalcined in the presence of 10 ppm SO_2 produce higher levels of calcium conversion during reinjection than sorbents precalcined in the absence of SO_2.

The time-resolved reactivity during reinjection of sorbents precalcined with and without 10 ppm SO_2 for a constant time 1.0 s was measured (Figure 7). Again, the lowermost curve (Figure 7) reflects the constant calcium conversion value achieved during precalcination in the presence of SO_2. The uppermost curve shows the calcium conversion obtained at various residence times during reinjection at 1,000 °C with 3,000 ppm SO_2. These values are, again, consistently higher than calcium conversions demonstrated by sorbents precalcined in the absence of SO_2. The shape of the curve for SO_2-modified precalcined sorbents reinjected into the reactor suggests a significant influence of pore diffusion, while the curve for precalcined sorbents generated without SO_2 is indicative of product layer diffusion. This would imply that sintering has not yet removed the fine pore structure in the SO_2-modified precalcined sorbents.

The result of increasing the SO_2 available during precalcination to 50 ppm was examined next (Figure 8). In this case sorbents are precalcines with and without 50 ppm SO_2 at 1,000 °C for times of 0.6 to 2.0 s. As before, these sorbents are then reinjected into the flow reactor at 1,000 °C for 0.8 s with 3,000 ppm SO_2 to determine differences in calcium conversion. In contrast to the sorbents modified with 10 ppm SO_2, sorbents modified with 50 ppm SO_2 are less reactive during reinjection than sorbents precalcined in the absence of SO_2. Data prsented above (Figures 2 through 5) suggest that sorbents modified with 50 ppm SO_2 have a lower propensity to sinter than those precalcined without SO_2. It must be assumed, therefore, that the 7.7 percent calcium conversion achieved during precalcination with 50 ppm SO_2 is sufficient to block much of the sulfation during reinjection. Sorbents

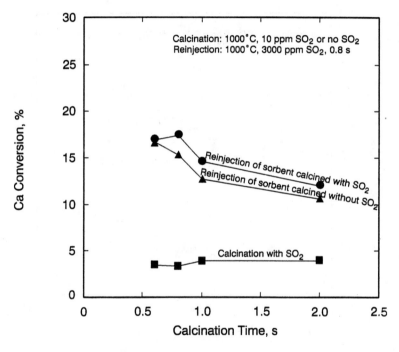

Figure 6. Calcium conversion achieved at a constant reinjection time by sorbents precalcined with and without 10 ppm SO_2.

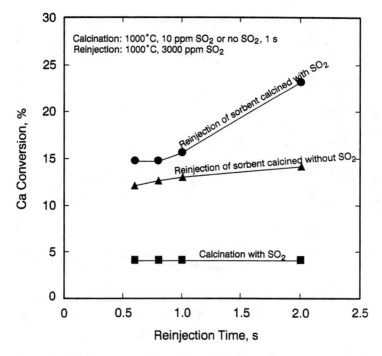

Figure 7. Calcium conversion achieved during reinjection by sorbents precalcined at a constant time with and without 10 ppm SO_2.

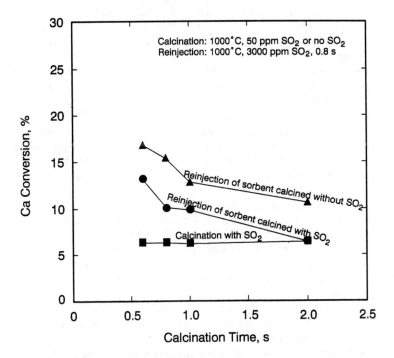

Figure 8. Calcium conversion achieved at a constant reinjection time by sorbents precalcined with and without 50 ppm SO_2.

precalcined with 10 ppm SO_2 appear to produce better calcium conversions during reinjection than sorbents precalcined without SO_2.

Discussion

Contrary to the observations by other workers (5,16), the results presented in this paper suggest that the presence of 300 ppm or less SO_2 during calcination significantly reduces the subsequent propensity of the sorbent to sinter. While it would be desirable to reconcile these apparently conflicting observations, differences in the experimental frameworks make this difficult. In one study (15) sulfate was introduced into reagent grade $CaCO_3$ by washing it in dilute sulfuric acid to produce partial dissolution, and subsequently drying it. These modified sorbents were then calcined in an electric furnace at 800 to 1,000 °C for 2 to 20 h. There are no similarities between experiments of the above work (15) and this work, and reasons for the differences in conclusions cannot be determined.

In another work investigating the effect of SO_2 (16), $CaCO_3$ was calcined at 980 °C for 0.5 with and without 180 ppm SO_2 present. The sorbent calcined without SO_2 had a mean pore diameter of 64 Å and virtually no pore volume above 100 Å. The sorbent calcined with SO_2 had a mean pore diameter of 155 Å and 80 percent of its pore volume above 100 Å. Using the distributed pore model, the conclusion was made (16) that the buildup of product layer would not produce the observed changes, and that they must be the result of enhanced sintering in the presence of SO_2. Lacking pore size distribution data from this work, the two studies cannot be directly compared.

The results reported here are fully consistent with those reported earlier (18) in which higher porosities were observed for sorbents calcined in the presence of SO_2 than for sorbents calcined in the absence of SO_2. The hypothesis set forth in that work suggests that the presence of $CaSO_4$ product on pore faces interferes with the exchange of migrating reactant species and lattice vacancies at the pore/sorbent interface. This in turn would retard the solid state diffusion process of sintering. This hypothesis still appears to be reasonable.

The results of this work suggest that precalcining sorbents in the presence of small amounts of SO_2 does not have a practical application in high temperature sorbent injection processes. Although it has been demonstrated that sorbents precalcined with 10 ppm SO_2 are more reactive during reinjection than sorbents calcined in the absence of SO_2, neither the modified nor the unmodified precalcined sorbent is as reactive with SO_2 as is the raw $Ca(OH)_2$. Raw $Ca(OH)_2$ sorbents typically have produced calcium conversions greater than 20 percent when injected at 1,000 °C for 1.0 s with 3,000 ppm SO_2.

Conclusions

Based on structural characteristics, $Ca(OH)_2$ sorbents precalcined to CaO in the presence of 300 ppm or less SO_2 retain greater amounts of surface area and porosity than sorbents precalcined in the absence of SO_2. It is believed that this behavior results from the ability of small amounts of $CaSO_4$ product on pore faces to interfere in the solid state diffusion process of sintering. After quantification, it would seem reasonable to include the effect of SO_2 on sintering in the empirical equation (5).

Sorbents precalcined with 10 ppm SO_2 are more reactive with SO_2 during reinjection into the reactor than sorbents precalcined in the absence of SO_2. Increasing SO_2 concentration during precalcination to 50 ppm reverses these results. Since no precalcined sorbent has been shown to be more reactive with SO_2 than is raw $Ca(OH)_2$, this process is not believed to have a practical application.

Acknowledgments

This research was supported by the U.S. Environmental Protection Agency's Air and Energy Engineering Research Laboratory (EPA Contract 68-02-4701). The authors wish to recognize the assistance of Monsie Gillis, Kevin Bruce, and Laura Beach of the Acurex Corporation, and George Gillis of EPA/AEERL.

Literature Cited

1. Overmoe, B.J.; Chen, S.L.; Ho, L.; Seeker, W.R.; Heap, M.P.; Pershing, D.W.; "Boiler Simulator Studies on Sorbent Utilization for SO_2 Control"; Proc. 1st Joint Symp. Dry SO_2 and Simul. SO_2/NO_x Control Technol. 1; EPA-600/9-85-020a (NTIS PB85-232353), 1985; 15-1-15-17.
2. Beittel, R.; Gooch, J.P.; Dismukes, E.B.; Muzio, L.J.; "Studies of Sorbent Calcination and SO_2-Sorbent Reactions in a Pilot-Scale Furnace"; ibid., 1985; 16-1–16-33.
3. Bortz, S.; Flament, P.; "Recent IFRF Fundamental and Pilot-Scale Studies on the Direct Sorbent Injection Process"; ibid., 1985; 17-1–17-22.
4. Slaughter, D.M.; Silcox, G.D.; Lemieux, G.D.; Newton, G.H.; Pershing, D.W.; "Bench-Scale Evaluation of Sulfur Sorbent Reactions"; ibid., 1985; 11-1–11-5.
5. Silcox, G.D.; Kramlich, J.C.; Pershing, D.W.; "A Mathematical Model for the Flash Calcination of Dispersed $CaCO_3$ and $Ca(OH)_2$ Particles"; *Ind. Eng. Chem. Res.* **1989**, *28*, 155-160.
6. Milne, C.R.; Pershing, D.W.; Kirchgessner, D.A.; "An Experimental and Theoretical Study of the Fundamentals of the SO_2/Lime Reaction at High Temperatures"; Proc. 1st Comb. FGD and Dry SO_2 Control Symp. 3; EPA-600/9-89-036a (NTIS PB89-172167), 1989; 7-1–7-20.
7. Milne, C.R.; Silcox, G.D.; Pershing, D.W.; Kirchgessner, D.A.; "Calcination and Sintering Models for Application to High Temperature, Short-Time Sulfation of Calcium-Based Sorbents"; *Ind. Eng. Chem. Res.* **1990**, *29*, 139-149.
8. Mai, M.C.; "Analysis of Simultaneous Calcination, Sintering, and Sulfation of Calcium Hydroxide Under Furnace Sorbent Injection Conditions"; Ph.D. Thesis, University of Texas, Austin, TX, 1987.
9. Borgwardt, R.H.; Bruce, K.R.; Blake, J.; "EPA Experimental Studies of the Mechanisms of Sulfur Capture by Limestone"; Proc. 1st Joint Symp. Dry SO_2 and Simul. SO_2/NO_x Control Technol. 1; EPA-600/9-85-020a (NTIS PB85-232353), 1985; 6-1–6-20.

10. Anderson, P.J.; Horlock, R.F.; Avery, R.G.; "Some Effects of Water Vapor During the Preparation and Calcination of Oxide Powders"; *Proc. Brit. Ceram. Soc.* **1965**, *3*, 33-42.
11. Ulerich, N.H.; O'Neill, E.P.; Keairns, D.L.; A Thermogravimetric Study of the Effect of Pore Volume—Pore Size Distribution on the Sulfation of Calcined Limestone"; *Thermochim. Acta* **1978**, *26*, 269-275.
12. Borgwardt, R.H.; "Calcium Oxide Sintering in Atmospheres Containing Water and Carbon Dioxide"; *Ind. Eng. Chem. Res.* **1989**, *28*, 493-500.
13. Borgwardt, R.H.; Bruce, K.R.; "Effect of Specific Surface Area on the Reactivity of CaO with SO_2"; *AIChE J.* **1986**, *32*, 239-246.
14. Borgwardt, R.H.; Roache, N.F.; Bruce, K.R.; "Method for Variation of Grain Size in Studies of Gas-Solid Reactions Involving CaO"; *Ind. Eng. Chem. Fund.* **1986**, *25*, 165-169.
15. Glasson, D.R.; "Reactivity of Lime and Related Oxides: XVI. Sintering of Lime"; *J. Appl. Chem.* **1967**, *17*, 91-96.
16. Newton, G.W.; "Sulfation of Limestone in a Combustion Environment"; Ph.D Thesis; University of Utah, Salt Lake City, UT, 1986.
17. Lange, H.B.; "Experimental Study of Gas Use to Improve the Limestone Injection Process for SO_2 Control on Utility Boilers"; Gas Res. Instit. Topical Report, GRI-87/0099, 1987.
18. Kirchgessner, D.A.; Jozewicz, W.; "Enhancement of Reactivity in Surfactant-Modified Sorbent for SO_2 Control"; *Ind. Eng. Chem. Res.* **1989**, *28*, 413-418.
19. Borgwardt, R.H.; "Sintering of Nascent Calcium Oxide"; *Chem. Eng. Sci.* **1989**, *44*, 53-60.
20. Kirchgessner, D.A.; Jozewicz, W.; "Structural Changes in Surfactant-Modified Sorbents During Furnace Injection"; *AIChE J.* **1989**, *35*, 500-506.
21. Bortz, S.J.; Roman, V.P.; Yang, R.J.; Flament, P.; Offen, G.R.; "Precalcination and its Effect on Sorbent Utilization During Upper Furnace Injection"; Proc. 1986 Joint Symp. on Dry SO_2 and Simul. SO_2/NO_x Control Technol. 1, EPA-600/9-86-029a (NTIS PB87-120465), 1986; 5-1-5-20.

This paper has been reviewed in accordance with the U.S. Environmental Protection Agency's peer and administrative review policies and approved for presentation and publication.

RECEIVED April 5, 1991

WATER MANAGEMENT

Chapter 5

Strategies for Photochemical Treatment of Wastewaters

Richard A. Larson, Karen A. Marley, and Martina B. Schlauch

Institute for Environmental Studies, Environmental Research Laboratory, University of Illinois, Urbana, IL 61801

Interest in the use of light energy to decompose pollutants in water has been rather sporadic, but attention has increased rapidly in recent years. Light can be understood as a chemical reagent that can be used to bring about a wide variety of selective transformations, some of which are virtually impossible to achieve using conventional reactants. A further advantage is that light is even free of charge when it comes from the sun. Although not all organic pollutants absorb light, or react once they do, many are susceptible to photodecomposition by one pathway or another. Knowledge of the chemical mechanisms of photochemical reactions is advantageous for the design of effective photochemical treatment systems for polluted waters.

Sunlight and Artificial Light

In addition to visible light and infrared (IR) radiation (heat), the sun emits radiation in the 290-400 nm (ultraviolet; UV) region that reaches the earth's surface. Approximately four percent of the total energy contained in sunlight occurs in the UV region. This radiation has a high potential for inducing chemical reactions due to its elevated energy content relative to visible and IR radiation. Of course, as the first law of photochemistry states, light must be absorbed by a system in order for chemical reactions to have a chance to occur. As a first approximation, then, molecules must have at least a little absorption at 290 nm in order to be affected by sunlight. Complications of this issue are addressed later.

Distinctions are sometimes made between the long-wavelength "UV-A" (400–320 nm) and the more energetic, short-wavelength "UV-B" (320–290 nm) radiation that is more strongly absorbed by many pollutants and biomolecules. In both cases, however, the fundamental mechanisms of photochemical damage are similar, although different receptor molecules (chromophores) are involved. UV radiation, because of its potent chemical energy content, has a

0097–6156/91/0468–0066$06.00/0

potential for causing direct damage to biochemically important molecules that absorb it and also to destroy dissolved pollutants in water under some conditions. The intensity of solar UV irradiance at the earth's surface varies greatly with season, time of day, latitude, ozone layer thickness, altitude, and cloud cover. The characteristics of sunlight have been reviewed by Finlayson-Pitts and Pitts (*17*).

Light can also be provided by artificial lamps; a wide variety are available, ranging from simple tungsten-filament bulbs that emit strongly in the visible region, to mercury arcs that produce high intensity UV light of even more energetic wavelengths than are found in sunlight. (This short-wave UV of less than 290 nm is sometimes called "UV-C.") Good general references on this subject are Jagger (*24*), Bickford and Dunn (*3*) and Gies et al. (*18*).

In the laboratory, the most widely used light sources are mercury lamps, which emit a variety of spectral lines, typically at 254, 313, 366, 404, 436, 546 and 578 nm (see Figure 1). The relative intensity of these lines depends on the pressure of the mercury vapor in the bulb; a low-pressure lamp emits almost pure 254 nm light, a medium-pressure lamp produces all the lines, and a high-pressure lamp gives, in addition, some output at wavelengths between the lines and thus has some characteristics of thermal (blackbody) emission.

A thermal source of light, which can be filtered to match sunlight rather well, is provided by the xenon arc lamp. The spectrum of the xenon lamp contains energy at every wavelength above 290 nm in both the UV and the visible region of the spectrum. These lamps are available in very high-power models, but are more expensive than mercury lamps.

Typical domestic "daylight" or "cool-white" fluorescent bulbs contain mercury vapor at very low pressure. Electrically excited mercury atoms transmit energy to the phosphor coating of the lamp tube and cause it to emit broad-spectrum light. The output spectra of these lamps show broad emission bands in the visible region; superimposed on these bands are the mercury emission lines. Thus, these lamps can be good, if low-power, sources of UV. Germicidal lamps contain mercury but not phosphors; they are rich in 254 nm (UV-C) light, which is transmitted by their quartz envelopes, although they also radiate about 15% of their output power in the longer-wavelength UV-B, UV-A, and visible regions. Special lamps with enhanced output in the longer-wavelength UV-A region (black lights) are available from Westinghouse (FS) and General Electric (BLB). These lamps are useful because of their relatively high output of short-wave UV-B and virtual absence of visible or infrared radiation. Some "sunlamps" emit very high levels of UV at less than 290 nm, whereas others, such as those used in commercial tanning salons, cut off sharply at around 300 nm; thus these latter lamps are useful sources of long-wavelength UV-A.

Metal halide lamps, such as those used to provide lighting in greenhouses, emit very high levels of visible and long-wavelength UV radiation. Tungsten bulbs are extremely inefficient UV sources; most of their energy is emitted in

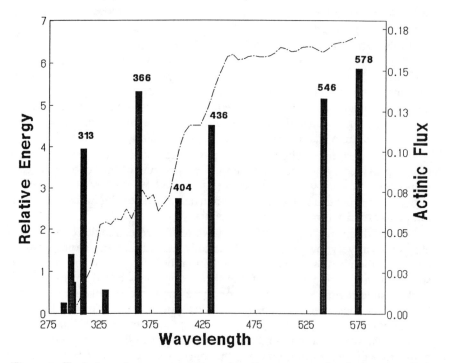

Figure 1. Emission of mercury arc lamp (medium-pressure) relative to sunlight energy distribution from 275–600 nm.

the long-wavelength visible and near infrared regions, with their greatest output around 900 nm.

When light is passed through glass or clear plastic, it is not completely transmitted. Typically these transparent media behave as cut-off filters, transmitting most of the light above some wavelength and having sharply diminishing transmittance below it. Figure 2 shows some typical transmittance spectra for some commonly used glasses and plastics. These spectra can vary greatly depending on the thickness of the medium and the presence or absence of processing impurities. Utilization of wavelengths below 280 nm requires quartz vessels. It should be remembered that such short-wave UV-C radiation, when it impinges on oxygen in the air, produces ozone, which must be removed from the treatment area if humans could be exposed.

Chromophores and Excited States

Photochemically damaging events are initiated by the uptake of the electronic energy of a photon by a light-absorbing molecule. In the UV region of the electromagnetic spectrum, the energy of such photons is sufficient to break covalent bonds, although it is unusual for their energy to be absorbed that efficiently. Usually, the absorbed energy converts the target molecule in its ground state to an electronically excited state, whose excess energy manifests itself in a different and often quite unstable electron configuration. Typical UV-absorbing groups in organic molecules commonly found in the environment are combinations of $C=C$, $C=O$, $C=N$, and $N=N$ multiple bonds. For example, an isolated $C=O$ group weakly absorbs light around 280 nm, just outside the solar region; but the absorbance maximum of a similar group conjugated with other double bonds is shifted to longer wavelengths. Thus acetophenone, $PhCOCH_3$, absorbs at 320 nm.

The initial excited state, a short-lived singlet having fully paired electrons, may be deactivated by fluorescence (the emission of a photon having a longer wavelength than the exciting radiation) and return to the ground state; it may react with neighboring molecules (although this is not common with singlets since their lifetimes are normally too short for them to diffuse over many molecular diameters); or it may undergo internal rearrangement to a longer-lived triplet excited state. The triplet state is much more likely to react chemically with surrounding molecules. Reaction, however, is only one of the possible fates of an excited triplet state; typically, less than one in ten of the excited molecules will undergo reactions. (Another way of saying this is that the quantum yield for chemical reaction is less than 0.1; in fact, often the quantum yield is orders of magnitude smaller than this, e.g., in the case of parathion the quantum yield in sunlight is 1.5×10^{-5}).

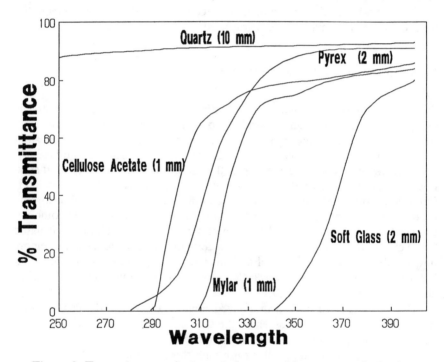

Figure 2. Transmittance of selected glasses and "transparent" plastics.

Kinetics of Photochemical Reactions

The rate of a photochemical reaction (at low concentrations, say less than 10^{-5} M of a light-absorbing organic molecule) is proportional to the degree of light absorption by the chromophore and the quantum yield of reaction for the process being observed. The degree of light absorption by a compound depends on the degree of overlap between the absorption spectrum of the compound and the emission spectrum of the incident light source, (e.g., the sun). An excellent computer program, GCSOLAR, is available for calculating rates of photolysis of organic compounds in sunlight when the absorption spectrum is known and the quantum yield of disappearance is either known or can be estimated [Zepp and Cline (48)]. Experimental methods for monitoring the rates of organic photochemical reactions have been reviewed by Zepp (47).

Direct Photolysis

Energy absorption is the primary prerequisite for a photochemical reaction, and only those contaminants exhibiting significant light absorption overlapping with solar spectral output can undergo direct photolysis in sunlight. During direct photolysis, the chemical absorbs the incident radiation directly and this can be followed by degradation [Zepp and Cline (48)]. A chemical such as a polycyclic aromatic hydrocarbon that has significant absorption in the solar region may photolyze rapidly even if the quantum yield is small [Mill et al. (34)]. Many light-absorbing compounds found in wastewaters are quite susceptible to chemical reaction after light absorption. For example, hexachlorocyclopentadiene, a synthetic compound used in the production of flame retardants and pesticides, absorbs light at wavelengths up to 390 nm. When exposed to sunlight in dilute solution, it disappears with a half-life of about four minutes [Chou et al. (10)]. Reaction products include chlorinated ketones, carboxylic acids, condensation products, and unidentified, water-soluble materials.

Organic Sensitizers and Energy Transfer

Many environmental pollutants are poor solar light absorbers (wavelengths greater than 290 nm). However, in addition to direct photoreactions, even molecules that do not absorb sunlight wavelengths can be induced to react in the presence of light by indirect means in which the light energy is trapped by an intermediate species. The use of sunlight, oxygen, and dissolved compounds capable of absorbing solar energy for the treatment of contaminated waters was first proposed nearly fifteen years ago [Acher and Rosenthal (1)].

The technique suggested by these authors, sensitized photooxidation, is a process in which light is absorbed by a sensitizing dye or other compound, raising it to a higher-energy electronically excited state. The excited dye is then capable of transferring some of its excess electronic energy to acceptor molecules to accomplish a chemical reaction. Acceptors may be either dissolved inorganic species, for example molecular oxygen, or organic materials. Methylene blue, one widely used sensitizer, is highly effective in transferring energy from its excited state to oxygen; the product of this reaction is singlet oxygen, often abbreviated $^1O_2$, a form of oxygen that is more reactive than ordinary molecular oxygen toward such compounds as olefins, cyclic dienes, and polycyclic aromatic hydrocarbons. A number of studies of wastewaters using methylene blue as sensitizer have been reported; Acher and Rosenthal (1) added methylene blue at 5–30 ppm concentrations to samples of sewage, exposed the samples to sunlight, and followed the losses of chemical oxygen demand (COD) and detergents. Declines in COD of more than half (from 380 to 100 ppm) were reported, and detergents declined by more than 90% (12 ppm to less than 1 ppm). More recently (12), an experimental plant using methylene blue and sunlight for wastewater disinfection began to operate in Livingston, Tennessee. The plant was built by Tennessee Technological University as part of a research effort funded by the U.S.-Israel Binational Agricultural Research and Development Fund. Preliminary findings on this project have been reported [Eisenberg et al. (12)].

Another approach to photodecomposition of pollutants, using riboflavin as a photosensitizing agent, has been studied in our laboratory [Larson et al. (31)]. Riboflavin (Vitamin B$_2$) occurs in natural waters and has been implicated in abiotic aquatic redox processes such as the photogeneration of hydrogen peroxide and the photooxygenation of amino acids and organic free radicals [Mopper and Zika (37)]. In general, it has been found that flavins (compounds containing the isoalloxazine ring system) are photochemically reactive toward electron-rich aromatic compounds. Riboflavin produces both singlet oxygen ($^1O_2$) and superoxide radical (O_2^-) when irradiated with visible or solar ultraviolet (UV) light in the presence of oxygen [Joshi (25)]. Hydroxyl radicals have also been postulated in some of its reactions [Ishimitsu et al. (23)]. In addition, excited flavins take part in energy-transfer processes and hydrogen atom abstraction [Chan (7)]. Other possible mechanisms for light-induced pollutant decomposition include direct electron transfer between dissolved pollutants and photochemically activated redox agents, or formation of hydrated electrons by photoionization of substrates that could then attack reducible solutes such as oxygen or electron-poor organic compounds in water.

In our experiments, riboflavin, when initially added at 5 μM to solutions of added phenols or anilines (structurally related to common environmental contaminants) greatly accelerated the rate of their loss in the presence of mercury arc light. For example, aniline's half-life toward direct photolysis under our conditions was about three hours, and in the presence of riboflavin its initial concentration was halved in about two minutes. Similarly, phenol was virtually inert to direct photolysis, but its original concentration was also halved in about two minutes. Although riboflavin is quite unstable in sunlight, under most conditions in water it is converted to lumichrome, which also has good sensitizing characteristics. The sensitized photolysis rates increased in the absence of oxygen, suggesting a mechanism involving direct energy or electron transfer between flavin excited states and acceptor molecules.

To determine whether riboflavin could be successfully used in a treatment process for removal of pollutants from contaminated waters, further experiments were performed in systems more closely resembling actual polluted environments [Larson (29)]. One such sample was obtained from Central Illinois Power Service from the site of a former coal gasification plant at Taylorville, Illinois. Analysis of this sample showed that the majority of the compounds in it are aromatic compounds (hydrocarbons and a few phenols), and are accordingly electron-rich compounds of the type expected to be treatable by riboflavin. The groundwater was exposed to light from the mercury arc lamp both with and without riboflavin present. The results indicated that several classes of compounds were partially removed by light alone, but that riboflavin significantly improved the rates of loss. Work is continuing on the products of these reactions.

Electron-transfer mechanisms have also been postulated to account for observations of the reductive degradation of methoxychlor sensitized by hydroquinone, used as a model compound for natural organic matter [Chaudhary et al. (8)]. These authors showed that the pesticide formed adducts with the sensitizer that could be explained by invoking an intermediate phenoxy radical.

Simple ketones can also act as sensitizing agents, although in these cases the mechanism does not usually involve singlet oxygen production but rather a free-radical pathway in which dissolved substrates are attacked by the excited state of the carbonyl compound. In an example of this type of process, Burkhard and Guth (6) demonstrated that atrazine and related herbicides were partly degraded in the presence of 0.15 M acetone.

Metal Ion Photochemistry

We have also initiated studies in which ferric and ferrous salts have been used to promote photodecomposition of dissolved compounds [Schlauch (41)]. Iron salts are known to effectively promote some photoreactions. For example, in dilute $FeCl_3$ and $Fe(NO_3)_3$ solutions, the insecticide parathion was photodegraded to paraoxon and p-nitrophenol [Bowman and Sans (5)].

We discovered that some iron salts were very significant promoters of photodestruction of the triazine herbicides atrazine, ametryn, prometon, and prometryn. All these triazines are very resistant to direct photolysis, having almost no absorption of solar ultraviolet light, and they are also only very slowly attacked by riboflavin. However, addition of either ferrous or ferric salts, even in the absence of other photosensitizers, led to extremely rapid triazine loss. For example, adding ferric perchlorate or ferric sulfate greatly increased the rate of triazine degradation (see Figure 3); the reactions proceeded more rapidly at low pH and low buffer concentrations.

There are several possible photochemical mechanisms by which iron salts could generate reactive species that could cause the photodestruction of triazines. Iron(III) complexes, either with water or other ligands, have significant absorption in the visible and solar UV regions [Balzani and Carassiti (2)]. A hydroxyl radical (OH·) could be produced by direct electron transfer (charge transfer) between an excited ferric ion and water molecule [Weiss (45)]; this process is especially favored in $Fe(OH)_2^+$, a complex that is stable around pH 4.5 in water [Faust and Hoigné (16)].

An alternative possibility for the ferrous ion is based on the cation photoionization process. In this case, a hydrated electron would be formed; in the presence of oxygen, superoxide anion (O_2^-) and H_2O_2 could be generated and ultimately a hydroxyl radical could be produced by the Fenton reaction [Walling (44)] of the H_2O_2 with reduced iron. Thus, either potent reductants or oxidants could be formed, depending on conditions. Ferrous ion does not absorb solar wavelengths very effectively, but shorter-wavelength irradiation produces potent reductants that can even convert CO_2 to formaldehyde [Borowska and Mauzerall (4)].

Free Radical Reactions, Advanced Oxidation Processes

If oxidation is initiated by some source of more reactive free radicals, the degradation of some compounds can occur in a very short time [Malaiyandi et al. (32); Glaze (19)]. Advantage is taken of these techniques in the so-called "advanced oxidation processes," in which combinations of UV, ozone, and hydrogen peroxide are exploited. When used in water in the presence of UV-C light, these systems initiate degradation of organic compounds due to the heterolytic cleavage of the O-O bond of either ozone or H_2O_2 that produces the hydroxyl radical (OH·). This radical is extremely reactive with almost all organic compounds. Polar organic substances are very susceptible to OH· attack and undergo oxidative degradation [Walling (44); Koubek (28)]. These techniques show considerable promise for removal of virtually all organic matter from contaminated waters, though at a considerable cost for energy input.

Malaiyandi et al. (32) found that the total organic carbon (TOC) content of distilled water samples was reduced by about 88% and of tap water by 98% when a combination of UV-C light and ozone was used. They also determined

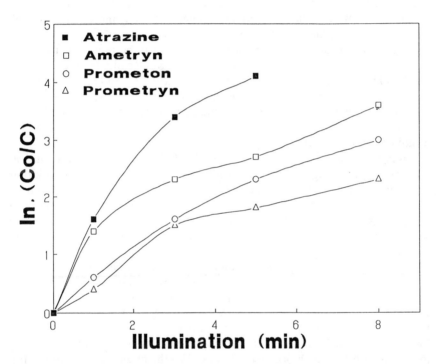

Figure 3. Photolysis of triazine herbicides (5 μM) in the presence of 260 μM ferric perchlorate in distilled water. Light source: Pyrex-filtered medium-pressure mercury lamp. [Data from Schlauch (41)].

that ozone treatment of up to 8 hours without UV was much less effective than H_2O_2 plus UV-C light for the removal of non-polar organic compounds in water. They concluded that small amounts of H_2O_2 in water in the presence of UV-C could be a very simple and efficient method of obtaining purified water.

The use of UV-C light and ozone for pesticide degradation in farm-generated pesticide wastewater was investigated by Kearney et al. (26). The species 2,4-D (1086 ppm) and atrazine (4480 ppm) [sic, the generally accepted solubility of atrazine is about 35 ppm] in aqueous solutions were greater than 80% degraded in 2–3 hours with UV (254 nm) and ozone. The product analysis revealed that most products identified were considerably more biodegradable and less toxic than the parent materials.

High-intensity short-wavelength UV-C lamps can be used to catalyze the formation of OH· from H_2O_2. It reacts with organic contaminants, theoretically oxidizing chemicals such as hydrocarbons to CO_2, and halogenated materials to halides. Oxidation of herbicides with UV and ozone has been shown to be successful when the influent organic carbon concentration is in the ppm to low ppb range [Nyer (38)]. The hydroxyl radical is an intermediate in this treatment process.

In natural waters, the principal source of HO· production is probably nitrate photolysis [Hoigné et al. (22)]. Deliberate addition of nitrate to solutions containing selected environmental contaminants has been shown to increase the rates of their photolysis in the presence of light [Kotzias et al. (27)], probably by this mechanism. Nitrate, however, has been reported to take part in some photochemical nitrations which can lead to mutagenic products [Suzuki et al. (43)].

Exposure to UV-C light from a low-pressure mercury lamp kills or renders bacteria incapable of reproduction by photochemically altering DNA as well as other targets in the cell. Thus, the use of UV alone for sterilization of water is possible. Oliver and Carey (39) used UV-C light for wastewater disinfection and found that 99% of the indicator organisms could be killed with fairly low doses. The effluent of UV-treated water was found to be much less toxic to aquatic life than a chlorinated effluent. Whitby et al. (46) also found that UV-C light treatment led to reduction in numbers of indicator bacteria from secondary wastewater effluents.

Surface Photochemistry

In recent years considerable work has been performed on photodegradation of organic pollutants by suspensions of semiconductor oxides such as TiO_2 and ZnO. These substances undergo light-triggered promotion of an electron from a ground-state valence-band to an excited-state conduction band, producing excited-state electrons and positively charged "holes" that can initiate chemical reactions in the presence of sufficient concentrations of acceptor molecules, such as those adsorbed to the surface. For example, conduction band electrons can be transferred to molecular oxygen to produce O_2, whereas the valence-band holes may accept electrons from such donors as HO^-, Cl^-, or HCO_3^- to produce the free radicals $\cdot OH$, $\cdot Cl$, and $HCO_3 \cdot$ respectively.

Both these oxides can be excited by solar UV-A wavelengths. Simple chlorinated alkanes can be completely mineralized to CO_2 and Cl by the use of TiO_2 in a 0.1% suspension [Pruden and Ollis (40)]. Matthews (33) reviewed the decomposition of a wide variety of organic compounds by suspended TiO_2, indicating that in their laboratory environment the rate of mineralization did not vary greatly with structural type; nitrobenzene was converted to CO_2 at about 40% of the rate for 2,4-dichlorophenol. These data are consistent with an intermediate oxidant such as $\cdot OH$. Draper and Crosby (11) reported that the common herbicide molinate could be readily photodecomposed under field conditions by adding ZnO (ca. 8 mg/L) to rice-paddy basins.

Photoreductive Dehalogenation

Recent work by Epling and co-workers has shown that sodium borohydride can act as a potent photoreductive agent when used in conjunction with 254 nm light. The dehalogenation reactions of chlorinated biphenyls [Epling and Florio (13)], brominated biphenyls [Epling et al. (14)], and chlorinated dibenzodioxins [Epling et al. (15)] have been shown to be strikingly enhanced in the presence of light. In most cases, the dehalogenation proceeds cleanly and sequentially to afford the corresponding hydrogenated compounds; thus, for example, a mixture of Cl_5-Cl_8 biphenyls was converted in good yield to a mixture of Cl_0-Cl_6 biphenyls by irradiation for 60 min in the presence of borohydride.

Photodechlorination of chlorobenzenes via electron transfer from *N,N*-dimethylaniline has been described by Chesta et al. (9).

Effects of Natural Water Constituents

Light-absorbing substances in water can inhibit photochemical reactions by absorbing light energy that could otherwise be used to degrade target molecules. Both dissolved and particulate materials can decrease the light intensity in water. Some very clear marine waters transmit the majority of the light in the 300–600 nm spectral region even at 10 meters depth [Smith and Baker (42)]. In "natural" freshwaters, however, the principal dissolved light-absorbing species are the polymeric humic and fulvic acids. These compounds are characterized by very similar absorption spectra which consist of curves that monotonically increase toward shorter wavelengths [Zepp and Schlotzhauer (49)], and accordingly remove UV radiation quite selectively, often within a few centimeters of the surface. In extremely eutrophic waters, significant absorption in the visible is sometimes observed due to release of algal pigments and decomposition products. The only other common dissolved absorbing species is nitrate ion, which absorbs a small amount of the shortest solar UV wavelengths.

Of course, turbid waters containing suspended sediments or other solid materials can have very low light transmittance. However, some suspended particles are UV-transparent, and it is conceivable that in these instances photolysis rates could even be increased due to the increased path length brought about by the scattering characteristics of the particulate matter [Miller and Zepp (35)].

The dissolved organic substances found in natural waters, such as humic and fulvic acids, can also influence photolysis rates by either promoting or retarding photochemical and photophysical processes. Miller et al. (36) showed that light attenuation is a retarding effect while mediation of indirect photoprocesses can be an accelerating effect. Zepp et al. (50) reported that humic materials sensitize photoreactions of several organic chemicals. In waters containing surface-active agents, rates of photolysis may either be accelerated or inhibited by the association of target compounds with the micelle-like aggregates of the surfactants.

The reaction of OH· with carbonate, CO_3^{2-}, is also an important factor. In natural freshwaters, OH· is consumed within a few microseconds by dissolved organic carbon and by carbonate [Hoigné and Bader (20,21)]. The reaction of OH· with carbonate is very rapid, although somewhat lower than the rate constant for the reaction of OH· with most organic compounds. However, the concentration of CO_3^{2-} in drinking water and wastewater is usually greater than that of the organic compounds found in these waters [Hoigné and Bader (21)]; thus, the scavenging of OH· by carbonate becomes the rate-determining reaction. Carbonate radical itself is not particularly reactive with most classes of dissolved organic compounds [Larson and Zepp (30)] and its fate in natural waters is at present uncertain.

Conclusions

Light can promote the decomposition of dissolved pollutants by a variety of mechanisms. Although some classes of organic chemicals absorb sunlight strongly and are rapidly photolyzed in natural waters or wastewaters, for many some means of indirect photodegradation must be resorted to. Classical sensitized photolysis with its singlet oxygen mechanism is probably of limited importance in water, but sensitizers such as riboflavin may be active toward compounds with which they can form light-absorbing complexes. Some dissolved and suspended species can produce highly reactive oxidizing free radicals such as HO, but the utility of these methods may be limited by high concentrations of scavengers of these radicals in the test water. Further studies on the use of all these treatment systems in actual polluted water samples is necessary.

Acknowledgments

We thank the Illinois Hazardous Waste and Information Center and the U.S. Geological Society for financial support (Contract No. HWR87032 and Grants 14-08-0001-G1298 and 14-08-0001-G1560, respectively) and David Ellis, Bruce Faust, Jürg Hoigné, Gary Peyton, and Richard Zepp for helpful discussions.

Literature Cited

1. Acher, A. M. and Rosenthal, I. Dye-sensitized photooxidation: a new approach to the treatment of organic matter in sewage effluents. *Water Res*, 11:557–562, **1977**.
2. Balzani, M. and Carassiti, V. *Photochemistry of coordination compounds*. Academic Press, London, 1970.
3. Bickford, E. D. and Dunn, S. *Lighting for plant growth*. Kent State Univ. Press, Kent, OH, 1972.
4. Borowska, Z. and Mauzerall, D. Photoreduction of carbon dioxide by aqueous ferrous ion: an alternative to the strongly reducing atmosphere for the chemical origin of life. *Proc Nat Acad Sci US*, 85:6577–6580, **1988**.
5. Bowman, B. T. and Sans, W. W. Stability of parathion and DDT in dilute iron solutions. *J Environ Sci Health*, B15:233–246, **1980**.
6. Burkhard, N. and Guth, J. A. Photodegradation of atrazine, atraton, and ametryne in aqueous solution with acetone as a photosensitiser. *Pestic Sci*, 7:65–71, **1976**.
7. Chan, H. W. Photo-sensitised oxidation of unsaturated fatty acid methyl esters: the identification of different pathways. *J Am Oil Chem Soc.*, 54:100–104, **1977**.
8. Chaudhary, S. K., Mitchell, R. H., West, P. R., and Ashwood-Smith, M. J. Photodechlorination of methoxychlor induced by hydroquinone. Rearrangement and conjugate formation. *Chemosphere*, 14:27–40, **1985**.
9. Chesta, C. A., Cosa, J. J., and Previtali, C. M. Photoinduced electron transfer from *N,N*-dimethylaniline to chlorobenzene: the decomposition rate constant of the radical anion of chlorobenzene. *J Photochem Photobiol*, 45:9–15, **1988**.
10. Chou, S., Griffin, R. A., Chou, M. M., and Larson, R. A. Photodegradation products of hexachlorocyclopentadiene (C-56) in aqueous solution. *Environ Toxicol Chem*, 6:371–376, **1987**.
11. Draper, R. B. and Crosby, D. G. Catalyzed photodegradation of the herbicides molinate and thiobencarb. In Zika, R. G. and Cooper, W. J., Eds.,

Photochemistry of Environmental Aquatic Systems, pages 240–247, ACS Sym Ser, No. 327, American Chemical Society, Washington, DC, 1987.

12. Eisenberg, T. N., Middlebrook, E. J., and Adams, V. D. Sensitized photooxidation for wastewater disinfection and detoxification. *Water Sci Technol*, 19:1255–1258, **1987**.

13. Epling, G. A. and Florio, E. Enhanced photodehalogenation of chlorobiphenyls. *Tetrahedron Lett*, 27:675–678, **1986**.

14. Epling, G. A., McVicar, W., and Kumar, A. Accelerated debromination of biphenyls by photolysis with sodium borohydride. *Chemosphere*, 16:1013–1020, **1987**.

15. Epling, G. A., Qiu, Q., and Kumar, A. Hydride-enhanced photoreaction of chlorodibenzodioxins. *Chemosphere*, 18:329–332, **1989**.

16. Faust, B. C. and Hoigne, J. Photolysis of Fe(III)-hydroxy complexes as sources of OH radicals in clouds, fog, and rain. *Atmos Environ*, **1990**. in press.

17. Finlayson-Pitts, B. J. and Pitts, Jr., J. N. *Atmospheric chemistry: fundamentals and experimental techniques*. John Wiley, 1986.

18. Gies, H. P., Roy, C. R., and Elliott, G. Artificial suntanning: spectral irradiance and hazard evaluation of ultraviolet sources. *Health Phys*, 50:691–703, **1986**.

19. Glaze, W. H. Drinking-water treatment with ozone. *Environ Sci Technol*, 21:224–230, **1987**.

20. Hoigné, J. and Bader, H. Beeinflussung der oxidationswirkung von ozon und OH radikalen durch carbonat. *Vom Wasser*, 48:283–304, **1977**.

21. Hoigné, J. and Bader, H. Ozonation of water: "oxidation-competition values" of different types of waters used in Switzerland. *Ozone Sci Eng*, 1:357–372, **1979**.

22. Hoigné, J., Faust, B. C., Haag, W. R., Scully, Jr., F. E., and Zepp, R. G. Aquatic humic substances as sources and sinks of photochemically produced transient reactants. In Suffet, I. H. and MacCarthy, P., Eds., *Aquatic Humic Substances*, pages 363–381, ACS Sym Ser, No. 219, American Chemical Society, Washington, DC, 1989.

23. Ishimitsu, S., Fujimota, S., and Ohara, A. The photochemical decomposition and hydroxylation of phenylalanine in the presence of riboflavin. *Chem Pharm Bull*, 33:1552–1556, **1985**.

24. Jagger, J. *Introduction to Research in Ultraviolet Photobiology*. Prentice-Hall, Englewood Cliffs, NJ, 1967.

25. Joshi, P. C. Comparison of the DNA-damaging property of photosensitised riboflavin via singlet oxygen and superoxide radical mechanisms. *Toxicol Lett*, 26:211–217, **1985**.

26. Kearney, P. C., Zeng, Q., and Ruth, J. M. Destruction of herbicide wastewaters. In Krueger, R. F. and Seiber, J. N., Eds., *Treatment and Disposal of Pesticide Wastes*, pages 195–209, ACS Sym Ser, No. 259, American Chemical Society, Washington, DC, 1984.

27. Kotzias, D., Parlar, H., and Korte, F. Photoreaktivität organischer chemikalien in wäßrigen systemen in gegenwart von nitraten und nitriten. *Naturwissenschaften*, 69:444, **1982**.

28. Koubek, E. Photochemically induced oxidation of refractory organics with hydrogen peroxide. *I&EC Proc Res Dev*, 14, **1975**.

29. Larson, R. A. *Sensitized photodecomposition of organic compounds found in Illinois wastewaters*. Technical Report RR-045, Illinois Hazardous Waste and Information Center, Champaign, IL, 1990.

30. Larson, R. A. and Zepp, R. G. Reactivity of the carbonate radical with aniline derivatives. *Environ Toxicol Chem*, 7:265–274, **1988**.

31. Larson, R. A., Ellis, D. D., Ju, H., and Marley, K. A. Flavin-sensitized photodecomposition of anilines and phenols. *Environ Toxicol Chem*, 8:1165–1170, **1989**.

32. Malaiyandi, M., Sadar, M. H., Lee, P., and O'Grady, R. Removal of organics in water using hydrogen peroxide in presence of ultraviolet light. *Water Res*, 14:1131–1135, **1980**.

33. Matthews, R. W. Photo-oxidation of organic material in aqueous suspensions of titanium dioxide. *Water Res*, 20:569–578, **1986**.

34. Mill, T., Mabey, W. R., Lan, B. Y., and Baraze, A. Photolysis of polycyclic aromatic hydrocarbons in water. *Chemosphere*, 10:1281–1290, **1981**.

35. Miller, G. C. and Zepp, R. G. Effects of suspended sediments on photolysis rates of dissolved pollutants. *Water Res*, 13:453–459, **1979**.

36. Miller, G. C., Zisook, R., and Zepp, R. G. Photolysis of 3.4-dichloroaniline in natural waters. *J Agri Food Chem*, 28:1053–1056, **1980**.

37. Mopper, K. and Zika, R. G. Natural photosensitizers in sea water: riboflavin and its breakdown products. In Zika, R. G. and Cooper, W. J., Eds., *Photochemistry of Environmental Aquatic Systems*, pages 174–190, ACS Sym Ser, No. 327, American Chemical Society, Washington, DC, 1987.

38. Nyer, E. K. Treatment of herbicides in ground water. *Ground Water Monit Rev*, 54–59, Summer **1988**.

39. Oliver, B. G. and Carey, J. H. Ultraviolet disinfection: an alternative to chlorination. *J Water Pollut Contr Fed*, 48:2619–2624, **1976**.

40. Pruden, A. L. and Ollis, D. F. Degradation of chloroform by photoassisted heterogeneous catalysis in dilute aqueous suspensions of titanium dioxide. *Environ Sci Technol*, 17:628–631, **1983**.

41. Schlauch, M. B. *Sensitized Photodecomposition of Triazine Herbicides*. Master's thesis, University of Illinois, Urbana, 1989.

42. Smith, R. C. and Baker, K. S. Penetration of UV-B and biologically effective dose-rates in natural waters. *Photochem Photobiol*, 29:311–323, **1979**.

43. Suzuki, J., Sato, T., and Suzuki, S. Hydroxynitrobiphenyls produced by photochemical reaction of biphenyl in aqueous nitrate solution and their mutagenicities. *Chem Pharm Bull*, 33:2507–2516, **1985**.

44. Walling, C. Fenton's reagent revisited. *Accts Chem Res*, 8:125–131, **1975**.
45. Weiss, J. J. Electron transfer processes in the mechanism of chemical reactions in solution. *Ber Bunsenges Phys Chem*, 73:131–135, **1969**.
46. Whitby, G. E., Palmateer, G., Cook, W. G., Maarschlakerweerd, J., Huber, D., and Flood, K. Ultraviolet disinfection of secondary effluent. *J Water Pollut Contr Fed*, 56:844–850, **1984**.
47. Zepp, R. G. Experimental approaches to environmental photochemistry. In Hutzinger, O., Ed., *Handbook of Environmental Chemistry, Vol. 2B*, pages 19–41, Springer Verlag, Berlin, 1982.
48. Zepp, R. G. and Cline, D. M. Rates of direct photolysis in the aqueous environment. *Environ Sci Technol*, 11:359–366, **1977**.
49. Zepp, R. G. and Schlotzhauer, P. F. Comparison of photochemical behavior of various humic substances in water. III. Spectroscopic properties of humic substances. *Chemosphere*, 10:479–486, **1981**.
50. Zepp, R. G., Schlotzhauer, P. F., and Sink, R. M. Photosensitized transformations involving electronic energy transfer in natural waters: role of humic substances. *Environ Sci Technol*, 19:74–81, **1985**.

RECEIVED April 5, 1991

Chapter 6

Disposal of Toxic Wastes by Using Concentrated Solar Radiation

J. L. Graham[1], B. Dellinger[1], D. Klosterman[1], G. Glatzmaier[2], and G. Nix[2]

[1]Environmental Sciences Group, University of Dayton Research Institute, 300 College Park, Dayton, OH 45469–0132
[2]Solar Energy Research Institute, 1617 Cole Boulevard, Golden, CO 80401

Laboratory and small-scale field studies have been conducted which illustrate that toxic organic wastes can be destroyed using concentrated solar radiation. Solar energy is a unique resource which provides large quantities of radiation ranging from the IR through to the visible and UV. In addition to the heat of combustion of a waste, the IR can be used to supply thermal energy to drive destructive photochemical reactions induced by the visible and UV portion of the spectrum. Laboratory studies of the gas phase destruction of various hazardous organic compounds using a simultaneous exposure to high temperature and simulated solar radiation equivalent to 300 times natural sunlight have shown that these materials can be destroyed with destruction factors $> 10^6$ in pyrolytic and oxidative environments within 5-10 seconds at temperatures below 800°C. Field tests with 500 - 1,300 suns have yielded destruction factors for 1,2,3,4-tetrachlorodibenzo-p-dioxin of $> 10^6$ at temperatures as low as 750°C. Research also indicates that products of incomplete combustion are formed with lower yields, and are destroyed at lower temperatures than in conventional incinerators.

For most of the history of the chemical process industries, the consequences of environmentally unacceptable disposal practices were not widely appreciated. The tragic result of this unfortunate attitude is a cleanup problem of extraordinary magnitude.

The seriousness of the threat to both the environment and to human health prompted the U.S. Congress to adopt the Resource Conservation and Recovery Act (RCRA) in 1976. In part, RCRA mandates the proper handling and disposal of hazardous wastes. Furthermore, regulations enacted under RCRA place the burden of liability on all parties associated with the waste in a "cradle-to-grave"

0097–6156/91/0468–0083$07.75/0

approach to waste disposal. The potential for crippling liability litigation has made the proper disposal of industrial wastes a significant economic issue for plant design and operation, and research into effective disposal practices has become economically attractive.

Despite the considerable long term risk involved with landfills, this technique has continued to be the method of choice for the disposal of the vast majority of the 290 million tons of hazardous wastes currently being produced each year in the United States (1). Research in the design of landfills has concentrated on determining the requirements for the long term stability of a disposal site, developing new liner materials and installation techniques, and developing methods for leak detection and monitoring.

Landfilling has remained popular primarily because of its relatively low short-term cost. However, despite the improvement in site design, landfilling remains little more than long term storage of wastes and does not constitute true disposal. New hazardous waste landfill sites have also become nearly impossible to construct as public opposition has become increasingly intense. So, as existing storage sites reach capacity, the cost of landfilling will inevitably increase. Furthermore, as storage capacity is reached, it will be increasingly important to reserve these facilities for the containment of hazardous materials which cannot be processed by alternative technologies.

Even before RCRA was enacted there was growing interest in using alternative disposal methods which result in the permanent destruction of hazardous organic wastes. Of particular interest was controlled high temperature incineration.

Regulations promulgated under RCRA require that for most wastes an incinerator must demonstrate destruction factors of greater than 10^4 for those principal components in the feed designated as hazardous. For materials which are particularly hazardous a far more stringent destruction factor of greater than 10^6 is required. To meet these specifications conventional incinerators must operate under extraordinary conditions. Temperatures are often in excess of 1000°C in atmospheres which are often highly acidic and abrasive. Consequently, incinerators have proven very expensive to operate and maintain. There is also considerable concern over whether they actually destroy wastes to environmentally acceptable levels in routine operation. Therefore, as in the case of new landfills, public opposition to the siting of new incinerators has become quite intense.

The difficulties encountered by incineration have prevented this technology from successfully addressing the hazardous waste problem as efficiently as had been hoped. In fact, only about 5% of the wastes eligible for incineration are actually being burned and only about 67% of the existing incinerator capacity is being utilized (2). Therefore, while it has been generally accepted that incineration is the ultimate disposal method for organic wastes, there is interest in developing new technologies, preferably based on incineration, which can effectively and economically destroy these materials. The recent studies discussed in the following pages have shown that using concentrated sunlight as the principal energy source in an incinerator can significantly increase the efficiency of this disposal technology.

Theory

The global reactions which would take place in a solar powered incinerator can be described by an energy versus reaction coordinate diagram as illustrated in Figure 1. The curves in this figure represent the lowest energy pathways for the global reaction of hazardous waste to products.

The lower curve in Figure 1, labeled S_o, represents those molecules which are in their ground electronic state, or those molecules which would decompose via purely thermal reactions such as would take place in a conventional incinerator. It has been shown that for these global reactions to take place requires 30 - 100 kcal/mol of thermal energy to overcome the ground state activation energy barrier (3). It is known, however, that if organic molecules are exposed to light of an appropriate energy they can be promoted to an electronically excited state. Considering the spectral distribution of sunlight, this would most likely be the lowest available singlet state as given by the curve S_1, from which the activation energies are typically on the order of 10 kcal/mol (4). Therefore, reactions from this state should proceed far more rapidly than comparable ground state reactions. Unfortunately, the lifetime of this state can be quite short, possibly only a few nanoseconds, so only very fast reactions can occur prior to the molecule leaving this state.

Fortunately, excited singlet states may not immediately revert to the ground state. Instead, they can relax to a meta-stable condition called a triplet state (curve T_1) from which the reaction energies are often as low as 2 kcal/mol (5). More importantly, the lifetime of this state can be quite long, typically three to six orders of magnitude longer than the comparable singlet state. Therefore, reactions from an excited triplet state usually have an ample opportunity to occur before the molecule returns to the ground state.

What we propose, then, is to use highly concentrated solar energy in an incinerator-type device. In this application, the near UV photons made available by the solar spectrum, which make up less than 10% of the suns's energy, would be used to promote a significant portion of the waste feed into electronically excited states, from which the thermal energy provided by the visible and IR

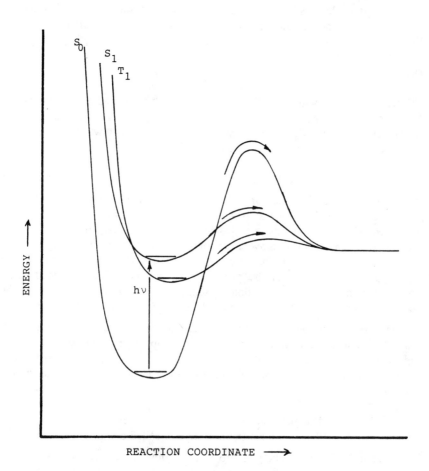

REACTION COORDINATE ⟶

THERMAL/PHOTOLYTIC DECOMPOSITION

Figure 1. Energy versus reaction coordinate diagram showing the possible initiation reaction pathways which would take place in a solar powered incinerator.

portion of the solar spectrum and the heat of combustion of the waste would be used to efficiently destroy the waste. This combined process of thermal and high temperature photo reactions should destroy the wastes far more efficiently than conventional, purely thermal technologies.

Experimental

To conduct a detailed study of the simple model proposed above, a dedicated instrument called the Thermal/Photolytic Reactor System (TPRS) was developed. As the schematic of the TPRS given in Figure 2 illustrates, this system consists of a thermally insulated enclosure which houses a fused quartz sample inlet chamber and high temperature reactor. The reactor is similar in form to a long spectroscopic absorption cell and lies within a small tube furnace which is sealed at either end with flat quartz windows. A conduit, which passes completely through the housing, allows the reactor to be illuminated by the radiation source. Adjacent to the reactor assembly is a second thermally controlled enclosure which houses a cryogenic trap that collects the material surviving the exposure in the reactor and holds them until all of the sample has passed through the system. For analysis of these products, the TPRS is fitted with an in-line programmed temperature gas chromatograph (GC, Varian Model 3700). Finally, the radiation source is a 1000 W xenon arc lamp (Kratos, SS-1000X) which is modified using air mass filters to simulate natural sunlight.

In its present configuration, the TPRS can be used to study nearly any gas phase system with simulated solar exposures up to about 300 times natural sunlight (where one sun = one air mass 1 solar constant = 0.1 W/cm^2), exposure times to 15 seconds, temperatures to 1000°C, and with any non-corrosive atmosphere.

In a typical experiment, 2 μg of a test sample is used for each analysis. Gases and volatile liquids are prepared in 1 liter bulbs filled with helium from which 250 μl aliquotes are withdrawn and injected into the system. Low vapor pressure liquids are injected directly. Solid samples are dissolved in a suitable solvent (typically cyclohexane) from which 2 μl aliquotes are drawn and deposited on a special quartz probe. The solvent is allowed to evaporate, and the probe is sealed within the sample inlet chamber.

For the tests with a mixture of methylene chloride and chlorine a sample of 7.95 x 10^{-4} mol/l chlorine and 2.65 x 10^{-4} mol/l methylene chloride was prepared in an atmosphere of dry helium contained in a 1 liter flask. For each analysis, 100 μl aliquotes of this stock sample were introduced into the TPRS giving 3.98 x 10^{-5} mol/l chlorine and 1.33 x 10^{-5} mol/l methylene chloride in the reactor. The mean exposure time for each test was 10 seconds in an atmosphere of dry air. A control sample was also prepared in an identical manner without the chlorine component.

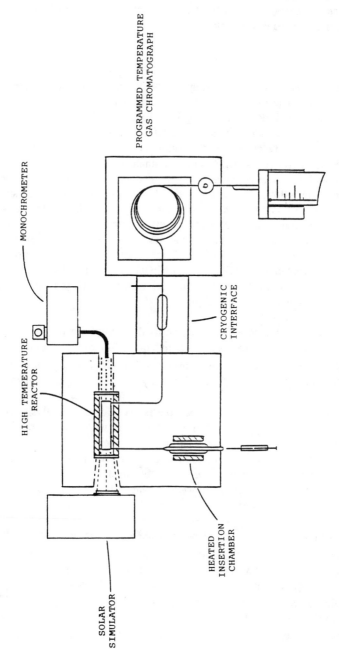

Figure 2.   Schematic of the thermal/photolytic reactor system (TPRS).

For each analysis, the insertion chamber is maintained at 300°C which is sufficient to transport most organic compounds in the gas phase without promoting thermal decomposition. So, within this chamber the sample is vaporized, mixed with flowing carrier gas, and is swept through the reactor. For studies in an oxidative atmosphere, dry air is used as the carrier gas, and for pyrolysis, helium is used. The resulting average concentration of sample in the reactor is typically on the order of 10^{-5} mol/l.

Initially, the reactor temperature is set fairly low (typically 300°C) so that the sample is not degraded. This allows calibration of the chromatographic response for the sample. The exposure time is controlled by the total volume flow rate of gas through the system. This flow was adjusted as the reactor temperature is changed to correct for expansion effects. Exposures of 5 and 10 seconds are typically used.

For capturing the organic material leaving the reactor, the cryogenic trap is cooled to -150°C. Typically, 10 minutes are allowed for the sample to be completely swept from the inlet and through the reactor. During this trapping sequence the exhaust from the trap is vented to the atmosphere. On completion of sample trapping, the system is purged with helium, the trap vent is sealed (thereby directing the flow of exhaust to the GC), and the system pressure is raised from ambient to 5 psig.

During the pressurization process the GC is cooled to -80°C to make it ready to receive the sample. The chromatographic column used for most of the analysis was a 0.32 mm x 15 m fused silica column with a 1 μm thick dimethyl silicone liquid film (J & W Scientific, DB-1). The column inlet is set up to include a 10:1 split ratio. Finally, a flame ionization detector (FID) is used for solute detection.

Once the GC is prepared to receive the sample, the cryogenic trap is heated to 300°C over a ten minute period thereby volatilizing the trapped material and passing it onto the GC column. After this transfer is complete the GC is heated to -5°C at 5°C/min, then to 260°C at 15°C/min.

The conversion (destruction) of parent compounds is calculated by normalizing the chromatographic response obtained at elevated temperatures with the response measured under non-destructive conditions. Furthermore, data at each temperature is obtained in pairs; first with the lamp off, giving a purely thermal exposure, then with the lamp on, giving a simulated solar exposure. The resulting data are summarized in a "decomposition profile" in which the data are plotted as weight percent remaining (on a logarithmic scale) versus exposure temperature. The difference between the data pairs are quantified by the ratio of the fraction remaining from the thermal exposure to the fraction remaining from the solar exposure. This value is called the solar enhancement ratio.

Results and Discussion

Based on previous experience with research on conventional incineration (6) three areas of key importance were identified for immediate study; the impact of the solar radiation on the overall thermal stability of target compounds, the effectiveness of the process under oxygen starved conditions, and the effect on the formation of products of incomplete combustion (PICs). The behavior of the process under various radiant flux levels was considered as an important scale-up issue.

Stability: Oxidation

The first goal of the experimental program was to determine if a simultaneous exposure to high temperature and simulated sunlight would affect the destruction of organic compounds as compared to an identical purely thermal exposure. For these tests, a series of aromatic hydrocarbons were selected which could be readily promoted to an excited singlet state with the simulated sunlight, but which were not likely to intersystem cross to a triplet state. This provided the simplest possible photochemical system.

Thermal and thermal/photolytic data was obtained on azulene, benzo(e)pyrene, and naphthalene exposure for 5 seconds in air. Data representative of this class of compounds can be illustrated by benzo(e)pyrene and naphthalene as shown in Figures 3 and 4, respectively. Specifically, little or no enhancement of the destruction is observed. Similar results were found for azulene. These tests suggest that, as in conventional low temperature photochemistry, the lifetime of the excited singlet state is simply too short to permit the molecules to react even with a much reduced activation energy.

The data presented in Figure 3 also illustrates an important point with respect to the experimental method. The TPRS is designed to allow the de-coupling of the reaction temperature and radiant flux. Specifically, the temperature of the reactor is controlled by the small tube furnace, while the radiant flux is established by the solar simulator. The success of this design has been demonstrated in several tests, of which the data for benzo(e)pyrene is an example. As Figure 3 illustrates, the lamp on/lamp off data pairs show that the level of conversion, and therefore the temperature, remains constant whether the reactor is illuminated or not. This would not be the case if the presence of the light significantly influenced the temperature of the reactor.

The next series of tests involved compounds which were good candidates for intersystem crossing to an excited triplet state. As it turns out, it is far easier to identify compounds which do intersystem cross to a triplet state than those that do

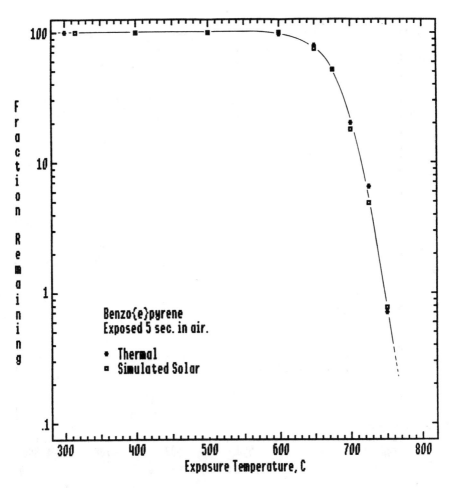

Figure 3. TPRS data for benzo(e)pyrene exposed to 0 and ~307 suns (simulated) for 5.0 seconds in air.

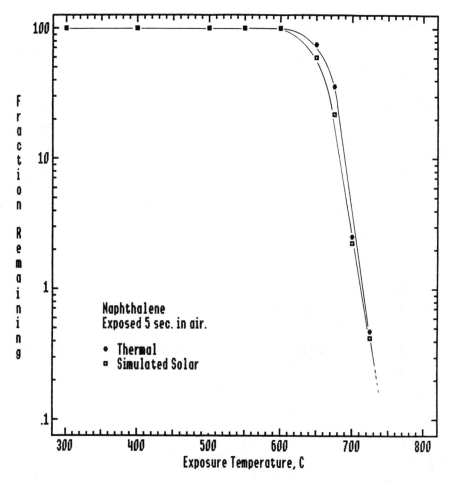

Figure 4. TPRS data for naphthalene exposed to 0 and ~ 307 suns (simulated) for 5.0 seconds in air.

not. That is because nearly any heteroatom substitution on a hydrocarbon molecule enhances the rate of intersystem crossing (4). Therefore, while the parent compound of a family of molecules, such as naphthalene, may not intersystem cross to a triplet state, most of the derivatives of that compound will.

Thermal and thermal/photolytic data was obtained on numerous compounds in this class including chlorine substituted benzenes, halogen, halogen substituted naphthalenes, 1-nitronaphthalene, aromatic ketones, 1,2,3,4-tetrachlorodibenzo-p-dioxin (TCDD), and 3,3',4,4'-tetrachlorobiphenyl (TCB). The data shown in Figure 5 for 1-nitronaphthalene exposed 5 seconds in air is an example of the results obtained for these compounds.

Comparing Figures 4 and 5 illustrates that the addition of a single nitro group to the naphthalene molecule results in a significant enhancement of the destruction. Specifically, even at the lowest temperature studied (300°C) nearly 70% of the starting material was destroyed, and the destruction efficiency is enhanced as much as 20 times during the simulated solar exposure as compared to the thermal exposure.

As additional examples, data for TCB is shown in Figure 6, in which an average enhancement of 3 was observed, and TCDD in Figure 7, in which an enhancement of 6 was found. Similar results were found for the other compounds in this class with enhancements varying from 3 in the case of TCB, to 160 for xanthone, with typical values in the range of 5 to 15.

Stability: Pyrolysis

Studying the decomposition of organic compounds under pyrolytic conditions is important for two reasons. First, pyrolysis is a simpler system to study as compared to oxidation as the number of possible initiation reactions is limited. Second, it is thought that uncontrollable oxygen deficient regions within a conventional incinerator play an important role in limiting the performance of these units. To evaluate the solar incineration process under pyrolytic conditions several of the test compounds examined in the oxidative thermal stability tests were re-examined in an atmosphere of flowing helium.

The data for TCDD exposed for 10 seconds in helium, shown in Figure 8, is typical of the results of these tests and illustrates that a significant enhancement (about tenfold) is still observed under these conditions. Similar results were found for the other test compounds with enhancements varying from 2, in the case of TCB, to 65, for xanthone, with typical values in the range of 5-10.

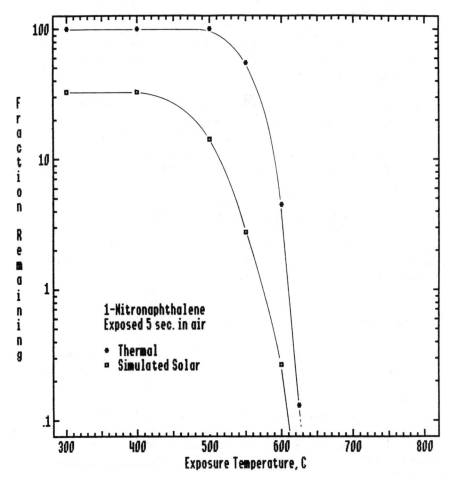

Figure 5. TPRS data for 1-nitronaphthalene exposed to 0 and ~307 suns
(simulated) for 5.0 seconds in air.

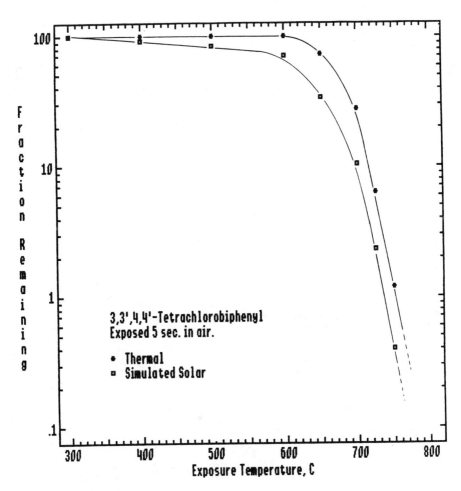

Figure 6. TPRS data for 3,3',4,4'-tetrachlorobiphenyl exposed to 0 and ~ 307 suns (simulated) for 5.0 seconds in air.

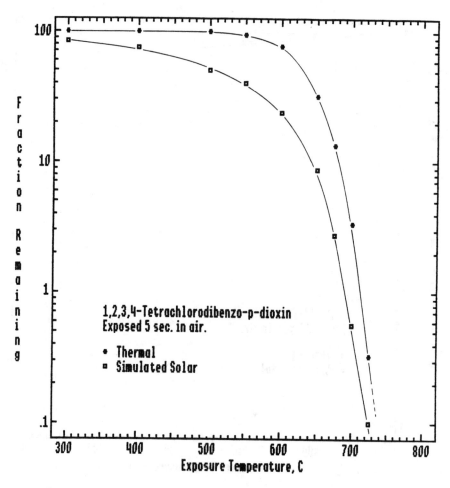

Figure 7. TPRS data for 1,2,3,4-tetrachlorodibenzo-p-dioxin to 0 and ~307 suns (simulated) for 5.0 seconds in air.

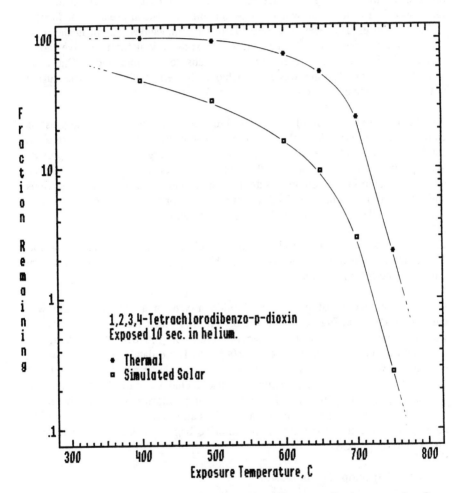

Figure 8. TPRS data for 1,2,3,4-tetrachlorodibenzo-p-dioxin exposed to 0 and ~ 307 suns (simulated) for 10 seconds in helium.

Stability: Non-Absorbing Wastes

Since real wastes are almost always a mixture of materials, the question arises: what about components which do not absorb sunlight? In conventional photochemistry, molecules can be indirectly promoted to excited triplet states through the use of energy transfer sensitizers. Unfortunately, these tend to be fairly specific in nature and would not be suitable for general use as an additive to a waste mixture. However, it is possible that the molecules present in the waste which do decompose as a result of the solar exposure will form reactive radical species which will in turn react with the non-absorbing compounds. One could even conceive of blending wastes to achieve this effect similar to the current practice of blending wastes to achieve desired energy content.

As a model case a mixture of methylene chloride was decomposed in the presence chlorine. In this case, the molecular chlorine photo-dissociates on exposure to the sunlight and the chlorine atoms then react with the organic molecules present. While it is impractical to suggest that chlorine gas be injected into a full-scale incinerator, chlorinated compounds are almost present in hazardous waste streams and the possibility of utilizing internal sources of chlorine could be considered. In any event, the use chlorine in this case is as an exploratory tool.

Figure 9 summarizes the results from control tests with methylene chloride. These data show that the decomposition of this simple molecule is not effected by the simulated solar radiation.

Data from a mixture of methylene chloride mixed on a 1:3 molar ratio with chlorine is shown in Figure 10. These data illustrate that the methylene chloride was destroyed to below the detection limit of 0.0982% throughout the temperature range studied (i.e., 400 - 800°C). These data suggest that organic compounds which do not absorb solar photons can be indirectly destroyed through solar induced radical attack. Figure 10 also includes a curious plateau in the thermal decomposition profile. This suggests an equilibrium process between the thermal dissociation of the chlorine and reformation of the methylene chloride. At higher temperatures the methylene chloride begins to decompose through a direct thermal process.

Products of Incomplete Combustion

With respect to a qualitative analysis of PICs, examining the chromatograms from the TCDD pyrolysis tests is informative. Note that in the decomposition profile presented in Figure 8 there is a region between 400 - 500°C where there is significant destruction of the TCDD under photolytic conditions, but there is no thermal destruction in this temperature region as illustrated by the thermal curve. Therefore, the reactions represented in this region are entirely photolytic and there is no thermal component to this destruction. Similarly, in the region of 650

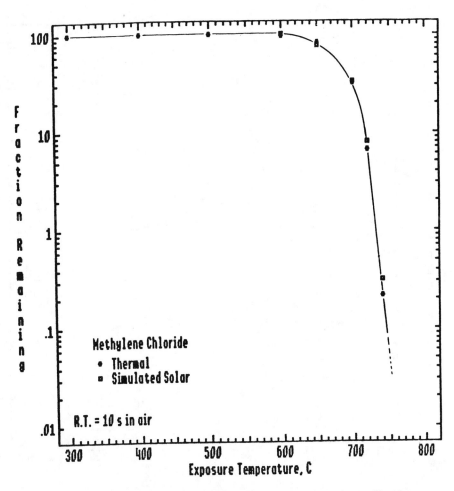

Figure 9.   Thermal and thermal/photolytic decomposition profiles for methylene chloride exposed for 10 seconds in air.

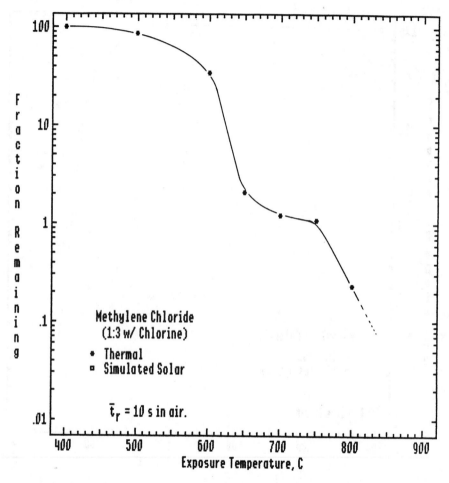

Figure 10. Thermal and thermal/photolytic decomposition profiles for methylene chloride mixed 1:3 with chlorine exposed for 10 seconds in air. Photolytic destruction was >99.9% at all temperatures.

- 700°C there are data on the thermal curve which represents a comparable level of conversion, but, of course there is no photolytic component to these reactions. Chromatographic data from these comparable levels of thermal and thermal/photolytic conversion are shown in Figure 11. This figure illustrates that PICs begin to appear at relatively low levels of thermal conversion and grow rapidly as the decomposition proceeds. In contrast, the simulated solar process produces relatively few products and in low yields.

A more quantitative study of PICs was part of a series of tests on TCB exposed to 95 suns for 10 seconds in helium. In these tests, numerous PICs were observed including two dichlorobiphenyls, three trichlorobiphenyls, and two tetrachlorobiphenylenes.

The data shown in Figures 12 and 13 illustrate the extremes in PIC production behavior observed. Specifically, the data in Figure 12 illustrates that a dichlorobiphenyl isomer forms at a lower temperature, reaches a lower maximum yield, and is destroyed at a lower temperature than its thermal counterpart. This behavior is typical for the PICs which are much lower in molecular weight than the parent TCB. The data given in Figure 13 for a tetrachlorobiphenylene isomer illustrates the other extreme in that this PIC was formed in significant yield from the thermal exposures, yet is almost completely absent from the solar data. This is typical of PICs which are heavier than the parent compound. PICs of intermediate molecular weight, the trichlorobiphenyls in this case, showed intermediate behavior.

Clearly, the effect of the solar exposure on the production of PICs is not directly proportional to the effect on the parent compounds. This suggests the presence of the intense solar radiation impacts the entire decomposition process, not just the initial decomposition of the feed material.

Radiant Flux Dependence

Early in this research program simple kinetic models were constructed to describe the solar incineration process (7). By expressing these models in terms of the enhancement ratio, the following equation can be derived:

$$R(x) = \left(\frac{f_r(0)}{f_r(n)}\right)^{x/n}$$

where $R(x)$ is the enhancement ratio with x suns, $f_r(0)$ is the fraction remaining after a thermal exposure, and $f_r(n)$ is the fraction remaining after an exposure to n suns.

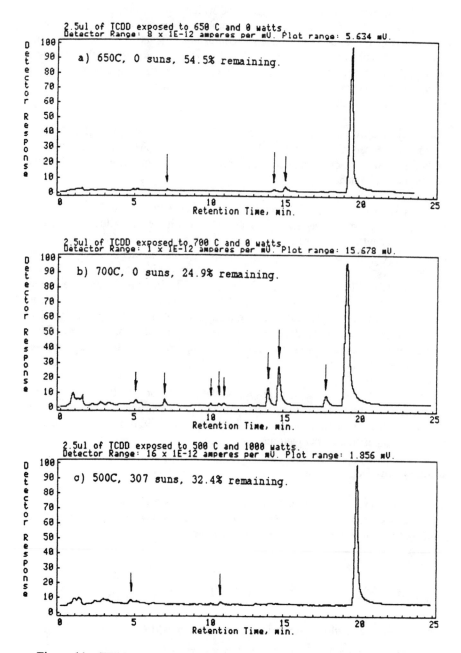

Figure 11. TPRS chromatograms for 1,2,3,4-tetrachlorodibenzo-p-dioxin exposed to 0 and ~307 suns (simulated) for 10 seconds in helium.

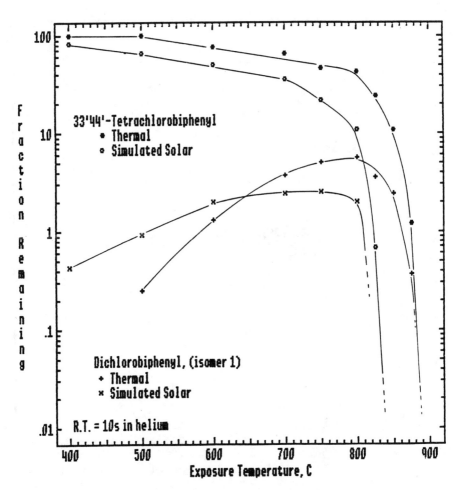

Figure 12. TPRS data for 3,3',4,4'-tetrachlorobiphenyl exposed 0 and ~95 suns (simulated) for 10 seconds in helium showing data for one of two dichlorobiphenyl isomers observed.

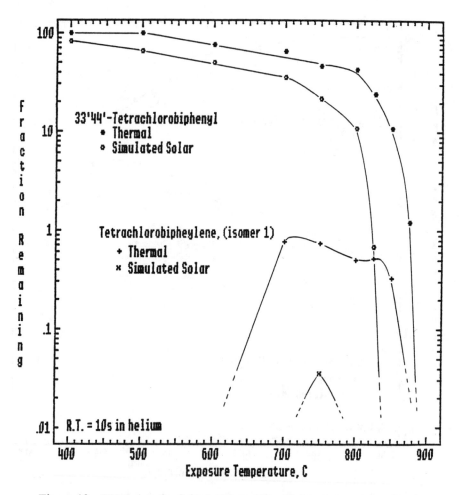

Figure 13. TPRS data for 3,3',4,4'-tetrachlorobiphenyl exposed to 0 and ~395 suns (simulated) for 10 seconds in helium showing data for one of two tetrachlorobiphenylene isomers observed.

The significance of this relation is twofold. First, it allows the enhancement ratio to be calculated at any radiant flux level once it has been measured at one known level, and it can be done so without having to find all of the numerous rate parameters needed to describe the system kinetically. Second, this equation shows that the enhancement should increase exponentially with increasing radiant flux.

To test this model the decomposition of TCB, TCDD, and xanthone were examined at a fixed temperature while varying the radiant flux from 0 to 307 suns. Furthermore, these tests were conducted both in air and helium. The data shown in Figure 14 for xanthone exposed for 5 seconds in air illustrates typical results and that an exponential dependence was indeed observed. Furthermore, values predicted by the line of best fit through these data (as found by a least squares analysis) closely match values predicted by the model.

The impact of this result can be illustrated by applying this model to a complete set of laboratory data to predict the performance of a system operating at 500 - 1,000 suns. Figure 15 shows the results of predicting the decomposition of TCDD exposed for 5 seconds in air under these conditions. Also shown on this figure are some of the results from field tests conducted with a large bench-scale reactor installed at the White Sands Solar Furnace (8). This reactor was similar to the laboratory system in that a quartz vessel was used so that both the thermal and UV energy of the sunlight could participate in the destruction of the test compounds, but differed in that the sunlight was the sole source of heat thereby offering some information as to how a full-scale system might perform. This also meant that the temperature in the field system was strongly coupled to the intensity of the sunlight, and hence to the intensity of the UV exposure. In any event, these data were obtained from an exposure to ~530 - 560 suns of actual sunlight, and 590 - 630 suns with the short wavelengths (<390 nm) removed. These data suggest that the predictions based on laboratory data may actually be conservative.

Conclusions

From the research conducted thusfar, several conclusions can be made with regard to the global performance of a solar incinerator as compared to a similar conventional system.

- It has been demonstrated that a simultaneous exposure to concentrated sunlight and high temperature can significantly increase the destruction efficiency of an organic waste as compared to a similar purely thermal exposure

- The photochemical reactions appear to occur primarily from excited triplet states

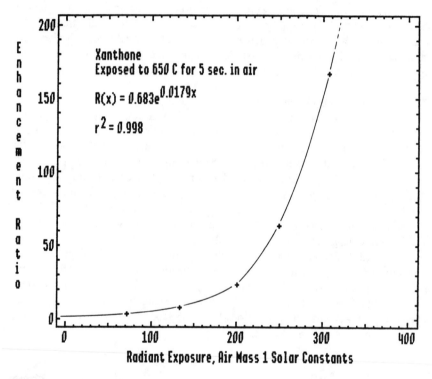

Figure 14. Enhancement ratio as a function of radiant intensity for xanthone exposed for 5 seconds in air.

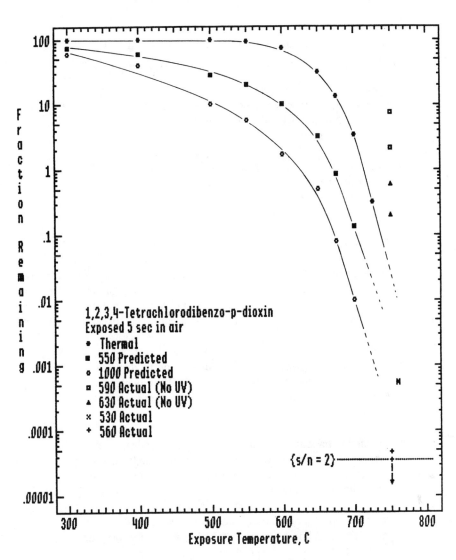

Figure 15. Actual and predicted decomposition profiles of TCDD exposed to 0 - 1000 suns. Note that (s/n = 2) denotes the detection limit for the field tests.

- A solar incinerator should be less sensitive to transient upsets than a conventional system as the solar incineration process remains effective under oxygen starved conditions

- Components of a waste which do not absorb solar energy directly may still benefit through indirect, solar induced destruction processes

- A solar facility should also operate more cleanly in that the solar process produces products in fewer numbers, in lower yields, and destroys them at lower temperatures than a conventional system

- Modeling the effect of increasing radiant flux suggests that a solar incinerator operating with 1,000 to 1,500 suns of radiant energy should operate 100 to 100,000 times more efficiently than its thermal counterpart

Future Research

The work described above establishes a foundation on which to build more thorough studies. This future work will include a continuation of the research which is applied in nature, as well as working toward establishing a sound, fundamental understanding of these high temperature photochemical reactions. This will include:

- Investigating the effect of the solar process on mixtures of wastes which are of varying complexity. Of particular interest is the effect of the process on materials which are not directly amenable to solar enhancement

- Conducting pressure dependence studies to investigate the possibility of inducing reactions from excited singlet states by increasing the rate of thermal energy transport from the bath gas to the target molecule

- Obtaining spectroscopic properties on test compounds in the gas-phase at high temperatures including absorption spectra, excited state lifetimes, and quantum yields

- Conducting detailed kinetic and mechanistic investigations of model systems including complete product identification to better quantify the solar destruction process

- Closely examining the near UV portion of the simulated solar spectrum and determining how well it compares to a "standard" spectrum to obtain a better measure of the accuracy of the simulation

- Conducting experiments under intense monochromatic radiation, i.e., using lasers, to provide a more precisely defined radiant exposure

Literature Cited

1. Zegel, William C., "An Overview of Hazardous Waste Issues," *Journal of the Air Pollution Control Association*," **1985**, Vol *35*, No. 1, , pp. 50-54.
2. Oppelt, E. T., "Incineration of Hazardous Waste, A Critical Review," *Journal of the Air Pollution Control Association*, **1987**, Vol. *37*, No. 5, pp. 558-586.
3. Dellinger, B. et al., "Determination of the Thermal Stability of Selected Hazardous Organic Compounds," *Hazardous Waste and Hazardous Materials*, **1984.**,Vol. *1*, No. 2, pp. 137-157.
4. Turro, N. J., *Modern Molecular Photochemistry*, the Benjamin/Cummings Publishing Co., Inc., Menlo Park, CA, **1978**.
5. McGlynn, S. P., Azumi, T., and Kinoshita, M., *Molecular Spectroscopy of the Triplet State*, Prentice-Hall, Inc., Englewood Cliffs, NJ, **1969**.
6. Graham, J. L., Hall, D. L., and Dellinger, B., "Laboratory Investigation of the Thermal Degradation of a Mixture of Hazardous Organic Compounds - I," *Environmental Science and Technology*, **1986**, Vol. *20*, No. 7, pp. 703-710.
7. Graham, J. L. and Dellinger, B., "Solar Thermal/Photolytic Destruction of Hazardous Organic Wastes," *Energy*, **1987**, Vol. *12*, No. 3/4, pp. 303-310.
8. Glatzmaier, G. C., Nix, R. G., and Mehos, M. S., "Solar Destruction of Hazardous Chemicals," *Emerging Technologies for Hazardous Waste Treatment, American Chemical Society*, Division of Industrial and Engineering Chemistry, Spring Symposium, **1989**, Atlanta, GA.

RECEIVED April 5, 1991

Chapter 7

Oxidation Technologies for Groundwater Treatment

R. E. Heeks[1], L. P. Smith[2], and P. M. Perry[2]

[1]Xerox Corporation, 800 Phillips Road, Webster, NY 14580
[2]H & A of New York, 189 North Water Street, Rochester, NY 14604

Xerox Corporation has pilot tested three UV / Oxidation processes for the treatment of contaminated groundwater containing chlorinated and non-chlorinated organic solvents. The technologies pilot tested included the ULTROX system developed by ULTROX International, the perox-pure process of Peroxidation Systems, Inc. and the Rayox process by Solarchem Environmental Systems. The three processes use a combination of ultraviolet light and hydrogen peroxide to oxidize organic solvents in water. The ULTROX system includes ozone as part of the treatment. Data gathered during pilot testing demonstrated that these processes are effective in the destruction of organic contaminants in groundwater. These results are discussed in regard to applicability to the groundwater remediation at the Xerox Facilities in Webster and Blauvelt, New York.

Over the past few years, Xerox Corporation has been aggressively investigating innovative technologies for treatment of contaminated groundwater. Those technologies that result in the destruction of contaminants rather than the transfer from one physical state to another such as the case with air stripping or activated carbon were of particular interest. As a result, chemical oxidation technologies involving combinations of ultraviolet radiation, hydrogen peroxide and ozone were evaluated as a method for groundwater treatment. The evaluation of oxidation as an emerging technology for treatment of contaminated groundwater involved field trails of the ULTROX system developed by ULTROX International , the perox-pure process by Peroxidation Systems, Inc. and the Rayox process by Solarchem Environmental Systems.

Each of these groundwater treatment systems was tested at the Xerox Facility in Webster, New York referred to as the Salt Road Site. The perox-pure system was also tested at the Xerox Facility in Blauvelt, New York. It is the intent of this paper to describe the results of testing of these units by Xerox for the treatment of contaminated groundwater.

0097–6156/91/0468–0110$06.75/0
© 1991 American Chemical Society

History

State of the art groundwater remediation activities have been evaluated and tested by Xerox Corporation over the past several years at the Xerox facilities in Webster and Blauvelt, New York as well as other sites not discussed in this paper. Activities at both sites have involved extensive hydrogeologic investigations resulting in the installation of over one hundred monitoring wells along with groundwater interim remediation systems which continue to operate today.

The Xerox Facility referred to as the Salt Road site is a 400 acre complex of manufacturing buildings and open space located in the northeast section of the Xerox manufacturing facility in Webster, New York. The Salt Road facility manufactures xerographic toner, developer and photoreceptors. During additional construction in the mid-1980's, it was discovered that underground containment tanks for Trichloroethene (TCE) and Toluene had leaked and contaminated surrounding soils. Subsequent investigations indicated that the groundwater quality in the vicinity had also been affected. Notification and coordination with the U.S. Environmental Protection Agency (EPA) and the New York State Department of Environmental Conservation (NYSDEC) has led to the implementation of a groundwater recovery and treatment system based on air stripping technology as an interim measure.

The Blauvelt site is located in Blauvelt, New York and at one time was used by Xerox as a refurbishing center for copier machines. The refurbishing process involved the use of such solvents as TCE and toluene. Today, the facility is used for the storage of copier parts. Groundwater contamination was discovered at the Blauvelt site in 1980. As in the case of Salt Road, the source of contamination at this site was leakage from underground tanks containing chlorinated solvents and toluene. Subsequent groundwater remediation activities at this site have included the installation of a groundwater recovery and treatment facility using air stripping followed by activated carbon as a polishing step.

Both the Salt Road and Blauvelt site have had extensive involvement with State and Federal regulatory agencies. This involvement has led to the negotiation of a Consent Order for the Salt Road site with EPA. A Consent Order under NYSDEC was recently signed for the Blauvelt site.

The investigation and remedial activities including the testing of UV / Oxidation systems have been performed in close consultation and cooperation with the regulatory agencies.

Remedial Measures To Date

The aerial and vertical dimensions of the contaminant plumes as well as chemical profile have been determined at the Salt Road and Blauvelt sites. Table I presents a profile of the groundwater quality at these sites.

Table I

Range Of Organic Chemical Constituent Concentrations Entering Groundwater Treatment

Contaminant	Salt Road (ppb)	Blauvelt (ppb)
Trichloroethene (TCE)	374-10600	1490-2380
1,2 Dichloroethene (Cis and Trans) (1,2 DCE)	138-8730	3020-6860
1,1,1 Trichloroethane (1,1,1 TCA)	ND-466	2440-4270
Toluene	ND-1980	ND-134
Vinyl Chloride	ND-470	ND
Tetrachlorethene (PCE)	ND-57	4540-7760
1,1 Dichloroethene (1,1 DCE)	ND-5	ND-165
1,1 Dichloroethane (1,1 DCA)	ND-334	ND-220
Methylene Chloride	ND	ND-177

ND - NOT DETECTABLE

Interim remediation measures implemented at the Salt Road site included the removal and off-site disposal of 500 cubic yards of contaminated soil from the source area. These activities were conducted shortly after the contamination was discovered. Since that time, over 100 monitoring wells of various depths have been installed at the Salt Road site along with 13 recovery wells and an air stripper for groundwater treatment. At the Blauvelt site, 25 monitoring wells as well as 10 recovery wells have been installed as part of groundwater remediation efforts.

Continued interim remediation measures at the Salt Road and Blauvelt sites have included the pilot-testing of UV/Oxidation technologies. Oxidation techniques for the treatment of contaminated groundwater were selected by Xerox for evaluation based on their potential to destroy specific chemical compounds, thereby minimizing contaminant concentrations in effluent discharges. The ULTROX and Peroxidation units were evaluated over a period of approximately ten (10) months each. The Solarchem Rayox process was not identified until the end of the testing sequence. Therefore, relatively few test points were obtained and it should not be considered that the unit was fully optimized in operation during this test period. The results of these units are discussed in detail.

Oxidation As An Emerging Technology

The oxidation technologies pilot tested at the Xerox facilities have involved the use of combinations of ultraviolet radiation, hydrogen peroxide and in one unit, ozone. The basic principal of oxidation technologies is the generation of the hydroxyl radical which promotes the destruction of organic compounds. Hydrogen peroxide, and ozone under the influence of ultraviolet radiation have greater oxidizing capabilities when compared to other common oxidants because of their potential to generate the hydroxyl radical in the presence of water. *(Ref. 1,2,3,4,5, 6).* Approaches to utilizing this principle for the purification of water have been embodied in the equipment provided by the three vendors studied at Xerox as described in this paper.

General Theory

When the oxidation of organic constituents in water is carried out to completion, the by products generated are carbon dioxide and water. If halogens are present, these are converted to the corresponding inorganic halides. Although ultraviolet light alone does have some destructive power, the hydroxyl radical is a superior oxidizing agent only to be surpassed by fluorine when compared to common chemical oxidants as is shown in Table II.

Table II
Oxidation Potential Of Oxidants

Relative Oxidation Power Chlorine = 1.0	Species	Oxidative Potential (volts)
2.23	Fluorine	3.03
2.06	Hydroxyl Radical	2.80
1.78	Atomic Oxygen (singlet)	2.42
1.52	Ozone	2.07
1.31	Hydrogen Peroxide	1.78
1.25	Perhydroxyl Radical	1.70
1.24	Permanganate	1.68
1.17	Hypochlorous Acid	1.59
1.15	Chlorine Dioxide	1.57
1.10	Hypochlorus Acid	1.49
1.07	Hypoiodous Acid	1.45
1.00	Chlorine	1.36
0.80	Bromine	1.09
0.39	Iodine	0.54

The literature *(Ref. 1,4,5)* identifies paths or steps by which hydrogen peroxide or hydrogen peroxide and ozone in the presence of UV energy form hydroxyl radicals which attack organic compounds in water. An example is the reaction of formic acid with hydrogen peroxide which has been catalyzed by UV.

General Theory

$$UV$$

$$H_2O_2 \text{-------} > 2\, {}^{\bullet}OH$$

$$254\,nm$$

$$HCOOH + {}^{\bullet}OH \text{--------} > H_2O + HCOO^{\bullet}$$

$$HCOO^{\bullet} + OH^{\bullet} \text{--------} > H_2O + CO_2$$

As with other chemical reactions, UV/Oxidation processes are dependent upon a number of reaction conditions which affect both performance and cost. These variables include the following.

- Type and concentration of organic contaminant
- Light transmittance of the water
- Type and quantity of dissolved salts such as iron and carbonates
- UV dosage
- Residence time

- Ozone and/or peroxide dosage
- pH

While the UV/Oxidation process is based on known chemistry, the equipment to carry out this process is continuously undergoing development. The challenge of the UV/Oxidation equipment producers today is to utilize the information that is being generated to maximize the power and reliability of the process while minimizing costs for long term use in purification of effluents with low contaminant concentrations.

Specification

The Xerox specification for a UV / Oxidation process provided to each vendor was 95 percent reduction of TCE at a through put rate of 250 gpm and 99 percent reduction of TCE at a throughput rate of 125 gpm. Additionally, an up time target of 90 percent calculated on a quarterly basis would, if not met, require penalties from the vendor unless the down time is under Xerox control. The validity of specification depends upon incoming groundwater quality and quantity conforming to the Xerox projection.   Additionally, specific preventative maintenance actions must be taken to maintain equipment up time.

Equipment Description

The UV / Oxidation processing units tested are described in the following paragraphs. Each process requires a stock feed of 50 percent hydrogen peroxide in water. The concentration of hydrogen peroxide, in the case of the ULTROX unit, for example was in the order of 0.002% during the processing. Peroxide is stored in polyethylene or aluminum tanks and is fed by chemical metering pumps directly into the incoming water stream. ULTROX uses ozone as a source of hydroxyl radical production in addition to hydrogen peroxide. Ozone for the ULTROX unit was produced on site by an ozone generator process involving a conventional spark gap method. The three processes use UV lamps of varying power with the principle wavelength being 254 nanometers. Some of the more unique characteristics of these devices are discussed in the following descriptions.

Peroxidation Systems. The Peroxidation unit tested was the perox-pure LV60 (Figure 1) (7). The reactor is rectangular with an 80 gallon capacity. There are four UV lamps stacked horizontally within the reactor. These lamps are protected from the water by quartz sleeves. The UV lamps can be controlled individually to provide variations in the number of lamps as well as power to the lamps. The LV60 unit used in this test contains four lamps, with a potential to produce 15 kW of energy per lamp.

Incoming water was mixed with hydrogen peroxide and entered into the bottom of the oxidation chamber. It then flowed upward over the UV lamps to exit from the top of the unit. The design of the flow pattern assures adequate mixing is obtained.   For the Xerox Salt Road application, Peroxidation Systems included a sand filter in an attempt to remove precipitated iron from incoming groundwater. The complete unit has alarms, automatic shutdown and controls to allow automatic

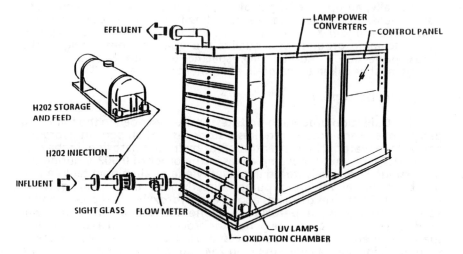

Figure 1. Peroxidation systems.

operation. The Peroxidation Systems LV60 was tested at both the Salt Road and Blauvelt sites.

Ultrox International. The ULTROX International model 725 unit was tested at the Salt Road site (Figure 2) (3). This unit has a capacity of 725 gallons. The reactor chamber contains 72 UV lamps, each having a 65 watts capacity. The lamps are placed vertically and protected by quartz sleeves.

This unit includes ozone in the treatment process. A generator is used to produce ozone on-site from dry air. Ozone is fed into the groundwater in the reaction chamber. There is an exit port for off gassing to the atmosphere. This contains a heated catalyst placed in the line to destroy ozone prior to entering the ambient air. The unit is equipped with an ozone monitoring device which reads the level of ozone and can be adapted to sound an alarm when an unacceptable level of ozone is encountered in the ambient air. The unit is designed to operate automatically.

Solarchem Environmental Systems. Late in the testing program a decision was made to test the Solarchem Environmental Systems, Rayox pilot unit at the Salt Road site (Figure 3) (8). The device obtained was a cylindrical reaction chamber containing a vertical 6 kW UV lamp protected by a quartz sleeve. Within the reaction chamber there was a mechanical cleaner which transversed the quartz sleeve automatically at a predetermined time interval to maintain cleanliness of the sleeve. Reactor volume was approximately 15 gallons.

Solarchem also makes provision for the addition of chemicals to catalyze the decomposition of hydrogen peroxide into the hydroxyl radical. This fluid, called ENOX™, is fed into the incoming groundwater stream in a matter similar to the hydrogen peroxide. The additive was not considered by Solarchem to be useful for the Salt Road groundwater. Solarchem as well as Peroxidation Systems have explored a variety of UV wavelengths to optimize absorption for more complete or higher rate of destruction of specific chemicals.

Process Methodology

The goal of the pilot trials was to determine the capability of the selected oxidation process units to meet the Xerox specification when operated under optimized conditions. Additionally, the ability of the unit to maintain a high degree of performance with continued operation was also evaluated along with relative cost of operation. Information was obtained to scale up to a unit capable of 250 gpm throughput.

Performance of the UV/Oxidation units was evaluated by measuring contaminant concentrations in the groundwater influent and effluent. Initial phase of the trial period focused on the adjustment of the unit to achieve optimum treatment of the incoming ground water. Real time measurements were taken by collecting influent and effluent samples which were screened for volitile organic compounds using a portable gas chromatograph. At start up, samples were submitted daily to an

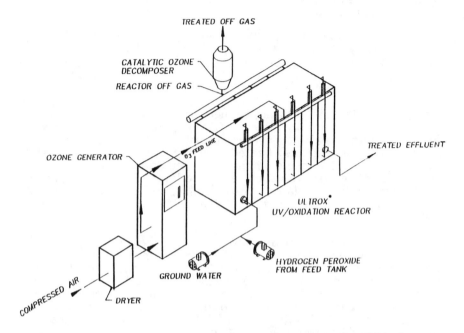

Figure 2. Isometric view of Ultrox International Model 725 unit.

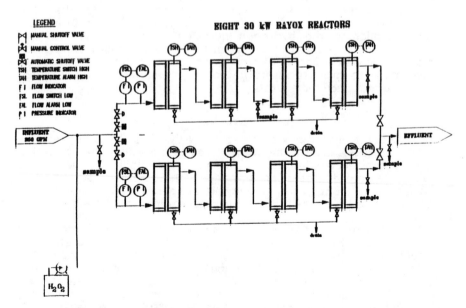

Figure 3. Solarchem Environmental Systems, Rayox pilot plant.

analytical laboratory for volatile organic compound analysis using EPA methods 8010/8020 to substantiate the screening results.

Establishment of optimum unit operation conditions was dependent upon proper adjustment of hydrogen peroxide flow, residence time of the water in the reaction chamber, ultraviolet radiation and in the case of the Ultrox unit, dosage of ozone. After ideal conditions were established, sampling for laboratory analysis was performed weekly to monitor the contaminant elimination efficiency. The test period for the Peroxidation Systems and Ultrox unit was approximately ten months each. As noted previously, the Solarchem unit was only evaluated for the period of about one week near the end of the test period.

Results

Results of the three UV/Oxidation processes pilot tested are summarized in Tables III through XI, where near optimized conditions are presented. The data are typical of performance achievable under routine operating conditions. In a few instances, particularly with saturated compounds, the influent stream contained a compound below the detection limit while in the effluent stream, concentration was slightly above the detection limit. These occurrences are not counted in the percent reduction averages. However, the data are presented in the Tables. The initials "ND" are used throughout this article to signify a compound was not present in quantity above the analytical method detection limit. Detection limits for influent ranged from 50 to 100 ppb and for effluent, from one to ten ppb. These ranges were affected by contaminant concentrations in the streams. It should be noted the pH of the incoming ground water was measured intermittently and found to be near seven with some excursions falling between six and eight.

Peroxidation Systems, Inc. Tables III and IV show the perox-pure LV60 trial results obtained at the Salt Road site. Optimized conditions were found to exist at a flow rate of 50 gpm which corresponds to 1.6 minutes residence time. Influent hydrogen peroxide level was optimized at 80 mg/l of 50% solution. The LV/60 contained four UV lamps, 15 kW each. Greatest reliability occurred when all four lamps were operated at their specified power of 15 kW. The LV60 as well as all the units tested was placed before an air stripper in the process line. At Blauvelt, some trials also were performed with the air stripper placed before the LV60 in the process.

As can be seen in the Tables, the perox-pure LV60 was very efficient in removing vinyl chloride, 1,2 DCE, TCE, toluene and PCE. The destruction of saturated compounds, 1,1 DCA and 1,1,1 TCA was achieved only to 73.0 percent and 17.7 percent, respectively. Difficulty in destruction of saturated compounds was common through out the trials for all units tested. However, owing to the chemical composition of the groundwater being purified, the overall total volatile organic reduction reached near 95 percent.

Table III. Perox-Pure LV60 Trial Results at Salt Road Site

Day	1		2		3		4		5	
	Influent (ppb)	Effluent (ppb)	Influent (ppb)	Effluent (ppb)	Influent (ppb)	Effluent (ppb)	Influent (ppb)	Effluent (ppb)	Influent (ppb)	Effluent (ppb)
Residence Time (Min.)	0.85		1.0		1.4		1.6		1.6	
Vinyl Chloride	163.0	ND	49.0	ND	ND	ND	ND	ND	198.0	ND
1,1-DCA	67.3	40.4	117.2	43.9	ND	36.0	250.5	31.2	105.0	30.5
1,2, DCE	1683.8	25.5	933.0	15.9	731.5	15.9	2750.0	6.9	1622.0	16.7
1,1,1-TCA	ND	182.0	75.7	76.8	85.1	74.8	127.5	85.7	ND	102.0
TCE	4200.0	968.0	789.5	27.2	6895.0	114.5	11550.0	73.5	3680.0	23.9
Toluene	434.0	13.9	227.5	1.2	272.5	1.7	642.5	ND	404.0	ND
PCE	309.0	21.2	ND	ND	ND	ND	ND	ND	242.0	4.4
Total	6857.1	1251.0	2191.9	165.0	7984.1	242.9	15320.5	197.3	6251.0	177.5
% Removal	81.8%		92.5%		97.0%		98.7%		97.2%	

Test Conditions:

- Sand and bag filters ahead of LV-60
- Quartz tubes cleaned and remained clean during test.
- Lamp #223 (Std) used. 15 kW
- Influent H_2O_2 approximately 80 mg/l. (50%)
- Four lamps on days 2 - 5.
  Two lamps on day one.

Table IV. Average Scores on Perox-Pure LV60 Trials at Salt Road Site

Contaminent	Average Influent (ppb)	Average Effluent (ppb)	% Removal ($\leq$)
Vinyl Chloride	82	ND	100.0
1,1 DCA	135.0	36.5	73.0
1,2 DCE	1544.1	16.2	99.0
1,1,1 TCA	96.1	79.1	17.7
TCE	5422.9	241.4	95.6
Toluene	396.1	3.4	99.1
PCE	110.2	5.1	95.4
Total	7786.4	381.7	95.1

During the trials of the perox-pure unit, the iron content of the water began to fluctuate with intermittent slugs of high concentrations coming into the system. In time, the quartz sleeves surrounding the UV lamps became coated with iron resulting in decreased efficiency of the lamps. This effect caused decreased performance of the unit. After cleaning of the sleeves, efficiency returned to its prior level of 95 + % reduction of total volatiles. However, after about three weeks the iron coatings repeated and again, performance decreased to an unacceptable level. Peroxidation Systems placed a sand filter in the influent line which appeared to alleviate the problem to a large extent. The conclusion from these trials was that the perox-pure LV60 is capable of removing an average of over 95% total volatiles from Salt Road groundwater at flow rates from 50 to 75 gallons per minute. However, the need to properly design and operate a pretreatment system was indicated for removal of iron from the water supply. The scaled up unit projected by Peroxidation Systems to meet the specification included the perox-pure   360 equipment which has a holding tank of 480 gallons capacity with energy input of twenty-four, 15 kW lamps. Peroxidation Systems also included the need for an adequate filtration system to purify the water prior to entry into the reaction chamber to minimize scaling of the quartz tubes surrounding the lamps.

At the Blauvelt site, the perox-pure LV60 system first was tested using influent groundwater after it had been processed through an air stripper to obtain highly purified water effluent. A second trial was conducted with the perox-pure unit placed prior to the air stripper. Tables V and VI describe the results of these trials.

Optimum conditions at Blauvelt included a residence time of 2.5 minutes, 4 UV lamps operating at their specified power 15kW and a flow rate of 50% concentration hydrogen peroxide at 40 mg/l. Although influent concentrations were greatly different using streams before and after processing through the air stripper, reduction efficiency of organic constituents was persistently in the range of 90%. As in prior studies of UV/Oxidation techniques, TCA demonstrated strong resistance to destruction with efficiency of removal of 28%.   The unsaturated compounds as found in other trials, demonstrated removal efficiencies near 100%. It should be noted the problem of iron coating the quartz sleeves was not pronounced at the Blauvelt site indicating low iron concentrations in the groundwater.   The perox-pure LV60 could accommodate adequate capacity to fulfill Blauvelt's requirement. Therefore, a perox-pure LV60 was leased by Xerox for use at its Blauvelt site.

Table V. Perox-Pure LV60 Trial Results at Blauvelt Site

Day	1		2		3		4		5		6	
	Influent (ppb)	Effluent (ppb)	Influent (ppb)	Effluent (ppb)	Influent (ppb)	Effluent (ppb)	Influent (ppb)	Effluent (ppb)	Influent (ppb)	Effluent (ppb)	Influent (ppb)	Effluent (ppb)
Residence Time (Min.)	2.5		2.5		2.5		2.5		2.5		2.5	
1,2,DCE	11.2	ND	9.1	ND	6.3	ND	16.1	ND	5.7	ND	3480.0	ND
1,1,1-TCA	5.9	3.7	3.6	2.8	2.8	2.6	8.6	7.6	1.9	ND	1980.0	1430
TCE	3.9	ND	2.5	ND	1.4	ND	6.7	ND	1.5	ND	1480.0	ND
PCE	15.7	ND	9.2	ND	15.8	ND	29.9	ND	4.5	ND	4990.0	ND
Total	36.7	3.7	24.4	2.8	26.3	2.6	61.3	7.6	13.6	ND	11930	1430
% Removal	89.9		88.5		90.1		87.6		100		88.0	

Test Conditions:

• Day 1 - 5 Ground water flowed through air stripper prior to the Peroxidation unit, LV60 used as polisher.
• Day 6 - Ground water flowed through LV60 prior to air stripper. Stripper used as polisher.
• Influent of H₂O₂ (50%) - 40 mg/l
• Four bulbs on during test. Standard # 223 bulbs used, 15 kW.

Table VI. Average Scores on Perox-Pure LV60 Trials at Blauvelt Site

Perox-Pure After Air Stripper

Contaminent	Average Influent (ppb)	Average Effluent (ppb)	% Removal (≤)
1,2 DCE	9.7	ND	100.0
1,1,1 TCA	4.6	3.3	28.2
TCE	3.2	ND	100.0
PCE	15.0	ND	100.0
Total	32.5	3.3	89.8

Perox-Pure Before Air Stripper

Contaminent	Average Influent (ppb)	Average Effluent (ppb)	% Removal (≤)
1,2 DCE	3480	ND	100
1,1,1 TCA	1980	1430	27.8
TCE	1480	ND	100
PCE	4990	ND	100
Total	11930	1430	88.0

Ultrox International. The greatest amount of data obtained during these trials was from the ULTROX International F-725 pilot unit. As noted previously, this device contained seventy-two 65 watt lamps as opposed to some of the competitive systems that used fewer but higher energy lamps. Tables VII and VIII demonstrate that optimized conditions for processing were near 24 minutes residence time or 30 gallons per minute at a hydrogen peroxide flow of 40 mg/l and an ozone flow of 58 mg/l. Some trials were run under the same conditions but using 14.5 minutes residence time relating to flow rate of 50 gpm. This appears to be a borderline condition for this unit because removal fell below 95 percent. On the tenth test day, hydrogen peroxide was reduced from 40 to 20 mg/l with other conditions remaining constant. Although prior trials indicated the hydrogen peroxide flow level of 20 mg/l to be excessively low, the data for day ten show little sensitivity to peroxide level in that particular test. It is considered desirable to reduce the hydrogen peroxide as far as possible to avoid obtaining the constituent in the effluent stream. Referring to Table IX it is seen that DCA and TCA follow the trend of low removal efficiency when under going UV/Oxidation treatment with removal efficiencies of 55.0 and 57.5 percent, respectively. The groundwater at Salt Road contains relatively small quantities of these saturated chlorinated hydrocarbons. Therefore, total removal efficiency in excess of 95 percent still can be achieved.

The USEPA, in a comprehensive evaluation of ULTROX equipment under its Superfund Innovative Technology Evaluation (SITE) Program, established the fully configured ULTROX system emitted no volatile organic compounds into the atmosphere. The EPA observed that the great majority of volatile organic compound destruction occurred in the water phase. A very small percentage of the volatile organic compound destruction, especially for saturated compounds such as TCA and DCA, occurred in the DECOMPOZON/D-TOX off gas control device. The EPA's findings are consistent with those found at Xerox. Using a unit without the D-TOX catalyst Xerox in its pilot runs detected trace quantities of organic materials in the effluent air stream on an intermittent basis. The full scale ULTROX system is to have the DECOMPOZON/D-TOX off gas device to ensure total volatile organic compound control. In the case of the Salt Road Application, the ULTROX machine will be placed in the line prior to a permitted air stripper presently in operation.

During the ULTROX trials a five micron filter was placed in the influent line to remove suspended particulate matter in the water. The ULTROX unit did not show a high tendency towards scaling of the quartz sleeves surrounding the UV lamps. After about five months running, some deterioration was noted. However, this was solved by use of a weak acid rinse. The unit performed on a high level of efficiency on a consistent basis.

Table VII. F-725 Pilot Unit Trial Results, Days 1–5

Day	1		2		3		4		5	
	Influent (ppb)	Effluent (ppb)	Influent (ppb)	Effluent (ppb)	Influent (ppb)	Effluent (ppb)	Influent (ppb)	Effluent (ppb)	Influent (ppb)	Effluent (ppb)
Residence Time Min.		14.5		14.5		14.5		14.5		14.5
Vinyl Chloride	ND	ND	ND	ND	ND	ND	ND	ND	ND	ND
1,1-DCA	ND	25.9	ND	10.1	ND	21.0	ND	14.9	ND	14.6
1,2-DCE	1220.0	ND	835.0	2.3	1030.0	4.61	733.0	28.9	681.0	23.5
1,1,1-TCA	61.3	21.5	66.9	18.1	52.5	37.2	16.3	20.4	14.3	13.7
TCE	2710.0	8.2	3930.0	23.4	1960.0	14.7	4290.0	211	2290.0	151.0
Toluene	ND	ND	ND	ND	71.3	ND	56.0	ND	29.5	ND
PCE	65.4	ND	ND	ND	ND	ND	ND	ND	ND	ND
Total	4056.7	63.4	4831.9	53.9	3113.8	80.51	5095.3	275.2	3014.8	202.8
% Removal		98.4		98.9		97.4		94.6		93.3

Test Conditions:

• 5 Micron bag filters ahead of F-725.
• Influent H_2O_2 approximately 40 mg/L except day 10.
• Ozone flow 58 mg/l

Table VIII. F-725 Pilot Unit Trial Results, Days 6–10

Day	6		7		8		9		10	
	Influent (ppb)	Effluent (ppb)	Influent (ppb)	Effluent (ppb)	Influent (ppb)	Effluent (ppb)	Influent (ppb)	Effluent (ppb)	Influent (ppb)	Effluent (ppb)
Residence Time Min.		14.5		24.0		24.0		24.0		24.0
Vinyl Chloride	20.4	ND	ND	ND	303.0	ND	ND	ND	60.4	ND
1,1-DCA	ND	13.8	78.7	38.6	97.5	27.8	52.1	31.5	72.3	37.4
1,2,DCE	958.0	33.6	1250.0	7.6	2260.0	ND	1470.0	4.1	1660.0	13.2
1,1,1-TCA	128.0	21.9	47.2	29.0	ND	24.2	ND	27.6	44.6	21.2
TCE	3970.0	211.0	1500.0	14.2	1790.0	6.8	1650.0	22.0	2190.0	39.9
Toluene	60.4	ND	118.0	ND	671.0	ND	337.0	ND	118.0	ND
PCE	ND	1.1	ND	ND	ND	ND	ND	1.0	22.8	2.0
Total	5136.8	281.4	2993.9	89.4	5121.5	58.8	3509.1	86.2	4168.1	113.7
% Removal		94.5		97.0		98.9		97.5		97.3

Test Conditions:

- 5 Micron bag filters ahead of F-725.
- Influent H_2O_2 approximately 40 mg/L except day 10.
  Day 10 influent H_2O_2 approximately 20 mg/L.

- Ozone flow 58.0 mg/l

Table IX. Average Scores of Trials with F-725 Pilot Unit

Contaminant	Average Influent (ppb)	Average Effluent (ppb)	% Removal ($\leq$)
Vinyl Chloride	38.3	ND	100.0
1,1 DCA	75.1	33.8	55.0
1,2 DCE	1209.7	11.8	99.0
1,1,1 TCA	53.9	22.9	57.5
TCE	2628.0	70.2	97.3
Toluene	146.1	ND	100.0
PCE	44.1	1.0	97.7
Total Volatiles	4195.2	138.7	96.7

Solarchem Environmental Systems. Optimum operating conditions for groundwater treatment with the Rayox unit at Salt Road involved the use of a 6 kW lamp with a two minute residence time in the reaction chamber and the addition of 24 mg / l of 50% hydrogen peroxide. Since this was a small pilot unit, flow through the unit was limited to 10 gpm. However, Solarchem stated sufficient data were available to scale up to a 250 gpm capacity unit.

The Rayox reactor was successful in destroying unsaturated organic compounds such as TCE and toluene. Destruction efficiencies were less for the saturated organic compounds, TCA and DCA, as has been the case with other reactors tested. As can be seen in Tables X and XI, removal of total contamination could be achieved to a level of up to 95.5 percent. Solarchem has stated that at other groundwater remediation sites, the Solarchem unit has achieved greater than 99.9 percent destruction of DCA and TCA using a larger unit than that employed at Salt Road. Competitive systems such as ULTROX and Peroxidation also have noted some success in destruction of saturated compounds such as DCA and TCA at other sites using different operating conditions.

Although sufficient time was not available to complete the testing of the Rayox unit, Solarchem personnel reported that twice the normal dose of UV energy was required for destruction of organic contamination in the Salt Road groundwater due to unusually high loading of inorganic salts. Use of a proprietary additive, ENOX, was considered not to be significantly beneficial to justify its use in these trials. In limited testing there was no evidence of stripping of contamination by volatization from the groundwater in the unit. Based upon this relatively short trial,

Table X. Rayox Reactor Trial Results

Day	1 Influent (ppb)	1 Effluent (ppb)	2 Influent (ppb)	2 Effluent (ppb)	3 Influent (ppb)	3 Effluent (ppb)	4 Influent (ppb)	4 Effluent (ppb)	5 Influent (ppb)	5 Effluent (ppb)
Residence Time (Min.)	2.0		2.0		2.0		2.0		1.0	
Influent H_2O_2 (50%) mg/l	24		24		24		0		30	
Vinyl Chloride	ND	ND	49.0	ND	ND	ND	68.1	14.4	ND	ND
1,1-DCA	ND	17.0	40.3	15.0	30.9	21.5	ND	11.2	ND	20.6
1,2-DCE	554.00	43.0	346.0	7.6	791.0	264.0	528.0	156.0	678.0	84.9
1,1,1-TCA	18.2	11.2	10.4	12.5	25.4	22.1	125.0	21.6	ND	13.1
TCE	1070.0	73.6	853.0	22.2	2050.0	225.0	3000.0	292.0	1880.0	177.0
Toluene	94.3	ND	27.2	ND	36.0	10.0	45.4	7.1	ND	ND
PCE	ND	ND	ND	ND	ND	ND	ND	ND	ND	ND
Total	1736.5	144.8	1276.5	57.3	2933.3	542.6	3766.5	502.5	2558.0	295.6
% Removal	91.7		95.5		81.5		86.7		88.4	

a scaled up unit was conceived to consist of eight separate reactors each with a single lamp. This unit would have a foot print of 10' X 16' in an area which would be relatively small compared to some competitive units. Solarchem produced impressive progress with the time that was alloted to them in demonstrating their approach to UV/Oxidation technology.

Table XI. Average Scores of Trials with Rayox Reactor

Contaminant	Average Influent (ppb)	Average Effluent (ppb)	% Removal (≤)
Vinyl Chloride	23.4	2.9	87.6
1,1 DCA	35.6	18.2	48.7
1,2 DCE	579.4	111.1	80.1
1,1,1 TCA	44.8	16.8	62.4
TCE	1770.6	158.0	91.1
Toluene	40.6	3.4	91.6
Total Volatiles	2494.4	310.4	87.4

Cost Estimate

Except for the Solarchem Rayox unit, sufficient information was generated to obtain cost estimates for the processing of 250 gpm flow rate of Xerox groundwater with 95% TCE removal. Information also was obtained to enable calculation of cost data at a flow rate of 25 gpm. Cost estimates were designed to be fully loaded, i.e., inclusive of all equipment, operating and maintenance cost.

Although variation among vendors does exist, average cost per thousand gallons of groundwater processed was determined at a flow rate of 250 gpm to be $1.57 . In reducing flow rate to 25 gpm and down sizing equipment, cost increased to beyond $5.00 per thousand gallons processed. This cost data emphasizes the merits of processing greater flow rates of water where fixed costs can be spread over large volumes of water processed per unit time.

Conclusions

Based upon the studies of the pilot units described, the following conclusions were drawn:

● UV/Oxidation technology is applicable on a commercial scale for the purification of groundwater containing the unsaturated compounds of TCE, DCE, PCE, Vinyl Chloride and Toluene.

● The processes evaluated were less efficient in the destruction of saturated compounds such as DCA and TCA. This is the subject of additional work by the equipment developers.

- The commercial units available rely upon similar technology. However, each application requires optimization of the process for destruction of specific organic compounds as well as adaptation to handle inorganic salts in the water that interfere with the process. Constituants in the groundwater that would not usually be considered toxic such as iron or bicarbonate, can influence the efficiency of a UV/Oxidation process to a significant extent.

- Cost per unit quantity of ground water processed can very significantly depending upon total volume per unit time. Some variation among different vendors should be anticipated. However, trials at Xerox suggest a cost range from $1.57 to near $5.00 for thousand gallons processed in progressing from a throughput rate of 250 down to 25 gallons per minute.

Discussion

Based upon the results of these trials, Xerox leased a Peroxidation, LV60 unit for its facility in Blauvelt, New York. The intent is to polish water after being processed by an air stripper to a high purity level for effluent to a storm sewer. At the Webster, Salt Road site an ULTROX system was ordered to accomodate full scale flow of 250 gallons per minute. The Salt Road groundwater will be processed first through the ULTROX unit to be followed by an air stripper. The intent is to provide highly purified water to a storm sewer system with levels of contamination being significantly below permitted both in the effluent air and water levels. Due to time constraints the Solarchem Rayox System was not fully optimized or performance tested in depth in this work. However, the results obtained were presented for completeness since sufficient information on the process was obtained to demonstrate this process will accomplish the cleaning of groundwater. The trials at the Xerox locations were specifically designed to purify water with characteristics present at those locations. The results presented in this paper are not intended to recommend or discourage the testing of any commercial system discussed.

Acknowledgement

The authors thank the persons listed below for their very significant contributions to this paper:

Frederick Bernardin
Peroxidation Systems, Inc.
P. O. Box 528
Clinton, Michigan 49326

Lee DiSalvo
H&A of New York
189 North Waster Street
Rochester, New York 14604

David Fletcher
ULTROX International
2435 South Anne Street
Santa Ana, California 92704

James C. MacKenzie
Xerox Corporation 0317-14S
Wilson Center for Technology
800 Phelps Road
Webster, New York 14580

R. D. Samuel Stevens
Solarchem Environmental Systems
40 West Wilmot Street
Richmond Hill, Ontario L4B1H8

ULTROX is a registered trademark of the ULTROX International Corporation.

Perox-Pure is a registered trademark of Peroxidation Systems Inc.

Rayox is a registered trademark of Solarchem Environmental Systems.

DECOMPOZON/D-TOX is a trademark of ULTROX International Corporation.

ENOX is a trademark of Solarchem Environmental Systems.

Literature Cited

1. William H. Glaze, Joon - Wun Kang & Douglas H. Chapin, Ozone Science & Engineering, Volume 9 pp. 335-352.

2. David B. Fletcher, Water World News, Volume 3, No. 3.

3. Norma M. Lewis, USEPA Site and Demonstration Bulletin, EPA / 540 / M5-89/012.

4. D. G. Hager and C. E. Smith, Proceedings of the Haztech International Conference and Exhibition, Institute for International Research, Denver, Colorado, pp. 215 - 231.

5. R. D. Samuel Stevens, "Application of Rayox for the remediation of chlorinated ground water," presented at the Hazmat Central Conference, Chicago, March, 1990.

6. Norma Lewis, Kirankumar Topudurti and Robert Foster, HMC, pp. 42-44, March/April 1990.

7. Carl Loven, Chris Giggy, Presented at AWWA Annual Conference and Exposition, Orlando, Florida June 19-23, 1988.

8. Dr. Samuel Stevens, Personal correspondence with Dr. Robert Heeks dated April 3, 1990.

RECEIVED April 5, 1991

Chapter 8

Multicomponent Ion-Exchange Equilibria in Chabazite Zeolite

S. M. Robinson, W. D. Arnold, and C. H. Byers

Oak Ridge National Laboratory, Oak Ridge, TN 37831–6044

Efficient design of ion-exchange columns, using Ionsiv IE-96 chabazite zeolite, for the decontamination of process wastewater that contains ppb levels of ^{90}Sr and ^{137}Cs requires a detailed study of binary and multicomponent ion-exchange equilibria. Experimental isotherms were acquired for Ca-Na, Mg-Na, Sr-Na, Cs-Na, Sr-Cs-Na, Ca-Mg-Ng, Sr-Ca-Mg-Na, Cs-Ca-Mg-Na, and Sr-Cs-Ca-Mg-Na comparing batch and column experimental approaches. Binary isotherms obtained by the batch technique were most successfully fitted with a modification of the Dubinin-Polyani equilibrium model. Prediction of the multicomponent equilibria from binary data will require more sophisticated modeling.

Fixed-bed ion exchange in which resins serve as the granular medium has been established for many years as an important technique for water purification and for the recovery of ionic components from mixtures. Inorganic media, such as porous, crystaline, aluminosilicate zeolites, have had limited application as ion exchangers in the past. However, zeolite molecular sieves have several characteristics which are unique compared to ion-exchange resins. They are porous crystalline aluminosilicates with a framework which consists of an assemblage of SiO_4 and AlO_4 tetrahedra. These tetrahedra are joined together in various regular arrangements through shared oxygen atoms to form an open crystal lattice containing pores of precisely uniform molecular dimensions with no distribution of pore size. The guest molecules must diffuse into the micropores before they can exchange with cations which are attached to the aluminum atoms in the framework of the zeolite. Therefore, zeolites exhibit both molecular sieve and ion exchange properties. They also tend to be cheaper than many organic resins and are resistant to thermal and radiation degradation. As

0097–6156/91/0468–0133$06.00/0

the field of applications for ion exchange broadens, a class of applications has developed where economic considerations, a high thermal and/or radiation flux, or the molecular sieve properties of zeolites make them more attractive than ion-exchange resins. Their molecular sieve properties give zeolites the advantage of selectivity sometimes not obtainable with adsorbers and resins. Because of their unique properties, growing interest has focused on molecular sieve zeolites for use in ion-exchange separations in water softening, pollution abatement energy production, agriculture, animal husbandry, aquaculture, metals processing, and biomedical areas (1).

Some of these considerations have led Oak Ridge National Laboratory (ORNL) to use chabazite zeolites for decontamination of process wastewater containing ppb levels of ^{90}Sr and ^{137}Cs. A typical characterization of the ORNL process waste stream is shown in Tables I and II. Treatability studies (2, 3) indicate that chabazite zeolites are highly selective for cesium and strontium while admitting high loadings of these metals. Thus they are suited to the removal of trace amounts of ^{137}Cs and ^{90}Sr from wastewater that contains relatively high concentrations of calcium and magnesium. These studies also indicate that the efficiency of the zeolite system depends strongly on the column design and operating conditions and that through optimization of the design of full-scale columns, one could halve the generation rate of loaded zeolite requiring disposal.

Models of multicomponent liquid ion-exchange systems were virtually nonexistent before the 1980s. Although a considerable effort has been made in this area, multicomponent models have not been developed to the point where they can be used in general industrial applications without using laboratory- or pilot-scale data to predict the equilibrium and mass transfer relationships (4, 5). A knowledge of multicomponent equilibrium is essential for modeling ion-exchange separation processes. While some studies have indicated significant progress in the field (6, 7), much remains to be elucidated, especially in the area of multicomponent systems or ion exchange in inorganic species.

Applications of zeolites for treatment of contaminated groundwater began to emerge over the last decade. The decontamination of radioactive waste solutions using zeolites and other inorganic ion exchangers have been studied since the 1950s. Unfortunately very little fundamental studies were done, and it has been difficult to make use of much of the literature because of the lack of standard procedures and theoretical bases (8). Predictive modeling for most multicomponent systems is complicated by competitive interactions among species. In the case of microporous materials such as zeolites, these interactions are further complicated by mutual interference in intraparticle mass transport as well as competition for available ion-exchange sites (9). Since mathematical models have not been available for column design of such systems, users have been restricted to designing columns based on pilot-plant tests and/or crude models. Neither of these approaches have been very successful for efficient column design.

The objective of this paper is to present experimental binary and multicomponent equilibrium data for synthetic mixtures that simulate ORNL's process wastewater. Binary equilibrium models are compared with experimental

Table I. Radiochemical Composition of ORNL Process Wastewater *(26, 27)*

Radionuclide	Concentration (Bq/L)	Concentration (N)
Gross alpha	5	-
Gross beta	6000	-
^{60}Co	25	2.0×10^{-14}
^{90}Sr	4000	1.7×10^{-11}
^{137}Cs	400	9.1×10^{-13}
^{106}Ru	10	2.3×10^{-15}

NOTE: Data are from refs. 26 and 27.

Table II. Chemical Composition of ORNL Process Wastewater *(26, 27)*

Component	Concentration (Mq/L)	Concentration (N)
Ca^{2+}	40	2.0×10^{-03}
Mg^{2+}	8	6.6×10^{-04}
Na^+	5	2.2×10^{-04}
K^+	2	5.1×10^{-05}
Si^{3+}	2	2.1×10^{-04}
Sr^{2+}	0.1	2.3×10^{-06}
Al^{3+}	0.1	1.1×10^{-05}
Fe^{2+}	0.1	3.6×10^{-06}
Zn^{2+}	0.1	3.1×10^{-06}
HCO_3^-	93	1.5×10^{-03}
SO_4^{2-}	23	4.8×10^{-04}
Cl^-	10	2.8×10^{-04}
NO_3^-	11	1.8×10^{-04}
CO_3^{2-}	7	2.3×10^{-04}
F^-	1	5.3×10^{-05}

NOTE: Data are from refs. 26 and 27.

data to assess the most appropriate model for the ion-exchange systems relevant to the problem at hand. The applicability of simple multicomponent relationships, based strictly on binary data, to the prediction of multicomponent data is examined. As noted above, zeolites have several characteristics which are unique compared to ion-exchange resins. These properties and their effects on ion exchange are noted throughout the text.

Theory

Binary ion exchange in zeolites may be represented by the following chemical reaction equation (10):

$$g_A B(z)^{g_B} + g_B A_{(s)}^{g_A} + \rightleftharpoons g_A B_{(s)}^{g_B} + g_B A_{(z)}^{g_A} \tag{1}$$

where g_A and g_B are the charges of the exchanging cations A and B, and the subscripts z and s refer to the zeolite and solution phases, respectively. The equivalent fractions of the exchanging cations in the solution and zeolite are defined by

$$A_s = g_A c_A / (g_A c_A + g_B c_B) \tag{2}$$

and

$$A_z = \frac{equivalents\ of\ exchanging\ cation\ A}{total\ equivalents\ of\ cations\ in\ the\ zeolite} \tag{3}$$

where c_A and c_B are the molalities of the ions A and B, respectively in the equilibrium solution.

The complexity and diversity of the mechanisms of single- and multicomponent ion-exchange equilibrium behavior have led to the development of a large number of equations, both theoretical and empirical in nature. Useful reviews of this area by Soldatov and Bichkova (11), Myers and Byington (12), and Shallcross et al. (13) are available. The most frequently used models are the following:

the binary Langmuir model,

$$q = \frac{q_s bc}{1+bc} = \frac{ac}{1+bc} \tag{4}$$

the binary Freundlich model,

$$q = kc^n \tag{5}$$

and the Dubinin-Polyani model,

$$q = q_s \exp\{-kR^2T^2[\ln(\frac{c_m}{c})]^2\}$$ (6)

In these equations, q and c are the equilibrium concentrations in the solid and liquid phases, respectively, q_s is the saturation concentration in the solid, and b, n, and k are coefficients fitted to the experimental data.

These models may be extended in a logical manner to describe multicomponent equilibrium. For instance, the multicomponent Langmuir model is

$$\frac{q_i}{q_{si}} = \frac{b_i c_i}{1+b_1 c_1+b_2 c_2+\dots b_m c_m}$$ (7)

the multicomponent Freundlich equation is

$$\frac{q_i}{q_{si}} = \frac{b_i c_i^{n_i}}{b_1 c_1^{n_1} + \dots b_m c_m^{n_m}}$$ (8)

and the Langmuir-Freundlich equation (4), which combines the form of the Langmuir equation with the parameters of the Freundlich equation for multicomponent mixtures, is

$$\frac{q_i}{q_{si}} = \frac{b_i c_i^{n_i}}{1 + b_1 c_1^{n_i} + b_2 c_2^{n_2} + \dots b_m c_m^{n_m}}$$ (9)

The coefficients, b_i and n_i, are obtained from binary isotherm data.

The multicomponent form of the Dubinin-Polyani equation (14) for liquids is

$$\frac{V_m(q_1+q_2+\dots)}{W_o} = \exp(-k\epsilon^2)$$ (10)

where

$$V_m = X_1 V_1 + X_2 V_2 + \dots$$ (11)

$$\frac{1}{\sqrt{k}} = \frac{x_1}{\sqrt{k_1}} + \frac{x_2}{\sqrt{k_2}} + \dots \tag{12}$$

$$and \quad \epsilon = -RT \ln\left(\frac{c_1 + c_2 + \dots}{c_m}\right) \tag{13}$$

The Ideal Adsorbed Solution Theory (IAST) Model is based on the Gibbs adsorption equation and only requires single-solute data to predict multicomponent equilibrium (4):

$$\frac{1}{q_1 + q_2 + \dots} = \frac{X_1}{q_1^0} + \frac{X_2}{q_2^0} + \dots \tag{14}$$

When the binary Freundlich adsorption isotherm equation is substituted in Eq 14, the following equation can be obtained for modeling purposes (16):

$$c_i = \frac{q_i}{\sum_{j-1}^{m} q_i} \left(\frac{\sum_{j-1}^{m} q_j n_j}{\frac{k_j}{n_j}}\right)^{\frac{1}{n_i}} \quad for \ i = 1 \ to \ m \tag{15}$$

Equilibrium equations usually used to model zeolites in the past fall into three groups: extension of pure gas isotherm equations, the Dubinin-Polyani potential theory, and thermodynamic methods (15). The simple gaseous model isotherms (Langmuir etc.) have been found to be accurate for some zeolite systems, but they are often not applicable. The Dubinin-Polyani isotherm equation has been found to apply to zeolites in many instances (7, 10). Predictions of liquid-phase equilibrium have been successful in a limited number of cases using the more rigorous classical thermodynamic approach and the generalized statistical model (5, 7). The thermodynamic approach has received much attention in recent years, but these rigorous models require refinement before they can be applied to systems containing several highly interactive components.

Unfortunately, for many mixtures none of the existing adsorption models are adequate (4). In these cases, researchers are limited to developing new theoretical models, modifying existing models to account for observed abnormalities, or developing empirical equations to fit experimental data. An empirical equation commonly used to describe multicomponent systems is

$$q = \frac{a_{io}c_i^{bio}}{a_i + \sum_{j=1}^{m} a_{ij}C_{ij}^{b_{ij}}} = f_1 \ (C_1, C_2, \ldots C_m) \tag{16}$$

Equation 16 includes some well-known relationships as special cases. For $b_{io} = b_{ij} = a_i = 1$, it becomes the Langmuir equation. The Freundlich equation is obtained when $a_i = 0$, and the Langmuir-Freundlich equation when $a_i = 1$.

Experimental

Chabazite zeolite is available both in natural (e.g., Tenneco Specialty Minerals, TSM-300) and synthetic (e.g., Union Carbide Ionsiv IE-90 or IE-96) form. Ionsiv IE-96, a pelletized form of Ionsiv IE-90 crystals in a clay binder, was used as the sorbent in these studies. Structurally, chabazite consists of stacked, double six-membered ring prisms, interconnected through four rings, in a cubic, close-packed array (10, 14). Repetition of the stacking of prisms produces 11 x 6.6 Å cylindrical cavities, joined to adjacent cavities through six eight-membered rings with 4.1 x 3.7 Å free diameters. The eight-membered rings have free apertures of 4.3 Å, and six-membered rings have 2.6 -Å diameters. The neutralizing cation mainly coordinates to water molecules in the eight-membered rings. It is coordinated only to the oxygen in the double six-membered ring. As a result, the six-membered rings are not usually active in ion-exchange.

The Ionsiv IE-96 was obtained from Union Carbide in the hydrated sodium form as 20- to 50-mesh (840 to 297 μm) irregularly shaped particles. Most of the material is 30 to 35 mesh (590 to 500 μm), and the laboratory measurements were made with fractions of this size fraction.

Simulated wastewater solutions were used in the experimental tests to avoid variability in feed composition. The solutions were prepared by dissolving various amounts of $SrCl_2 \cdot 6H_2O$, $CsCl$, $CaCl_2$, and $MgCl_2 \cdot 6H_2O$ in demineralized water. The solutions containing strontium and cesium were also spiked with ^{85}Sr and ^{137}Cs tracers, respectively. The calcium and magnesium concentrations in the simulated wastewaters were the same as the concentrations listed in Table II, but the strontium and cesium concentrations were increased to 0.002 and 0.001 $\underline{N}$, respectively, so that residual concentrations in the equilibrated solutions could be accurately measured with available analytical equipment.

The concentrations of cations in the solution phase were measured using the atomic absorption or gamma-counting analytical equipment. The concentrations of cations in the zeolite were calculated from mass balances using the initial and final solution concentrations. All batch tests were performed in triplicate at room temperature (23° C).

Ion-Exchange Capacity Tests. The total ion-exchange capacity of Ionsiv IE-96 was measured with $(NH_4)_2SO_4$, $CsCl$, $SrCl_2$, and $CsCl$-$SrCl_2$ solutions using the procedure developed by a vendor of zeolites. Initially, the zeolite was contacted three times with 20 mL/g of 2 $\underline{N}$ NaCl overnight to assure that it was in the sodium form. The zeolite was then washed three times by mixing with 20 mL/g of deionized water for 30 min. The washed zeolite was dried overnight at 110° C. The zeolite was then contacted with 20 mL/g of 0.2 $\underline{N}$ exchange solution overnight. The mixture was weighed, centrifuged for 30 min at 5000 relative centrifugal force, and the supernate was decanted and analyzed by atomic adsorption for sodium.

The results indicated that Ionsiv IE-96 has a cation-exchange capacity of 2.24 meq/g when the solution phase does not contain cesium and 2.60 meq/g when cesium is present. Such variations are not uncommon since zeolite ion exchange is complicated by ion sieve, steric, and electrostatic forces within the zeolite pores (22, 23, 24). These results indicate that certain sites in chabazite zeolite are available to cesium, but are not available for calcium, magnesium, or strontium exchange. Since some sites are not available for ion exchange with the latter cations, the effective exchange capacity of the zeolite is lowered. All experiments discussed in the remainder of this report are consistent with these findings.

Isotherm Tests. Several references have reported that the experimental method used in the study affects the isotherm for multicomponent systems. Crittenden and Weber (17) have indicated that multicomponent adsorption equilibria in fixed beds cannot be adequately predicted from batch tests. Several other researchers indicate that complex equilibrium behavior will occur if multicomponent feed solutions, multivalent cations, or high solution concentrations are used (1, 18-21). In these cases, researchers have noted that the equilibrium isotherm depends on the initial solution concentration, and the isotherm should be determined by varying the ion-exchanger dosage.

Because of the problems described in the literature, three different experimental methods were compared for use in this study: (1) batch tests in which the volume and concentration of the solution phase were held constant and solids concentration was varied, (2) batch tests in which the solution concentration was varied and the solids concentration was held constant, and (3) column tests in which the amount of solids was held constant and the solution concentration was varied. Binary Sr-Na and Cs-Na and ternary Sr-Cs-Na isotherms were obtained by each method. Some tests were repeated using the first experimental method at different initial solution concentrations (the relative concentration of cations was held constant) for several binary and multicomponent systems. The isotherms obtained by each experimental method for the Sr-Na system are shown in Figure 1. The strontium isotherms for the Sr-Na and Sr-Cs-Na systems obtained by experimental method (1) are shown in Figure 2, and the strontium isotherms for the Sr-Cs-Ca-Mg-Na system obtained by method (1) for two initial solution concentrations are shown in Figure 3.

Based on the results from these tests (discussed in detail in the Results and Discussion section), a standard method was adopted for obtaining isotherms for modeling purposes. A batch process was used in which the solution phase volume was held constant at 10 mL and the solids concentration was varied between 0.002 and 0.4 g. This is consistent with the methods used by previous researchers who noted experimental effects on isotherm results (1, 15, 19, 20).

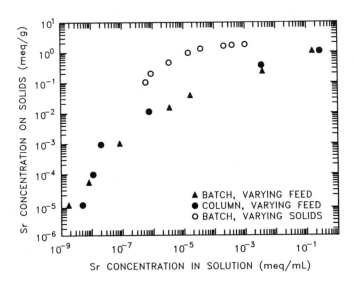

Figure 1. Effect of experimental method on the strontium isotherm for the Sr–Na system (▲, 0.5 g solids, 2.1 mL solution, 2×10^{-06} to 0.4 $\underline{N}$ Sr in feed; ●, 40 g solids, 166 mL solution, 2×10^{-06} to 0.4 $\underline{N}$ Sr in feed; ○, 0.005 to 0.4 g solids, 10 mL solution, 2×10^{-03} $\underline{N}$ Sr in feed).

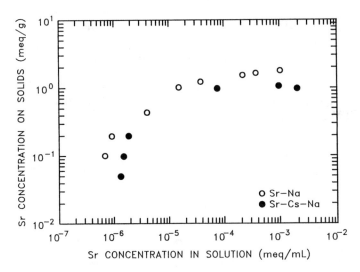

Figure 2. Strontium isotherms for binary and ternary systems (0.005 to 0.4 g solids, 10 mL solution; ○, 2×10^{-03} $\underline{N}$ Sr; ●, 2×10^{-03} $\underline{N}$ Sr, 1×10^{-03} $\underline{N}$ Cs).

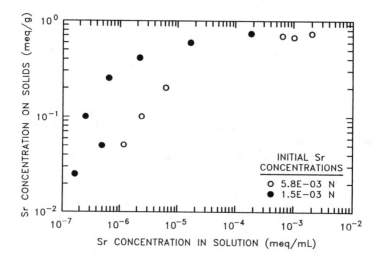

Figure 3.   Effect of feed concentration on the strontium isotherm for the Sr-Cs-Ca-Mg-Na system (0.005 to 0.4 g solids, 10mL solution, 2/2/1/0.8 Sr/Ca/Cs/Mg ratio in feed).

The solution and solids were contacted for 24 h to obtain each isotherm point. The initial concentrations of the exchanging cations in the solution phase were 2.0×10^{-03} $\underline{N}$ Sr, 1.0×10^{-03} $\underline{N}$ Cs, 2.0×10^{-03} $\underline{N}$ Ca, and/or 8.0×10^{-04} $\underline{N}$ Mg. The initial sodium concentration was zero in the solution phase. The resulting isotherms for the binary systems and a typical multicomponent system are given in Figures 4 and 5, respectively.

Triplicate samples were analyzed to obtain each isotherm point. Standard deviations were typically less than 1 %. The zeolite reached greater than 90% equilibrium in less than 1 h and at least 99.7% equilibrium after 5 h. Maximum zeolite loadings of 96% of the total ion exchange capacities were obtained in the equilibrium tests. Mass balances were performed on the solution phase to ensure that stoichiometry of exchange was established. Experimental errors were typically less than 10%. Therefore, the possibility of complications due to non-exchange adsorption is eliminated.

Results and Discussion

Impact of Experimental Conditions. Figure 1 is a typical example of how the experimental method affected the binary isotherms. There was no statistical difference between the isotherms obtained by the batch and column methods when the initial solution concentration was varied. However, there were differences between isotherms obtained by these two methods and those obtained by the batch method when the solids concentration was varied and the solution concentration was held constant. This held true for both monovalent-monovalent and divalent-monovalent ion-exchange. The data for the binary and ternary systems showed similar trends.

For two very strongly held cations, such as cesium and strontium, there was no significant difference between the binary and ternary isotherms as long as the same experimental method was used (See Figure 2). This held true when the total solution concentration was changed. However, the total solution concentration had a significant effect on the isotherms obtained for cations with different selectivities. In these cases, the multicomponent isotherms (such as Figure 3) followed the concentration valence theory which states that an exchanger will have a higher selectivity for higher valence cations when the total normality of the solution phase is decreased (*20, 21*). The isotherms for the different solution concentrations converged at the upper end of all graphs where the solution and solids-phase concentrations were highest. This held for all cations, regardless of whether they were strongly or weakly held.

These results indicated that column equilibrium can be accurately predicted from batch data. They also indicate that the initial concentration of cations, the total solution concentration, and the solid-to-liquid ratio will affect the equilibrium results for multicomponent systems. These results are consistent with earlier findings for non-ideal multicomponent ion-exchange systems (*1, 18-20*).

Binary and Multicomponent Isotherms. The binary isotherms shown in Figure 4 are "favorable" for uptake of the sorbates. The isotherms for the multicomponent Sr-Cs-Ca-Mg-Na system (Figure 5) are typical of all

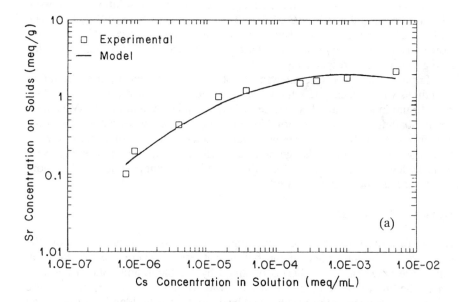

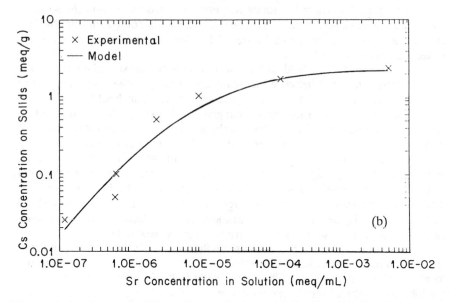

Figure 4. Isotherms for binary ion exchange. (a) Sr–Na; (b) Cs–Na. Data are from refs. 26 and 27.

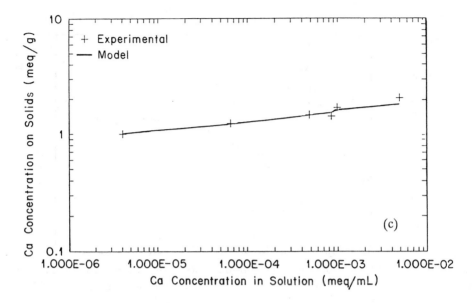

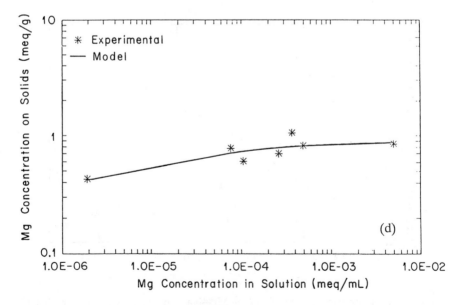

Figure 4. *Continued.* Isotherms for binary ion exchange. (c) Ca–Na; (d) Mg–Na. Data are from refs. 26 and 27.

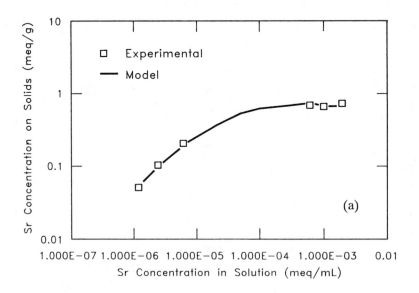

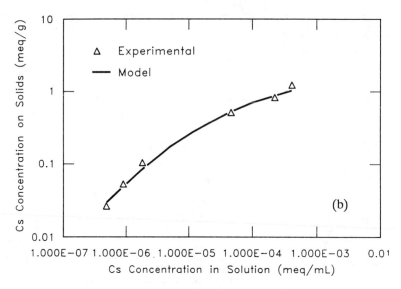

Figure 5. Isotherms for Sr–Cs–Ca–Mg–Na ion exchange. (a) Sr; (b) Cs. Data are from ref. 26.

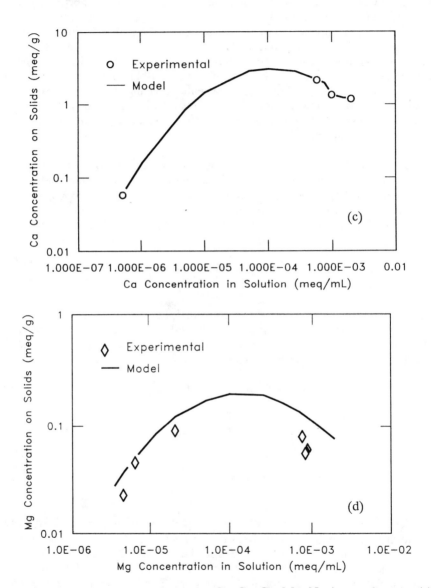

Figure 5. *Continued.* Isotherms for Sr–Cs–Ca–Mg–Na ion exchange. (c) Ca; (d) Mg. Data are from ref. 26.

multicomponent systems tested. The isotherm data indicate that Ionsiv IE-96 has the selectivity order $Cs > Sr > Ca > Mg$ and that the zeolite is selective for the entering cations over sodium for the solution composition range tested.

In the multicomponent systems, the shapes of the isotherms for the cation with the highest selectivity were similar to those of the corresponding binary isotherms. However, cation interactions affected the isotherms for the cations with lower selectivities, and these cations were displaced when the solution concentration increased relative to the available exchange sites. This was more pronounced with decreasing selectivity and resulted in a magnesium isotherm with a convex shape. Slight displacement of strontium and calcium occurred at the highest loadings tested.

Isotherm Models. Standard isotherm equations were used to model the data. The binary data were modeled using Equations 4-6. The Dubinin-Polyani equation was solved for k by substituting values for R, T, and c_m into Equation 6. Values for c_m , the solubility of the salt in water, were obtained from the literature (25). A modified version of the Dubinin-Polyani equation was also tested:

$$q = \exp \left\{ b_0 + b_1 \cdot \ln(c) + b_2 \cdot [\ln(c)]^2 \right\} . \tag{17}$$

The Dubinin-Polyani equation was solved for b_0, b_1, and b_2. This form of the equation is referred to as the modified Dubinin-Polyani model in the remainder of this report. If the systems are ideal, Equations 6 and 17 should be equal, resulting in the following relationships:

$$b_0 = \ln(q_s) - kR^2T^2[\ln(c_m)]^2 ,$$

$$b_1 = 2kR^2T^2\ln(c_m) ,$$

$$and \quad b_2 = -kR^2T^2 . \tag{18}$$

The constants for the binary isotherm equations were determined from the isotherm data by regression analyses of the linearized equations. The results indicate that all of the models except the Dubinin-Polyani equation are reasonable fits for the data. The modified Dubinin-Polyani model is the best equation for fitting the data for all four systems. The fact that Equations 6 and 17 are not equal for any of the four binary isotherms indicates that the solubility of salts in the liquid phase is not adequate to account for the nonidealities which occur during liquid ion-exchange. This is not surprising, since the Dubinin-Polyani equation was developed from the theory of pore filling for gases in microporous carbon adsorbents.

Turning to the multicomponent systems, predictions were made of the isotherms which had been evaluated experimentally. Predictions were based on binary data using the multicomponent Langmuir, Freundlich, Langmuir-Freundlich, and the IAST models. When the coefficients for the binary isotherms were substituted into Equations 7, 8, 9, and 15, the results did not accurately predict the experimental multicomponent data. These results indicate that the multicomponent data in this study cannot be predicted from binary data, and rigorous thermodynamic or empirical equations will be required to model the multicomponent data. Since thermodynamic models are not well developed for highly interactive multicomponent systems, empirical equations having the basic forms of Equations 7, 8, 9, 16, and 17 were fit to the equilibrium data. Equation 17 fit the data most accurately with the least number of coefficients. The resulting equation parameters are listed in Table III, and the modeling results are shown in Figures 4 and 5.

Summary

This study has generated isotherm data for binary and multicomponent systems containing Ca, Mg, Na, Sr, and Cs. The results indicate that the isotherms are dependent on experimental method and the total solution concentration. All data obtained in this study were accurately predicted by the modified Dubinin-Polyani model, but multicomponent data could not be predicted from binary data. These data will be used to design zeolite exchange columns for wastewater treatment at ORNL.

Notation

A equivalent fraction
a isotherm constant
b isotherm constant
c sorbate concentration, meq/mL
c_m solubility, meq/mL
g charge of cation
k isotherm constant
m number of solutes
n_i exponent of the i component, Freundlich Isotherm
q sorbent concentration, meq/g
q_o for the pure components, meq/g
q_s saturation limit, meq/g
R gas constant
T temperature
v_i partial molar volume of component
V_m molar volume of sorbate
W_o specific micropore volume of adsorbent
X mole fraction in adsorbed phase

Table III. Parameters for the Modified Dubinin-Polyani Model[1]

Isotherm	Coefficients			
	b_0	b_1	b_2	R^2
Binary Systems				
Sr Isotherm	-1.637	-0.6839	-0.05015	0.97
Cs Isotherm	-2.165	-0.6839	-0.06261	0.95
Ca Isotherm	1.833	0.2655	0.00961	0.96
Mg Isotherm	-0.4651	-0.1279	-0.01202	0.74
Sr-Cs-Na System				
Sr Isotherm	-4.952	-1.262	-0.0781	0.97
Cs Isotherm	-4.315	-1.134	-0.0733	0.99
Ca-Mg-Na System				
Ca Isotherm	-5.326	-1.506	-0.09814	0.99
Mg Isotherm	-15.81	-3.312	-0.1877	1.00
Sr-Ca-Mg-Na System				
Sr Isotherm	-1.592	-0.5100	-0.04096	0.99
Ca Isotherm	-0.6791	-0.1495	-0.01558	0.99
Mg Isotherm	-18.06	-3.674	-0.2017	0.91
Cs-Ca-Mg-Na System				
Cs Isotherm	-5.073	-1.287	-0.08128	0.97
Cs Isotherm	-3.219	-0.7620	-0.04545	0.99
Mg Isotherm	-29.40	-6.105	-0.3213	0.91
Sr-Cs-Ca-Mg-Na System				
Sr Isotherm	-4.742	-1.157	-0.07507	0.99
Cs Isotherm	-1.532	-0.5890	-0.05002	0.99
Ca Isotherm	-4.212	-0.8560	-0.04703	1.00
Mg Isotherm	-17.28	-3.345	-0.1818	0.93

[1] See Eq. 10.

Literature Cited

1. Sherman, J. D., AIChE Symp. Ser., *Ion Exchange Separations with Molecular Sieve Zeolites*, 1978, **74**, 179.
2. Robinson, S. M., and Parrott, Jr., J. R., *Pilot-Scale Demonstration of Process Wastewater Decontamination Using Chabazite Zeolites*, ORNL/TM-10836, Oak Ridge National Laboratory, Oak Ridge, TN, December 1989.
3. Robinson, S. M., et al., *The Development of a Zeolite System for Upgrade of the ORNL Process Waste Treatment Plant*, ORNL/TM-11426, Oak Ridge National Laboratory, Oak Ridge, TN, in publication.
4. Weber, Jr., W. J.; Smith, E. H.; "Simulation and Design Models for Adsorption Processes," *Environ. Sci. Technol.*, 21(11), 1987, 1040-50.
5. Wankat, P. C., *Large-Scale Adsorption and Chromatography*; Vol. 1, CRC Press, 1986.
6. Miller, G. W.; Knaebel, K. S.; Ikels, K. G. "Equilibria of Nitrogen, Oxygen, Argon, and Air in Molecular Sieve 5A," *AIChE Jour.*, 1987, 33(2), 194-201.
7. Richter, E. Schultz W.; Myers, A. L., "Effect of Adsorption Equation on Prediction of Multicomponent Adsorption Equilibria by the Ideal Adsorbed Solution Theory," *Chem. Eng. Sci.* 1989, **44**(8), 1609-1616.
8. Dyer, A; Keir, D. "Nuclear Waste Treatment by Zeolites," *Zeolites*, **4**, 1984, 215-217.
9. Liang, S.; Weber, W. J., Jr. "Parameter Evaluation for Modeling Multicomponent Mass Transfer, *Chem. Eng. Commun.*, 35, 1985, 49-61.
10. Breck, D. W., *Zeolite Molecular Sieves*, Wiley and Sons, New York, 1974.
11. Soldatov, V. S.; Bichkova, V. A. "Ion Exchange Selectivity and Activity Coefficients as Functions of Ion Exchange Composition," *Separ. Sci.*, 15, 1980 89-110.
12. Myers, A. L.; Byington, S. "Thermodynamics of Ion-Exchange: Prediction of Multicomponent Equilibria from Binary Data," In *Ion-exchange: Science and Technology*, Rodrigues, A. E., Ed.; NATO ASI Series E 107; Martinus Nijhoff Publishers; Boston, 1986, 119-145.
13. Shallcross, D. C., Hermann, C. C., and McCoy, B. J. "An Improved Model for the Prediction of Multicomponent Ion Exchange Equilibria," *Chem. Eng. Sci.*, 43(2), 1988, 79-288.
14. Ruthven, D. M., *Principles of Adsorption and Adsorption Processes*, John Wiley & Sons, New York, 1984.
15. Singhal, A. K., *Multicomponent Sorption Equilibria on Hydrocarbon Mixtures in Zeolite Molecular Sieves*, AIChE Symposium Series, 1978, 74(179), 36-41.
16. Crittenden, J. C. et al. "Prediction of Multicomponent Adsorption Equilibria Using Ideal Adsorbed Solution Theory," *Environ. Sci. Technol.*, 1985, 19(11), 1037-43.
17. Crittenden, J. C. and Weber, Jr., W. J. "Model for Design of Multicomponent Adsorption Systems," *ASCE Env. Eng. Div.*, 104(EE6), 1978, 1175-1195.
18. Ausikaitis, J. P., *Fundamentals of Adsorption*, AIChE, New York, 1984, 49-63.

19. Loizidou, M., and Townsend, R. P. "Ion-Exchange Properties of Natural Clinoptilolite, Ferrigrite, and Mordenite," *Zeolites*, **7(2)**, 1987, 153-159.
20. Barri, S. A. I., and Rees, L. V. C. "Binary and Ternary Cation Exchange in Zeolites," *J. Chromatogr.*, **201**, 1980, 21-34.
21. Cremers, A., "Ion Exchange in Zeolites," *Molecular Sieves-II*, J. R. Catzer, Ed., In ACS Symp. Series, **40**, Washington, DC, 1977, 179-93.
22. Chu, P.; Dwyer, F. G. *Inorganic Cation Exchange Properties of Zeolite ZSM-5*"; Stucky, G. D., Dwyer, F. G., Eds.; In ACS Symp. Series; **218**; Washington, DC, 1983, 59-77.
23. Barrer, R. M.; Klinowski, J., "Influence of Framework Charge Density on Ion-Exchange Properties of Zeolites"; *J. Chem. Soc. Faraday Trans.* I, **9**, 1972, 1956-63.
24. Franklin, K. R., et al., "Ternary Exchange Equalibria Involving H_3O^+, NH_4^+, and N_a^+ Ions in Synthetic Zeolites"; *New Developments in Zeolite Science and Technology*, Proceedings of the 7th International Zeolite Conference, Y. Murakami, A. Iijima, and J. W. Word, Eds.; Elsevier Science Pub. Co., Inc., 1986, 289-296.
25. Nemeth, B. A., *Chemical Tables*, John Wiley and Sons, New York, 1975, 350-365.
26. Robinson, S. M.; Arnold, W. D.; Byers, C. H. "Design of Fixed-Bed Ion-Exchange Columns for Wastewater Treatment"; Deutsch, B. P., Ed; In *HAZTECH International '90 Conference Proceedings*, Houston, Texas, 1990, 807.
27. Robinson, S. M.; Arnold, W. D.; Byers, C. H. "Design of Fixed-Bed Ion-Exchange Columns for Wastewater Treatment"; Post, R. G., Ed; In *Proceedings of the Symposium on Waste Management*, Tucson, Arizona, 1990, Vol. 2, 518.

RECEIVED April 5, 1991

Chapter 9

Decontamination of Low-Level Radioactive Wastewaters by Continuous Countercurrent Ion Exchange

Reginald Hall, J. S. Watson, and S. M. Robinson

Chemical Technology Division, Oak Ridge National Laboratory, Oak Ridge, TN 37831-6044

A mobile pilot-scale continuous countercurrent ion-exchange (CCIX) system is being operated at the Oak Ridge National Laboratory (ORNL) for the treatment of wastewaters that contain predominantly calcium, sodium, and magnesium bicarbonates and are slightly contaminated with ^{90}Sr and ^{137}Cs radioisotopes. A demonstration study is being conducted to evaluate the near-steady-state performance and feasibility of a pilot-scale CCIX column for the selective removal of strontium from wastewater. Test results show that the process removes strontium sufficiently from the wastewater to permit discharge while significantly reducing the volume of secondary waste generation. CCIX has the potential for effective use in several applications; however, it has not been utilized often by industries. The CCIX system could offer an economical alternative for decontamination of wastewaters containing trace amounts of contaminants prior to discharge into the environment.

This paper discusses (a) modeling of multicomponent equilibrium, (b) application of the Thomas model for predicting breakthrough curves from ion exchange column tests, (c) methods for scaleup of experimental small-scale ion-exchange columns to industrial-scale columns, and (d) methods for predicting effluent compositions in a CCIX system.

Alternative methods for reducing secondary waste generation in the treatment of wastewaters slightly contaminated with ^{90}Sr and ^{137}Cs radioisotopes are being developed at the Oak Ridge National Laboratory (ORNL) (1). The objective is to decontaminate process wastewater sufficiently for release to the environment while concentrating the radioactive materials (^{90}Sr and ^{137}Cs) into an immobilized solid waste form that can be safely stored with minimum surveillance. Large quantities of calcium are often found in waste streams that are only slightly contaminated with trace quantities of strontium and other radioactive materials. Calcium and magnesium compete with strontium and other radionuclides for sites

0097-6156/91/0468-0153$07.00/0

on cation-exchange resin, making the efficient removal of strontium from multicomponent mixtures of ions in wastewater extremely difficult (2). Ion-exchange resins can be selected which will easily separate strontium from monovalent ions found in wastewaters using a fixed- bed ion-exchange system. However, fixed-bed ion-exchange systems presently in use cannot efficiently separate strontium from chemically similar divalent cations such as calcium. Additionally, resin regeneration in fixed-bed ion-exchange systems produces secondary wastes that require subsequent treatment. Thus, finding a treatment process that will efficiently separate ^{90}Sr from calcium would reduce the volume of secondary waste generated and reduce disposal costs.

A pilot-scale continuous countercurrent ion-exchange (CCIX) system marketed by CSA, Inc., in Oak Ridge, Tennessee, was tested as a potentially economical alternative means of treating slightly contaminated process wastewaters. This system has the potential for efficiently removing strontium from the wastewater while simultaneously minimizing secondary waste generation by (a) separating calcium and strontium by regeneration with 1 $\underline{M}$ and 5 $\underline{M}$ NaCl solution, (b) precipitating strontium from the recyclable 5 $\underline{M}$ NaCL solution using a sodium carbonate solution followed by subsequent disposal of strontium-rich solid precipitates, and (c) discharging a nonradioactive calcium-rich stream.

One paramount feature of the CCIX process is the ability to selectively remove a preferred component, strontium, from many other ions, calcium, magnesium, etc., even at conditions where ion-exchange separation factors are near unity (3). To maximize production efficiency and minimize volumes of secondary wastes, a CCIX system is often preferred over a fixed-bed ion-exchange system since CCIX is more efficient in utilizing the resin-exchange capacity. Resin is continuously reacted and regenerated, thus minimizing the amount of resin required for a given amount of liquid processed (4).

Composition of ORNL Process Wastewater. The process waste system at ORNL is used to collect waste streams that are slightly contaminated with radioactivity, such as process wastewater from research laboratories, condensate from the evaporators, and surface water. ORNL process wastewater contains a number of trace radionuclides, as shown in Table I, and relatively large amounts of competing ions (representative of city water and local groundwater in Oak Ridge, Tennessee), as shown in Table II (5). The major chemical constituents are calcium, sodium, and magnesium bicarbonates, which are introduced by shallow drainage wells. The major radionuclides are ^{90}Sr and ^{137}Cs. The ^{90}Sr is the more hazardous contaminant because of its potential for entering the human and animal food chain. Discharge limits into the environment, as defined by the Department of Energy (DOE Order 5400.5), are 37 Bq/L for ^{90}Sr and 100 Bq/L for ^{137}Cs.

The concentrations of ^{90}Sr and ^{137}Cs in the feed to the process waste treatment system average 2000 and 200 Bq/L, respectively. However, concentration spikes have occurred occasionally due to drainages from construction sites, leaks in underground piping, and so forth. Feed compositions may vary between 500 and 8000 Bq/L ^{90}Sr and 100 and 1000 Bq/L ^{137}Cs.

Table I. Radiochemical Composition of Process Wastewater at ORNL

Radionuclide	Concentration	
	$(Bq/L)^a$	(mg/L)
Gross Alpha	5	—
Gross Beta	2500	—
^{60}Co	25	4.60E-11
^{90}Sr	2500	2.99E-08
$^{95}ZrNb$	50	4.84E-12
^{106}Ru	10	6.21E-12
^{137}Cs	200	4.80E-09

^{a} 1 Bz = 1 dis/s = 2.7 x 10^{-11} Ci.

Table II. Chemical Composition of ORNL Process Wastewater (pH = 7.7)

Cation	Concentration (mg/L)	Anion	Concentration (mg/L)
Ca	40	HCO_3	60
Mg	10	SO_4	18
Na	20	Cl	6
Si	2	NO_3	4
K	2	F	1
Sr	0.2	PO_4	0.2
Al	0.1		
Fe	0.05		
Zr	0.05		
Cu	0.02		
Ni	<0.02		
Cr	<0.02		
U	<0.001		

Experimental

A general understanding of the ion exchange process, where counter ions in the resin phase are displaced in stoichiometric proportions by counter ions initially in the liquid phase, was needed to develop mathematical models to predict column performance at various operating conditions. Four factors needed to predict the ion exchange column performance were (a) the equilibrium relation for the multicomponent system, (b) the bulk flow rate of feed solution through the column, (c) the ion exchange kinetics and mass transfer rates, and (d) column length and geometry. These parameters provide information needed to describe the breakthrough curves. Breakthrough curves for the multicomponent process wastewater feed solution were measured from continuous sampling of the effluent from the loading of process wastewater feed on a Dowex HGR-W2 cation resin in a bench-scale column. Likewise, the elution curves were measured by displacing the feed solution counter ions from the loaded resin using sodium chloride solutions.

Batch Equilibrium Tests. For the multicomponent process wastewater, the feed solutions with the major cations found in the ORNL wastewater (Sr, Ca, Mg, and Na) were equilibrated with an ion-exchange resin in the Na^+ form. All initial equilibration solutions contained 100 mg/L Sr, 71 mg/L Ca, 41 mg/L Mg, and 47 mg/L Na. Approximately 100 mL batches were allowed to mix for 24 hours with various quantities of resin at ambient temperature (23 °C) using a Burrell Model 75 wrist action automatic shaker (setting equal 6). Triplicate samples of the feed and supernate liquid were analyzed (within 10% accuracy) using an atomic absorption spectrophotometer to measure the equilibrium concentrations for strontium, calcium, magnesium, and sodium in the solution and the quantities of these cations that had been transferred to or from the resin. Concentrations in the solid resin phase were then determined from the changes in the solution concentrations. The multicomponent equilibrium data for both liquid and solid phases are given in Table III. The distribution coefficient (K_d) changes, depending on the ionic concentration of the reacting species (i.e., feed concentration). The separation factor (α) is the ratio of the distribution coefficient for each component using calcium distribution as the reference. Because the concentrations in the resin phase were determined from changes in the solution concentrations, the accuracy of the separation factors for each ion pair depends upon the distribution coefficient and were most accurate when approximately half of each ion was transferred from the solution to the resin. The Sr-Ca separation factor was of most interest. Therefore, most of the multicomponent data were taken under conditions more suitable for determining this coefficient. These conditions were less satisfactory for determining the separation factors for the other ions. Uncertainties in the individual separation factors can be estimated from the results given in Table III. Separation factors calculated from the individual experiments were averaged, and both the average value and standard deviation from the average are shown in Table III. A few individual data points were ignored in the averages because the concentrations or the concentration differences were too low to be measured accurately as indicated in Table III. Note that the Sr-Ca separation factors were the most

Table III. Multicomponent Equilibria Data for Simulated Process Wastewater

Strontium Equilibrium Data

Resin Wt. (g)	Liquid Phase Conc. (meq/L)	Solid Phase Conc. (meq/g-resin)	K_d Sr (mL/g-resin)	α $K_d/K_d(Ca)$
0.2958	0.0106	0.817	7.70E+04	1.96
0.1875	0.318	1.13	3.54E+05	1.62
0.1574	0.572	1.18	2.06E+03	1.51
0.0564	1.47	1.69	1.15E+03	1.45
0.6728	0.0006 [a]	0.361	6.12E+05	4.99 [b]
0.3433	0.0065	0.705	1.09E+05	1.62
0.2529	0.0301	0.948	3.15E+04	1.87
0.3145	0.0100	0.769	7.67E+04	1.99
			Average α = 1.72 ± 0.22	

[a] Concentration may be too low for accurate measurement.
[b] Value not used in average.

Calcium Equilibrium Data

Resin Wt. (g)	Liquid Phase Conc. (meq/L)	Solid-Phase Conc. (meq/g-resin)	K_d Ca (mL/g-resin)	α $K_d/K_d(Ca)$
0.2958	0.0305	1.19	3.92E+04	1
0.1875	0.701	1.53	2.18E+03	1
0.1574	1.13	1.55	1.36E+03	1
0.0564	2.47	1.95	7.90E+02	1
0.6728	0.0043	0.529	1.23E+05	1
0.3433	0.0154	1.03	6.70E+04	1
0.2529	0.0815	1.38	1.69E+04	1
0.3145	0.0292	1.12	3.86E+04	1

Continued on next page

Table III. (Continued)

Magnesium Equilibrium Data

Resin Wt. (g)	Liquid Phase Conc. (meq/L)	Solid Phase Conc. (meq/g-resin)	K_d Mg (mL/g-resin)	α $K_d/K_d(Ca)$
0.2958	0.0926	1.12	1.21E+04	0.309
0.1875	1.56	0.988	6.33E+02	0.291
0.1574	1.64	1.126	6.87E+02	0.504
0.0564	2.92	0.865	2.96E+02	0.375
0.6728	0.0119	0.505	4.24E+04	0.345
0.3433	3.28	0.0389 [a]	1.19E+01	1.78E-04 [b]
0.2529	0.0729	1.32	1.81E+04	1.07 [b]
0.3145	0.0771	1.06	1.38E+04	0.357
			Average α =	0.364 ± 0.075

[a] Concentration may be too low for accurate measurement.
[b] Value not used in average.

Sodium Equilibrium Data

Resin Wt. (g)	Liquid Phase Conc. (meq/L)	Solid Phase Conc. (meq/g-resin)	K_d Na (mL/g-resin)	α $K_d/K_d(Ca)$
0.2958	10.7	1.08	273	0.007
0.1875	7.44	1.12	387	0.177
0.1574	8.36	0.0001 [a]	480	0.352 [b]
0.0564	4.45	0.0065	960	1.215 [b]
0.6728	11.7	2.56	123	0.004
0.3433	8.82	2.02	224	0.003
0.2529	10.8	0.557	320	0.019
0.3145	11.4	1.02	261	0.007
			Average α =	0.036± 0.069

[a] Concentration may be too low for accurate measurement.
[b] Value not used in average.

accurate with a standard deviation of approximately 13% while the Mg-Ca and the Na-Ca separation factors include greater uncertainties and have standard deviations of approximately 20%. One should not attempt to extrapolate these results to solution compositions that are far beyond the range studied.

Column Tests. Bench-scale column tests were performed to obtain kinetic rate data for the multicomponent system from breakthrough curves using the Dowex HGR-W2 cation-exchange resin in the sodium form and ORNL process wastewater feed. The purpose of these tests was to obtain data to determine the feasibility of efficiently separating strontium from calcium using ORNL process wastewater feed in a fixed-bed column. The resulting data were then used to establish operating conditions for the pilot-scale CCIX tests.

Bench-scale Breakthrough Curves. A fixed-bed ion-exchange column (7 mm ID, 76 mm long) was packed with glass wool (to eliminate end effects) and then loaded with 3 mL (1.5 g) of Dowex HGR-W2 sodium-form resin [20 to 50 mesh (0.84 to 0.30 mm), resin capacity equals 2.2 meq/mL wet resin] (6). ORNL process wastewater with a composition of 10.5 mg/L Sr, 53 mg/L Ca, 10 mg/L Mg, and 10.5 mg/L Na was fed to the top of the column at a flow rate of 3 mL/min (1 bed volume per minute). Effluent samples were collected at the bottom of the column and analyzed at intervals to produce concentration profiles (breakthrough curves) for calcium, strontium, and magnesium. These breakthrough curves, shown in Figure 1, were obtained by plotting the mean throughput versus fractional breakthrough (i.e., effluent concentration divided by feed concentration). The distribution coefficient, K_d, is a volume-based (mL solution per mL resin) distribution approximately equal to the number of bed volumes that have been processed at 50% breakthrough. Distribution coefficients are often given in terms of the proportion of a metal adsorbed or not adsorbed (mL solution per gram of dry resin) by an ion-exchange resin (7). Distribution coefficients of 879 mL solution per mL resin (1730 mL solution per gram of dry resin) and 631 mL solution per mL resin (1245 mL solution per gram of dry resin) were measured for strontium and calcium, respectively, during continuous loading on the ion-exchange resin. Results from the breakthrough curves indicated that magnesium in the process wastewater tends to offer little competition for sites on the resin and is easily displaced by calcium.

Bench-scale Elution Curves. A test was performed to compare the separation factor for elution of strontium and calcium from the loaded resin. Strontium and calcium were removed by scrubbing (eluting) with 1 $\underline{M}$ NaCl solution or 5 $\underline{M}$ NaCl at a flow rate of 1 mL/min (one-third bed volume per minute) (8). The elution curves (Figure 2) were obtained by plotting the concentration in the effluent divided by the mean concentration on the resin. Breakthrough (initial) occurs almost instantly. Separation factors between monovalent and divalent ions will change greatly between dilute and concentrated solutions because of Donnan effects (9). Thus, elution with Na^+ ions should be done at high concentrations, and loading with Ca^{++} and Sr^{++} ions will be best at low concentrations.

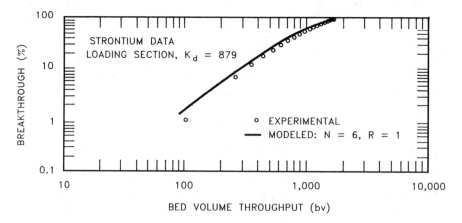

Figure 1. Breakthrough curve for bench-scale column using ORNL process wastewater and Dowex-HGR-W2 sodium-form resin.

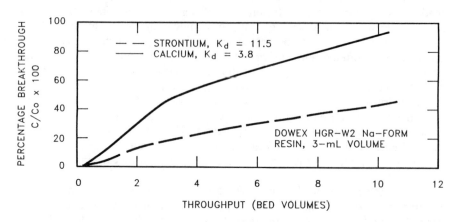

Figure 2. Elution curve for removal of strontium and calcium from Dowex HGR-W2 sodium-form resin using 1 N NaCl in a bench-scale column.

Pilot-Scale CCIX System. A pilot-scale CCIX system marketed by the CSA, Inc., in Oak Ridge, Tennessee, was purchased for the purpose of treating slightly contaminated wastewaters. A schematic of the CCIX column operated at ORNL to test the selective removal of strontium from wastewater is shown in Figure 3. The stream compositions given in Figure 3 are based on experimental results obtained during pilot testing (*10*). This system is mounted on a mobile trailer bed to allow easy transport to remedial sites. A Dowex HGR-W2 cation exchange resin (10% cross-linkage) was chosen to perform the separation between the multicomponents based on the resin's affinity for calcium and strontium relative to sodium, cost and availability, and high-throughput capability. The pilot CCIX column is sized to treat 3 gal/min (11 L/min). The column is 27 ft (8.2 m) high and composed of four subsections—loading, pulse, scrub, and strip/rinse—each connected in a closed loop separated by air-operated valves (*11*).

Resin [20 to 50 mesh (0.84 to 0.30 mm)] is initially charged into the CCIX column. During a loading cycle, the column loading section contains a fixed volume of resin [maximum volume is ≈ 12.3 L, i.e., the volume of a 5-ft (150-cm)-long, 4-in. (10-cm)-ID PVC pipe]. Process wastewater is intermittently blended with the scrub recycle in a 200-gal (756-L) blended feed tank. The blended feed is continuously fed to the column at a constant flow rate of 3 gal/min (11 L/min) in a countercurrent direction with respect to the resin flow during a loading cycle. The loading cycle usually requires 30 to 40 minutes. At the end of each cycle, a small volume (≈ 600 ml) of resin, loaded primarily with strontium and calcium, is pulsed from the top of the column loading section around the column loop as a freshly regenerated volume of resin is simultaneously introduced in the bottom of the loading section. Normally, there is a maximum interval of 10 seconds between the end of one loading cycle and the start of the consecutive loading cycle. This interval is the intermittent, semicontinuous operating cycle for the CCIX system.

Strontium and calcium are separated in the scrub section [a 9-ft (2.7-m) long, 2-in. (5-cm)-ID PVC pipe] using a 0.5 to 1.0 $\underline{M}$ NaCl solution countercurrent with respect to the resin flow. Approximately 80 to 100% of the strontium remains on the resin. The resulting solution is collected in the scrub recycle stream and blended with the process wastewater feed. Strontium loaded on the resin is ultimately removed in the strip section [a 14 ft (4.3 m) length, 2-in. (5 cm) ID PVC pipe] using a 3.5 to 5 $\underline{M}$ NaCl solution countercurrent with respect to the resin flow. After strontium has been removed, the resin is washed with clean rinse water prior to entering the bottom of the loading section. During the pilot demonstration, small batches (≈ 1 L) of the strontium-rich eluate collected in the strip waste tank were precipitated as the carbonate by addition of sodium carbonate to confirm operation feasibility. These precipitates were then filtered and ultimately disposed of as solid waste.

Modeling Experimental Results

The equilibrium relation, q_i^*, which is usually called the adsorption isotherm is, in general, a complicated nonlinear function of C_i, in which mutual interactions among different solutes are taken into account. The model representing this relation was first introduced by Langmuir (1916) to describe an ideal localized

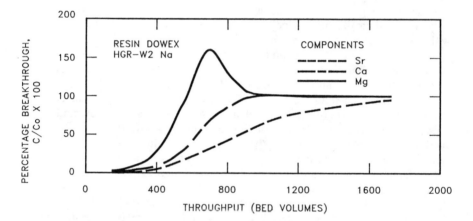

Figure 3. Schematic of pilot-scale CCIX column tested using ORNL process wastewater. Percentages are relative to blended feed composition.

monolayer sorption of gases on a solid surface (14-16). An appropriate form of the Langmuir equation for multicomponent systems expressed in terms of the weight of uptake by the resin and concentrations of solute in the fluid phase surrounding the resin is given by Equation 1,

$$q_i^* = \frac{Q'K_i'C_i}{1+\sum_{j=1}^{m} K_j'C_j} \quad for \ i=1 \ to \ m \tag{1}$$

where

q_i^* = equilibrium uptake of solute i by the adsorbent, meq/g resin,

Q' = asymptotic maximum solid phase concentration, meq solute/g resin,

C_i = equilibrium concentration solute i in liquid phase, meq/L,

K_i' = equilibrium constant for individual solute component i, L/meq,

m = number of solute components in the feed solution.

Modeling Equilibrium Data. The multicomponent equilibrium data were fitted to the Langmuir model using a nonlinear parameter algorithm (17). There are a number of different methods for obtaining nonlinear parameter estimates. The three most widely used computer-based routines include linearization, steepest descent, and Levenberg-Marquardt's compromise (18). Of these three, the Levenberg-Marquardt's compromise method was employed in this study, because of its ability to combine the best features of both the linearization (or Taylor series) method and the steepest descent method while avoiding their most serious limitations to successfully solve many nonlinear problems.

The linearization method uses the results of linear least squares in a succession of steps. This is shown by assuming that the proposed model is given by Equation 2,

$$Y^E = f(\eta_1, \eta_2, ..., \eta_\kappa; \theta_1, \theta_2, ... \theta_p) + \varepsilon \tag{2}$$

where

η_k = the independent variable,

θ_p = parameter estimates,

and choosing the initial parameter estimates, $\theta_1^*, \theta_2^*, ..., \theta_n^*$ for the parameters $1, 2, ..., n$. A Taylor series expansion of $f(\eta_j, \ )$ is carried out about the point θ^* (i.e., $\theta_1^*, \theta_2^*, ..., \theta_p^*$). The series is then truncated after the first derivatives,

$$f(\eta_j, \theta) = f(\eta_j, \theta^\circ) + \sum_{i=1}^{P} \left[\frac{\partial f(\eta_j, \theta)}{\partial \theta_i} \right]_{\theta = \theta_\cdot} (\theta_i - \overset{\circ}{\theta}_i), \tag{3}$$

where

$$\eta_j = \begin{vmatrix} \eta_{11} & \eta_{12} & \cdots & \eta_{1j} \\ \eta_{21} & \eta_{22} & \cdots & \eta_{2j} \\ \vdots & \vdots & \ddots & \vdots \\ \eta_{n,1} & \eta_{n,2} & \cdots & \eta_{n,j} \end{vmatrix} \tag{4}$$

Letting

$$\overset{\circ}{f}_j = f(\eta_j, \overset{\circ}{\theta})$$

$$\overset{\circ}{\Delta}_i = \theta_i - \overset{\circ}{\theta}_i$$

$$Z_i = \left[\frac{\partial f(\eta_j, \theta)}{\partial \theta_i} \right]_{\theta = \theta}$$

Equation 3 becomes

$$Y^E - \overset{\circ}{f}_j + \sum_i \overset{\circ}{\Delta}_i \overset{\circ}{Z}_{ij} + e_j \tag{5}$$

Equation 5 is linear in terms of θ and therefore the linear least squares theory may be applied. In vector notation, $b^p_\cdot$, the estimates of Δ° are obtained by solving the following equation:

$$b^P_\cdot = (Z'^\circ Z^\circ)^{-1} Z'^\circ (Y^E - f^\circ) \tag{6}$$

Since linear least squares theory is applied, $b^p_\cdot$ will minimize the sum of squares,

$$SS(\theta) = \sum_{j=1}^{n} \left[Y_j^E - f(\eta_j, \theta^\circ) - \sum_{i=1}^{P} \overset{\circ}{\Delta}_i \overset{\circ}{Z}_{ij} \right]^2 \tag{7}$$

with respect to $\overset{\circ}{\Delta}_i$, $i = 1, 2, \ldots, p$. New values of $\theta_{i(k+1)}$ are obtained by letting $\theta_{i(k+1)} = \theta_k + b_k$, where k is the number of iterations. This iterative process is continued until the solution converges, that is, until in successive iterations k, (k+1),

$$\left| \frac{\theta_i(k+1) - \theta_{ik}}{\theta_{ik}} \right| < \delta \qquad (8)$$

or

$$\left| \frac{SS(\theta_{k+1}) - SS(\theta_k)}{SS(\theta_k)} \right| < \delta_s, \qquad (9)$$

where δ and δ_s are some specified amount (e.g., 0.000001). At each stage of the iterative process, $SS(\theta)$ should be evaluated to assess whether a reduction in its value has actually been achieved.

The multicomponent equilibrium constants calculated in the regression model are given in Table IV. Distribution coefficients, calculated from the equilibrium constants based on the average feed composition for ORNL process wastewater, are in good agreement with the observed values obtained in column tests for Sr and Ca. However, the calculated distribution coefficients for Mg and Na differ significantly from the observed values. As shown in Figure 4, Sr and Ca data clearly resemble favorable "Langmuir-like" equilibrium isotherms while Mg and Na do not. The Langmuir model accurately represents Sr and Ca equilibria. However, the Langmuir relation can not adequately model the Mg "irregular-shaped" displacement (Figure 4c) which occurs as Ca displaces Mg on the loaded resin (see breakthrough curve Figure 1). The lack of fit generates gross errors in the nonlinear model exercise.

Table IV. Comparison of Predicted vs. Observed Distribution Coefficients

Feed Component	Equilibrium Constant (L/meq)	Predicted[a] K_d (mL/g resin)	Observed K_d (mL/g resin)
Sr	90.8	1850	1730
Ca	59.2	1206	1245
Mg	20.3	414	917
Na	0.207	4.21	569

[a]Distribution of coefficient calculated using equilibrium constants and initial feed concentrations.

Ion-Exchange Column Modeling. The experimental data obtained from the ion-exchange column runs were analyzed using a mathematical model originally developed by Thomas *(11-12)*. The Thomas Model assumes that a Langmuir isotherm is used and that ion-exchange mass transfer can be approximated by the second-order kinetic equation and modeled by: (a) T, a throughput parameter; (b) N, column height as transfer units; and (c) R, the separation factor. The value of N was determined from laboratory-scale experimental breakthrough

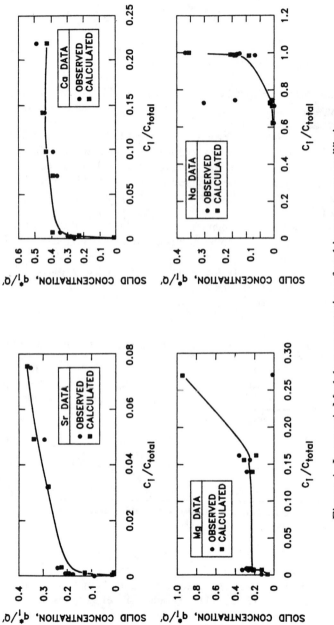

Figure 4. Langmuir Model representation of multicomponent equilibrium.

curves and used to predict scaleup for the pilot-scale CCIX system. For dilute solutions, the separation factor, R, is approximately constant and can be determined from the equilibrium constants.

The original Thomas model, which was designed for a single fixed-bed ion-exchange column, was extended by R. M. Wallace to predict the operation of several columns in series *(19)*. Wallace used a numerical analysis solution of the Thomas equation with input values of R, N, and K_d. In his calculations, a series of columns was loaded until a specified breakthrough occurred. At that time, a fresh column was added to the front (outlet end), and a column was removed from the back (inlet end) of the series. The columns again operated until a specified breakthrough occurred, and the sequence was repeated. This sequence of operations with a series of fixed beds simulates the behavior of a CCIX loading section. In the present study, the Wallace model was expanded to estimate the steady-state performance of the pilot-scale CCIX system *(20)*. The cyclic pulse operation in the CCIX column was modeled as several columns in series where the volume of each column was equivalent to the volume of freshly generated resin per pulse cycle. The model was used to predict distribution coefficients and performance of the CCIX system under various operating conditions.

Thomas Model. Thomas used experimental data to account for the equilibrium Langmuir relationship between solute A in solution and concentration A in the resin phase. The mass transfer between solute A and counter ion B originally on the resin was approximated by a second-order kinetic equation. Thomas assumed that the rate of adsorption of a single component on a resin can be represented by an expression suggested by the stoichiometry of the ion-exchange reaction:

$$nA + P_n B \rightleftharpoons B + nPA, \tag{10}$$

where A and B are the concentrations of the cations in the solution; PA and PB are the concentrations of the respective cations adsorbed on the resin, P, and n is the valence ratio of B to A. The general equation for the reaction kinetics of fixed-bed ion exchange column of length Z is:

$$\left(\frac{\partial Y}{\partial(NT)}\right)_N = X(1 - Y) - RY(1 - X) \tag{11}$$

and

$$\left(\frac{\partial X}{\partial N}\right)_{NT} = -X(1 - Y) + RY(1 - X), \tag{12}$$

where

X = dimensionless concentrations of the solute ion in the fluid phase,

Y = dimensionless concentrations of the solute ion in the solid phase,

N = represents the length of exchange column in transfer units,

NT = represents the time parameter for the system,

R = constant separation factor,

T = throughput parameter.

The variable X is defined as C/C_o, where C and C_o are the concentrations of the solute ion of interest in the effluent and feed, respectively. The variable Y is defined as q^*/q_s^*, where q^* is the actual concentration in the solid phase and q_s^* is the solute saturation concentration in the solid phase at equilibrium with fluid at the inlet concentration, C_o. The dimensionless expressions which are easily derived from material balances around the ion-exchange column are defined by:

$$N = \frac{K_d' (1 - \epsilon) Z \Omega_a}{\epsilon U_z}, \tag{13}$$

$$NT = \Omega_a \left(t - \frac{Z}{U_z} \right) \tag{14}$$

$$T = \frac{\epsilon U_z \left(t - \frac{Z}{U_z} \right)}{(1 - \epsilon) Z K_d'}, \tag{15}$$

$$r^* = \frac{1}{1 + K'C_o}, \tag{16}$$

$$\Omega_a = k_a \frac{1 + K'C_o}{K'}, \tag{17}$$

where ϵU_z is the flow rate through the column. K_d' is the distribution coefficient, (q_s^*/C_o), when $X = 1$; Ω_a is an expression for the mass transfer relation; t is the elapsed time at a point Z (i.e., length of the column bed) after a fluid particle has arrived at that point in the column having started at the entrance to the bed at time $t=0$ (21); and r^* is the variable separation factor determined from the equilibrium isotherm. The constant separation factor, R, as derived by Vermeulen and Hiester, is an explicit function of the variable equilibrium separation factor, r^* (22); R is conveniently used in the Thomas model to predict breakthrough. For a partial equilibrium isotherm, R is more linear (≈ 1) than r^*, which ranges between 0.62 to 0.83 for strontium data in the dilute wastewater feed.

For an ion exchange system, the throughput parameter, T, is often expressed as the ratio of the number of bed volumes of solution (V) that has passed through the bed, per unit volume of ion exchange bed (v). Using Equation 15, V can be shown to equal $\epsilon U_z S$ $(t\text{-}Z/U_z)$ and v is equal to SZ. When p_b is constant, the volume-based distribution coefficient is defined as $K_d = q_v/C_o$, where q_v is the concentration of the solute ion per unit volume of sorbent bed (sorbent plus void space), and C_o is the concentration in the feed. Equations 13 and 15 can then be expressed in terms of the volume-based distribution coefficient as

$$N = K_d K_a \left(\frac{v}{f}\right), \tag{18}$$

where

$f \quad = \epsilon U_z S$ which is the volumetric flow rate through the column,

$K_a \quad =$ the system mass transfer coefficient, Ω_a, and

$$T = \frac{(V/v)}{K_d}. \tag{19}$$

For an ion exchange system initially free of adsorbate, the boundary conditions are

$$X = \frac{C}{C_o} = 1 \ at \ N = 0 \ for \ all \ T \tag{20}$$

and

$$Y = \frac{q^*}{q_s^*} = 0 \ at \ T = 0 \ for \ all \ N. \tag{21}$$

When Equation 11 and Equation 12 are integrated, assuming the above boundary conditions, the solution obtained by Thomas is:

$$X = \frac{J(RN, NT)}{J(RN, NT) + [1 - J(N, RNT)] \exp [(R - 1)N(T - 1)]} \tag{22}$$

and

$$Y = \frac{1 - J(RN, NT)}{J(RN, NT) + [1 - J(N, RNT)] \, exp \, [(R - 1)N(T - 1)]} \quad (23)$$

Where J is a zero order Bessel function.

Modeling Breakthrough Curves. Breakthrough curves are S-shaped curves obtained by plotting T (or V/v) vs X on linear scales. Plotting these variables on logarithmic-probability scales eliminates the curvature of such plots and allows direct comparison with theoretical curves. For large values of RN, X is 0.5 when T equals 1 and is independent of RN. Therefore, K_d is approximately equal to V/v at the point where X is $\approx$ 0.5. Experimental data can be used to construct breakthrough curves of X vs T (or V/v). Values of R and N can be obtained from the experimental data by repeated iterations of Equations 22 and 23 or by comparing the experimental breakthrough curves (plotted on logarithmic-probability paper) with theoretical values available in many reference books *(23)*. These values can the be used to estimate performance at increased residence times and for columns in series *(24)*. The strontium data (plotted on logarithmic paper) obtained from the multicomponent breakthrough curves (Figure 1) were modeled as a binary system consisting of calcium and strontium to obtain the Thomas model parameters given in Figure 5.

Scaleup. The scaleup of small ion-exchange systems to industrial units can be performed once values of R, N, T, K_d, and K_a are obtained. These values can be used to predict the performance of columns at different operating conditions. The breakthrough curve at the new set of conditions will pass through the stoichiometric point ("center of mass," i.e., the value of X at T = 1 for which the area below and to the left of the breakthrough curve equals the area above and to the right of the breakthrough curve), but the slope of the breakthrough curve will change. The slope of the new curve at each X value will be proportional to that at the central X point and is determined by two factors: (a) how N changes with the new conditions and (b) how the slope ratio changes with N. It can be seen from Equation 18 that the value of N will vary with K_a, K_d, v, and f. During scaleup, the rate parameter, (i.e., the mass-transfer mechanism) cannot change. For linear isotherms (R = 1) and constant-sorption material, K_d is constant. Therefore, N becomes a function of v and f only.

The value of N $(N = 120)$ for the pilot-scale column was estimated for bench-scale data using Equation 18. The bench-scale column and pilot-scale CCIX columns had diameters of 7 and 102 mm ID, respectively. The columns had length-to-diameter ratios of 11 and 15, respectively. The feed flow rate was constant with a 1 min residence time in each column. Mass-transfer coefficients, K_a, and the value of K_d were assumed to be constant during scaleup. Direct modeling of the pilot-scale column breakthrough curves using Equations 22 and 23 revealed that the actual observed values of N were in the range of $30 \leq N \leq 50$. Thus, the scaleup equation predicted a higher value of N than was observed. The pilot-scale data suggest that a combination of controlling mass transfer

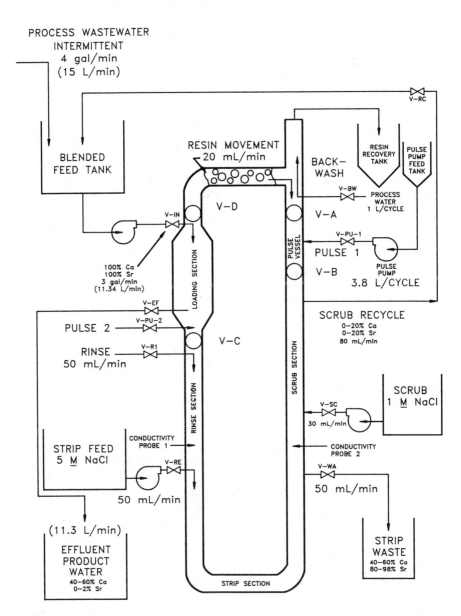

Figure 5. Modeling of bench-scale column breakthrough curve using Thomas Model.

mechanisms may occur. Consequently, more fundamental mass-transfer data are needed to accurately predict the CCIX column performance.

Wallace Model. Model parameters (R, N, T, and K_d) obtained from the Thomas equation were used in the Wallace model to predict effluent concentrations and steady-state performance for the pilot-scale CCIX system. A schematic of the Wallace Model representation of a CCIX loading section is illustrated in Figure 6. The CCIX column was modeled as four fixed-bed columns in series. The series of columns was loaded until a specified breakthrough occurred. At that time, a fresh column was added to the front (effluent outlet end), and a column was removed from the back (feed inlet end) of the series. This model would represent a case where the loading cycle is 30 min and length of resin pulse in the column loading section is equal to 15 in. (38 cm) per cycle. Each cycle adds a freshly regenerated column bed equal to 15 in. (38 cm). Column loading during each cycle is predicted in the mathematical model as illustrated in Figure 7.

Results of the Wallace model suggested that approximately 20 to 25 operating cycles (10 to 12h of semicontinuous operation for a 30-min loading cycle) were required to reach steady-state under conditions where the composition of the blended feed was constant and the resin entering the bottom of the loading section was thoroughly stripped (100% removal of strontium and calcium, leaving only sodium ions on the resin).

Summary and Conclusions

The CCIX system employs use of well-known, often-used water softening technique to perform a relatively difficult separation between strontium and calcium and many other ions (^{137}Cs, Mg, Na, etc.) found in low-level radioactive wastewaters. The pilot demonstrations showed the CCIX process to be more efficient than traditional fixed-bed ion-exchange systems. Data obtained from the near-steady-state operation of the pilot CCIX system for less than 50h of continuous operation have indicated that maximum removal efficiency (>99.7%) of the strontium from the blended feed stream can be achieved. The effluent stream from CCIX contained <0.005 mg/L total strontium and ≈3.4 Bq/L radioactive strontium, which is below the federal discharge limit of <11 Bq/L for radioactive strontium.

The maximum separation between Sr and Ca via CCIX was not achieved in the pilot demonstration. Optimization techniques will have to be performed to improve the separation from 40 to 60% removal of Ca to ≥90% removal in the strip waste stream. Continuous recycling of the 5 $\underline{M}$ NaCl regeneration stream used in the strontium removal needs to be demonstrated. Studies to date were conducted in bench-scale batch operations and have only partially confirmed the feasibility of the precipitation step. More fundamental studies are also needed to accurately model the CCIX system and predict optimum performance. Additional data are needed to accurately model the diffusion mechanisms, CCIX breakthrough curves, and predict optimum performance at different operating conditions. Results to date indicate that these additional studies are warranted.

CCIX LOADING SECTION

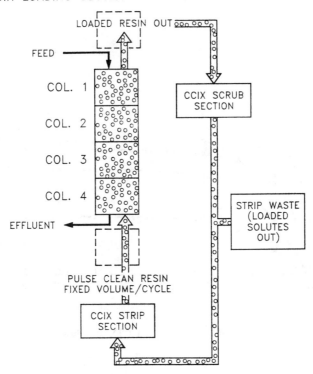

OPERATION SEQUENCE
1. FEED ENTERS DURING LOADING CYCLE; LOADING ENDS.
2. COLUMN 1 IS REMOVED FROM TOP AT END OF LOADING CYCLE.
3. OTHER COLUMNS ARE ADVANCED ONE POSITION.
4. NEW COLUMN IS ADDED AT BOTTOM POSITION.
5. REPEAT SEQUENCE.

Figure 6. Schematic of Wallace Model for a CCIX system.

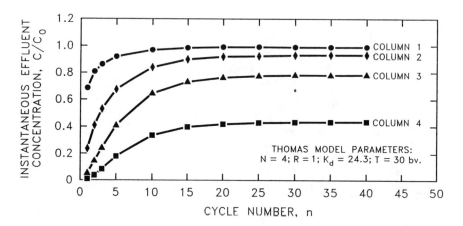

Figure 7. Wallace Model prediction of CCIX steady-state performance.

Acknowledgments

This research was sponsored by the Department of Energy under contract DE-AC05-840R21400 with Martin Marietta Energy Systems, Inc., at the Oak Ridge National Laboratory (ORNL), Oak Ridge, Tennessee. This report was prepared as a thesis and submitted to the faculty of the Graduate School of the University of Tennessee in partial fulfillment of the Master of Science in the Department of Chemical Engineering. The authors wish to acknowledge the assistance and support provided by J. H. Stewart of the Analytical Chemistry Division and J. D. Hewitt of the Chemical Technology Division at ORNL.

Literature Cited

1. E. D. Collins et al., "An Improved Ion-Exchange Method for Treatment of slightly contaminated Wastewaters," Proc. Am. Nuclear Soc. Int. Meeting on Low-, Intermediate-, and High-Level Waste Manage. and Decontamination and Decommissioning, Niagara Falls, NY (1986).
2. S. M. Robinson, *Initial Evaluation of Alternative Flowsheets for Upgrade of the Process Waste Treatment Plant*, ORNL/TM-10576, Oak Ridge National Laboratory (in-press).
3. R. E. Blanco and J. T. Roberts, "Separation of Lithium Isotopes by Batch or Continuous Ion Exchange with Decalso or Dowex-50 Media," Progress in Nuclear Energy, Series IV, 4, Technology Engineering and Safety (1961).
4. I. R. Higgins and M. S. Denton, "CSA Continuous Countercurrent Ion Exchange (CCIX) Technology," *Sep. Sci. Technol.*, 22 (2&3), 99 (1987).
5. S.M. Robinson et al., *Treatment of Radioactive Wastewaters by Chemical Precipitation and Ion Exchange,* presented at the Am. Inst. Chem. Eng. Spring National Meet., Houston, TX, March 29-April 2, 1987; AIChE Symp. Ser., 83, 52 (1987).
6. Dow Chemical Corporation, "Physical and Chemical Properties, Midland, MI.
7. A. Ringbom, *Complexation in Analytical Chemistry*, Interscience Publishers, John Wiley & Sons, New York, Chap. VI (1963).
8. R. E. Treybal, *Mass-Transfer Operations*, 34d Ed., University of Rhode Island, McGraw-Hill, New York, Chap. 11 (1980).
9. F. Helfferich, *Ion Exchange*, University of California, Berkeley, McGraw-Hill, NY (1962).
10. R. Hall et al., "Operation of a Mobile Pilot-Scale Continuous Countercurrent Ion Exchange System for Treatment of Low-Level Radioactive Wastewater," in "Proceedings of 62nd Annual Water Pollution Control Federation Conference," San Francisco, CA (1989).
11. I. R. Higgins and M. S. Denton, *Model 42 CCIX Pilot Unit Installation Operation and Maintenance Manual*, CSA Document No. 263-050, Rev. 0, CSA, Inc., Oak Ridge, TN (May 1988).
12. H. C. Thomas, *J. Am. Chem. Soc.*, 66, 1664 (1944).
13. I. Langmuir, *J. Am. Chem. Soc.*, 40, 1361, (1918).

14. J. N. Wilson, *J. Am. Chem. Soc.,* **62,** 1583 (1940).
15. E. Glueckauf, *Discuss. Faraday Soc.,* **7,** 12, (1949).
16. R. T. Yang, *Gas Separation by Adsorption Processes,* Univ. New York, Boston, 1987.
17. J. J. Moore, B. S. Garbow, and K. E. Hillstorm, *User Guide for MinPack-1,* Argonne National Laboratory, ANL-80-74.
18. N. R. Draper, and Smith, H., *Applied Regression Analysis,* John Wiley and Sons, 1981.
19. E. D. Collins, D. O. Campbell, L. J. King, J. B. Knauer, and R. M. Wallace, "Development of the Flowsheet Used for Decontaminating High-Activity-Level Water," in "The Three Mile Island Incident, Diagnosis and Prognosis," ACS Symp. Ser., **293,** 212 (1986).
20. R. Hall, J. S. Watson, and S. M. Robinson, *Pilot-Scale Demonstration of Strontium Removal from Low-Level Radioactively Contaminated Process Wastewater Using Continuous Countercurrent Ion Exchange (CCIX),* ORNL/TM-Draft, Oak Ridge National Laboratory (in preparation).
21. N. K. Hiester and T. Vermeulen, "Saturation Performance of Ion-Exchange and Adsorption Columns," *Chem. Eng. Prog.,* **48,** 505 (1952).
22. N. K. Hiester and T. Vermeulen, "Ion-Exchange Column Kinetics with Uniform Partial Presaturation," *J. Chem. Phys.,* **22,** 96 (1954).
23. N. K. Heister, T. Vermeulen, and G. Klein, *Chemical Engineers' Handbook,* 6th ed., J. H. Perry et al., Eds., McGraw-Hill, New York, Chap. 16, 1984.
24. H. K. S. Tan, "Programs for Fixed-Bed Sorption," *Chem. Eng.,* **91,** (26), 57 (1984).

RECEIVED April 5, 1991

Chapter 10

Economic Model for Air Stripping of Volatile Organic Chemicals from Groundwater with Emission Controls

R. M. Counce[1], J. H. Wilson[2], S. P. Singh[2],
R. A. Ashworth[3], and M. G. Elliott[3]

[1]Chemical Engineering Department, The University of Tennessee,
Knoxville, TN 37996–2200
[2]Oak Ridge National Laboratory, Oak Ridge, TN 37831–6044
[3]Engineering and Services Laboratory, Air Force Engineering and Services
Center, Tyndall Air Force Base, FL 32403

A spreadsheet-based computer simulation program has been developed to assist in the economic evaluation of VOC groundwater cleanup technology. This simulation consists of three general parts: (1) process design, (2) estimation of fixed capital and annual operating expense, and (3) operating lifetime analysis. Analysis of the estimated lifetime operating costs indicates that (1) the use of emission control increases the cleanup costs considerably and (2) the cost of remediation is largely dominated by annual operating costs.

This paper provides details of a spreadsheet-based computer simulation program designed to assist with the economic evaluation of VOC groundwater cleanup technology along with selected results for cleanup of jet fuel and TCE from groundwater. This simulation consists of three general parts: (1) process design, (2) estimation of fixed capital and annual operating costs, and (3) analysis of operating lifetime. The output from this simulation is an estimated unit processing cost (U.S. dollars/1000 gal) for remediation of contaminated groundwater. Cost estimating techniques are compatible with the procedures of the U.S. General Accounting Office. Usually, these unit processing costs are largely dominated by annual operating costs rather than by fixed capital investment. Much of the information provided in this paper is from a field study of air stripping with emissions control performed by the Oak Ridge National Laboratory (ORNL) for the United States Air Force (1).

Jet fuel contains a large percentage of hydrocarbon compounds that are easily removed by aeration. These compounds are called volatile organic compounds (VOCs) and have a Henry's coefficient of >1 to 2 m^3 atm/kmol. Benzene is a VOC, having an estimated Henry's coefficient of 4.7 m^3 atm/kmol at 20°C (2). Naphthalene is not considered to be a VOC, having a Henry's coefficient estimated to be 0.42 m^3 atm/kmol at 20°C (3) and are not efficiently removed in aeration devices designed for VOC removal.

0097–6156/91/0468–0177$10.00/0
© 1991 American Chemical Society

Stripping fuel components from water merely transfers them to another medium, usually air. Abatement of gaseous hydrocarbons by adsorption onto activated carbon has been utilized for many years. This process transfers the contaminants to another medium — activated carbon — which requires disposal or regeneration. Abatement by catalytic incineration offers a potentially very high destruction of fuel components resulting in CO_2 and H_2O.

The simulation described here is designed to provide the user with a method of estimating unit processing costs (U.S. dollars/1000 gal) for remediation of groundwater contaminated with VOC components. Two options are offered for air stripping — a traditional packed tower and a rotary air stripper. Also, two options are offered for emission control — activated carbon adsorption and catalytic incineration.

Background

Contamination of groundwater by VOCs is an important environmental problem. Sources of such contamination include accidental spills and leaking storage tanks and transport lines. Contamination of soil and groundwater from a leaking storage tank is illustrated in Figure 1. As shown in this figure, remediation of a contaminated aquifer is part of a larger problem of remediating the contaminated site, which includes the unsaturated zones. Site remediation usually involves containment of the contaminant(s). The National Oil and Hazardous Substances Contingency Plan (4) identifies three general categories of remediation activities: (1) initial, (2) source control, and (3) off-site measures. Pumping of groundwater can be useful for all three categories. Groundwater that ordinarily flows continuously through the site is an important agent for transporting soluble contaminants. Pumping activities usually involve depressing the groundwater table in the affected region by pumping from a system of wells. Contaminated water from such an operation requires treatment, as necessary, for its discharge; such treatment is the subject of this paper. A block diagram for the remediation of contaminated groundwater by air stripping with emission controls is presented in Figure 2.

A review of relevant technology was recently presented by Singh and Counce (5). Air-water contactors may be designed to give high removal of VOCs from water; contaminants of less volatility will be removed simultaneously to a lesser extent. The contaminated effluent air from the air-water contact is often a matter of concern. If this stream is not suitable for discharge, then treatment to restore acceptable air quality will be required. During the remediation activity, the concentration of contaminants in the groundwater will be reduced to an acceptable level for discharge. At some point in this activity, the effluent air quality may improve to the level that any emission control originally required is no longer necessary.

The objective of the activity described here is to provide a means of assessing the unit processing costs (U.S. dollars/1000 gal) of several configurations of an engineering system for the removal of VOC components from contaminated groundwater with costs estimated as a function of a number of key engineering and cost parameters. Some background information is also

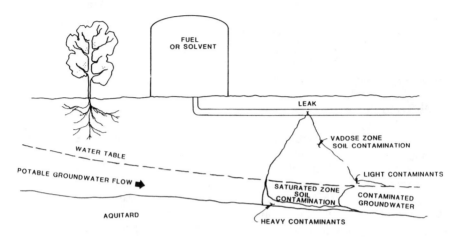

Figure 1. Subsurface behavior of spilled hydrocarbons.

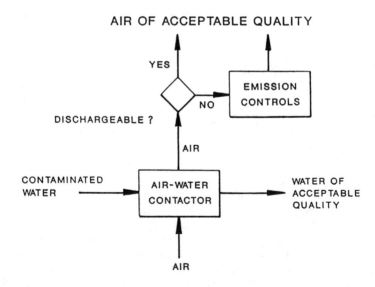

Figure 2. Block diagram of remediation of contaminated groundwater by air stripping with emissions control.

provided to give the reader a brief description of relevant technical information. The possible system configurations include an air stripper with and without emission controls. Two types of air strippers may be used: (1) a traditional packed tower for countercurrent gas-liquid contact and (2) a centrifugal contactor, also for countercurrent gas-liquid contact. The choice of gaseous emission control devices include: (1) none, (2) catalytic incineration, and (3) activated carbon adsorption.

Relevant Technical Information

AIR STRIPPING

Packed Air Strippers. Packed gas-liquid contactors with countercurrent gas and liquid flow provide a highly efficient means of stripping VOCs from groundwater. The design of packed towers is well developed in the chemical engineering literature (6-8). A typical packed tower is shown in Figure 3. The design process for strippers usually begins with known liquid flow rate and composition information, including that of the solute to be removed. The first step consists of the selection of the stripping medium; in this section, it is generally assumed that air is that medium. The air flow rate is selected such that an adequate "driving force" for this operation can be maintained. For conditions as typically existing in air strippers, this may be satisfied by choosing a stripping factor ≥1. The stripping factor is defined by

$$S = mG/L , \qquad (1)$$

where m is the equilibrium-phase distribution ratio, y/x, where y and x are mole fractions in the gas and liquid phases, respectively. The stripping factor also contains the ratio of gas-to-liquid superficial molar velocities, G/L. It is helpful to note that this ratio is identical to that of the molar flow rates of these two phases, since the tower cross-sectional area is not often known in the early stages of the design process. The optimum value of S is frequently found to be between 1.25 and 4.

Two general types of packing, random and structured, are useful for VOC removal by air stripping. Random packings come in a number of varieties, with standard saddles and slotted rings most commonly used for commercial applications. The commercial names may vary with the manufacturer, such as Flexirings® from Koch Engineering Company vs Pall rings from Chemical Processing Products Division of Norton Company and Flexisaddles® from Koch Engineering Company vs Intalox® saddles from Norton Chemical Process Products or Novalox® saddles from Jaeger Products, Inc. Random packing is sometimes referred to as "dumped" packing due to the usual method of placement in the tower. Random packings are available in a number of nominal sizes or diameters, d_p, of up to 89 mm and in materials of ceramic, plastic, or metal. There is a tendency for maldistribution of the liquid phase in towers filled with dumped packings, especially when the packing size and depth of packing in the bed are incorrectly chosen. This tendency is less when the

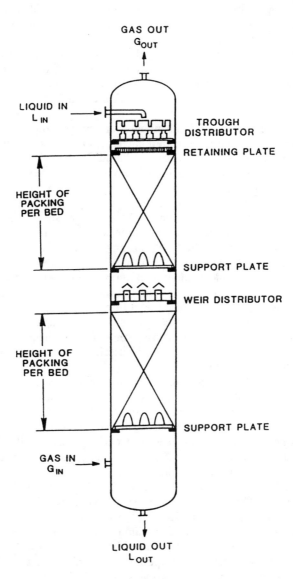

Figure 3. Typical packed tower.

ratio of column diameter to packing size, d_c/d_p, is >8 (6), with the "best" value occurring at 15. Eckert (9) recommends minimum d_c/d_p values of 30 for Raschig rings, 15 for standard saddles (other than Raschig rings), and 10 to 15 for slotted rings. In view of the tendency of the liquid and gas flows to segregate in towers filled with dumped packings, it is customary to redistribute the liquid at intervals, Z_p, which vary from 2.5 to 10 times the tower diameter; Eckert (9) recommends maximum Z_p/d_c values of 2.5 to 3 for Raschig rings, 5 to 8 for standard saddles (other than Raschig rings), 5 to 10 for slotted rings, or a maximum Z_p of 6 m, whichever is smaller. There is still speculation over the extent that good initial liquid distribution can affect the maximum Z_p/d_c ratio (10).

Structured packings offer advantages of low pressure-drop service and usually have excellent liquid distribution characteristics. It is critical, however, that the initial distribution of gases and liquids be done correctly in order to take advantage of the characteristics of these packings. The structured packings are manufactured as elements that are carefully fitted to the inside dimensions of the tower in an ordered, or structured, manner. These packings are commonly made of gauze (woven cloth), ceramic, sheet metal, or various plastics. The gauze packings are sometimes referred to as "high-efficiency" packing, an example being the Koch-Sulzer packing. The cloth nature of the surface promotes a capillary action so that the liquid covers all the available surface even at low liquid loading, apparently greatly enhancing liquid-phase transport (11-13). Similar packing elements fabricated of sheet metal or plastic do not appear to have the reduced liquid-phase resistance and near-constant interfacial area properties of the gauze-type packings. The gauze-type-structured packing is more expensive than either sheet-metal or plastic-structured packing or dumped packing; its use, however, can significantly reduce design height requirements.

Discussion of typical tower intervals, distributors, packing support, etc., is presented by Treybal (6) and Perry et al. (14). The design of these items is critical for efficient packed-tower operation (15 and 16). They are usually manufactured and distributed by the same commercial concerns that supply tower packing. The flow-through type of packings, such as Pall® rings, metal Intalox® saddles, and structured packings, requires more attention to the distribution of both gases and liquids than is required for older types of packings (bluff-body packings), such as Raschig rings, Berl saddles, and ceramic Intalox saddles. These older types of packings forced fluids to flow around them, causing a higher pressure loss, and provided more capacity to correct for maldistribution (17). Entrainment eliminators are not essential for VOC removal in the stripper, but they may be very important for equipment operating downstream from the stripper tower. Knitted wire mesh is especially effective for removing entrained droplets of liquid from gas streams, although many other devices are available. Further discussion on this subject is found in refs. 6 and 8.

Packed towers are usually circular in cross-section due to ease of construction and strength (6). The diameter of a tower for fixed gas and liquid rates is normally bounded by limits of operability. At a sufficiently small diameter, the tower will flood. At too large a diameter, the packing will not be sufficiently wetted for efficient mass transfer. Towers filled with dumped packing have operated with superficial liquid velocities as low as 0.18 mm/s; however, special liquid distribution systems are required (18). Usually, a tower with dumped packing is designed to operate at a given differential pressure drop per meter of packing with values for strippers of 0.25 to 0.5 in. H_2O/ft (200 to 400 Pa/m) of packed depth being common (6). Alternately, the tower may be designed by selecting a gas velocity as a fraction of the flooding gas velocities.

The height of packing required for a given separation may be conveniently calculated using the transfer unit concept,

$$Z = H_{toL} \, N_{toL} \, ,\qquad (2)$$

where Z is the height of packing required, and H_{toL} and N_{toL} are the height and number of overall liquid transfer units, respectively. The above relationship may also be written in terms of overall gas transfer units; the form chosen usually indicates where the principal resistance lies. The height of an overall liquid transfer unit is calculated in the procedures used herein from estimates of heights of individual gas (H_G) and liquid (H_L) transfer units,

$$H_{toL} = H_L + H_G/S \, .\qquad (3)$$

The correlations for H_L and H_G are based on a revised version of the Cornell equations, as presented by Fair et al. (8). The stripping factor, S, was defined in equation 1.

The equilibrium distribution ratio, m, is related to Henry's constant, H, by

$$m = y/x = H(\rho_L/M_L P_T) \, ,\qquad (4)$$

where ρ_L and M_L are the liquid density and average molecular weight, respectively, and P_T is the total pressure.

The Henry's constant determines the distribution of the solute between the gas and liquid phases and depends upon the nature of the solute and the temperature. Although the Henry's constant can be estimated from the vapor pressure of pure solute and the solubility of the solute in water, experimental determination is usually recommended. The experimentally determined expressions for the variation of Henry's law constant with temperature for benzene, which were used in this procedure, are from Ashworth et al. (2) and are given as

$$H = 1000 \, \exp(5.534 - 3194/T) \, [Benzene]\qquad (5a)$$

and

$$H = 1000 \exp(7.845 - 3702/T) \ [\textit{Trichloroethylene}] \quad . \qquad (5b)$$

Further information on estimates of H_L and H_G can be found in a review by Singh and Counce (5) as well as in Fair et al. (8).

The number of transfer units at conditions common to VOC removal is

$$N_{toL} = \int_{x_1}^{x_2} \frac{dx}{x - y/m} \ , \qquad (6)$$

where x_1 and x_2 are the solute concentrations in the inlet and outlet liquid, respectively.

The N_{toL} is estimated in the procedure used here by

$$N_{toL} = \frac{\ln[((x_2 - y_1/m)(1 - 1/S) \ / \ (x_1 - y_1/m)) + 1/S]}{(1 - 1/S)} \ , \qquad (7)$$

where y_1 is the solute concentration in the inlet gas. In its simplest form, the number of transfer units is the change in liquid-phase solute composition divided by the average solute driving force composition.

The design of packed towers usually includes a substantial safety factor to account for uncertainties in the data base. Bolles and Fair (19) state that the calculated height should be multiplied by 1.7 to achieve 95% confidence when using their correlations for H_{toL}.

The data base for estimation of mass transfer information is often not available for new packings. Advertisements for new packings often give overall coefficient information. These data, however, usually involve the absorption of CO_2 into caustic solutions in which liquid-phase resistance to mass transfer is minimal and, thus, are inappropriate for general use in the design of air strippers for VOC removal.

In general, inexpensive saddles and slotted rings, which have long been reliably used in industry, appear to be cost-effective for VOC strippers (20). Specialty random packings, however, are continually being developed and marketed, although these packings are usually much more costly than "standard" types of packing. The economics of the use of specialty random packings should be carefully investigated before they are specified and used. The random packing used for the performance and cost estimation in this manual is plastic Flexirings®. The major use of structured packings for VOC stripping appears to be to repack existing towers where greater efficiency or more capacity is required (20). Another use of structured packings may be to improve situations where minimum space exists for the stripper or where semivolatile components are being stripped.

Centrifugal Air Strippers. An alternative to the traditional packed column for countercurrent gas-liquid contact is the centrifugal contactor. A schematic of the centrifugal contactor is shown in Figure 4. The centrifugal vapor-liquid contactor is composed of two major components: the rotating packing and the stationary housing. The liquid phase is fed into the center of the rotating packing and flows outward due to the centrifugal force. After exiting the packing, the liquid phase impacts the housing wall and flows by gravity out of the unit. The vapor phase is introduced into the annular space between the packing and the housing and flows inward due to the pressure driving force. Seals are provided between the rotating packing and the housing to prevent the vapor phase from bypassing the packing. The high shear forces experienced by the liquid phase cause the formation of very thin films and rapid renewal of the interfacial surfaces. The rotation of the packing also causes considerable turbulence in the vapor phase. Both of these factors contribute to efficient mass transfer.

The concepts used to design conventional packed towers can be modified for the design of centrifugal vapor-liquid contactors. The latter design is based on tests of the HIGEE® centrifugal contactor marketed by Glitch, Inc., of Dallas, Texas. Characteristics of packing material used in these tests are shown in Table I. In designing a conventional packed tower, the diameter of the tower and the depth of packing are the two variables which need to be determined. Similarly, for the centrifugal vapor-liquid contactor, the cross-sectional area at the inner radius and a value of the outer radius are the two critical variables. An additional complexity arises in the design of the centrifugal vapor-liquid contactor because the cross-sectional area at the inner radius can be varied by changing either the radius or the axial length. This results in an iterative design process in which the inner radius, outer radius, and axial length are varied to arrive at an optimum design solution. The maximum OD is thought to be ~400 mm and the minimum 254 mm. The OD:ID ratio should be ~2.

Table I. Characteristics of Packing Material[a] used by Wilson et al. (*1*) in Tests of the HIGEE® Centrifugal Contactor

ROTOR	SPECIFIC SURFACE AREA ft^2/ft^3	VOIDAGE (%)
1	763	95.0
2	630	93.4

[a]Material: 85% nickel, 15% chromium (Sumitomo Electric Industries, Ltd.)

The cross-sectional area required at the inner radius is dependent on the desired hydraulic capacity. This design of these units should be closely coordinated with the manufacturer. The results of hydraulic capacity tests (*1*) are shown in Figure 5, along with the Sherwood correlation. These results indicate that the Sherwood correlation underestimated the limit of operability for the rotational speed in the case of the Sumitomo packing. However, there

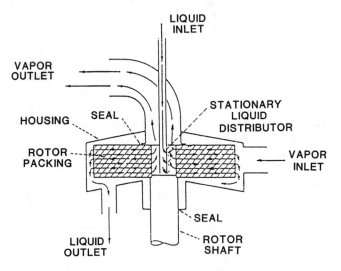

Figure 4. A schematic of centrifugal vapor-liquid contactor.

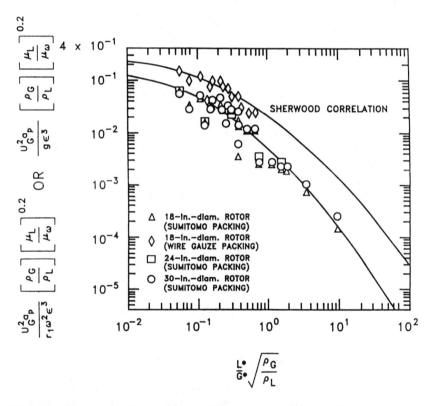

Figure 5. Comparison of limit of operability data with that predicted by the Sherwood Flooding Correlation.

was good agreement for the wire gauze packing. A second-order polynomial curve fit for the experimental data is also shown in Figure 5. The equation of this curve is:

$$\log Y = -2.274484 - 1.1367\log(X) - 0.168118[\log(X)]^2 , \tag{8}$$

with a coefficient of determination (r^2) of 0.80. The Y and X in this equation are the ordinate and abscissa, respectively, for Figure 5.

The pressure drop across the rotating packed torus of a centrifugal vapor-liquid contactor is modeled as a function of two terms. The first term accounts for the pressure drop due to rotation of the packing, and the second term accounts for pressure drop resulting from the flow of fluids through a porous media:

$$\Delta P_T = 1.392 \times 10^5 \rho_G \, \omega^2 (r_2^2 - r_1^2) + 1.50 \times 10^5 \frac{a_p}{e} \rho_G (r_2 - r_1) V_{G,avg}^2 . \tag{9}$$

The coefficient of determination (r^2) for the regression fit is 0.94. Although the approach outlined above is a rather simple representation of a complicated system, it does a reasonable job of describing the experimental data and is convenient to use.

The mass transfer results for the HIGEE® centrifugal contactor are based on a study by Wilson et al. (1). A variation of the transfer unit concept was developed by Singh (21) to represent the results of this study:

$$\pi(r_2^2 - r_1^2) = A_{toL} N_{toL} , \tag{10}$$

where

$$A_{toL} = \frac{3.37 \times 10^5}{a_p^2} (L^*/\mu_L \, a_p)^{0.6} (\rho_L^2 a_c/\mu_L^2 \, a_p^3)^{-0.15} . \tag{11}$$

The dimensionless groups are the Reynolds number and the Grashof number. Although the coefficient of determination (r^2) from regression analysis was only 0.61, the correlation usually predicts the area of a transfer unit (ATU) within ±20%, which is similar to what existing correlations predict for the conventional packed tower. It should be pointed out that the above correlation is based on the assumption that the rate of mass transfer is liquid-film controlled. The proposed correlation could be made more general by including the Schmidt number. However, since the Schmidt number in the experimental data remained constant, it may be considered to be part of the coefficient.

Power consumption for a centrifugal vapor-liquid contactor can be modeled using two distinct terms. The first term can be used to account for all the frictional losses and the second term to account for the power required to accelerate the liquid entering the packing torus to the rotational speed at the outer radius. The power required to accelerate the liquid can be described by a theoretical model (22). An operation for the overall power consumption was obtained by Wilson et al. (*1*). The experimental data for the region where the rotational speed was greater than its limit of operability gave the following equation:

$$P' = 1.21 + 0.000464 \, \rho_L r_2^2 \omega^2 \, Q_L .$$ (12)

The coefficient of determination (r^2) for this equation was 0.92. The correlation does a reasonable job in describing the power consumption over the operating conditions studied by Wilson et al. (*1*).

Emission Controls

Although air stripping is an effective technology for removing VOCs from groundwater, it simply transfers contaminants from one medium to another. In many cases, the removal of VOCs from the air may be necessary to reduce exposure to these chemicals. Emission control technologies have been used by industries that use solvents or other VOCs in production processes to either recover or destroy these compounds. The three most commonly used techniques for controlling VOCs from air streams are (1) adsorption onto activated carbon, (2) catalytic destruction, and (3) thermal destruction. This work focuses on VOC control by means of adsorption onto activated carbon and catalytic destruction.

Activated Carbon. Activated carbon can be produced from petroleum fractions, wood, coconut shell, or coal. It is then given treatment by superheated steam, which extends the pore network of the particles, giving activated carbon its large surface area.

The use of activated carbon to remove solvents from airstreams is well established and has been in use since the 1930s (*23*). The adsorption of the solvent molecules onto the activated carbon is mainly due to Van der Waal forces, with no chemical reaction taking place. The adsorption of the VOCs from the gas stream onto the activated carbon bed depends on (1) type of carbon, (2) relative humidity, (3) temperature, (4) concentration and type of VOC, and (5) regeneration steps used (*24*).

Several equations are available in the literature to describe the adsorption of compounds onto the activated carbon. These include the Braunauer-Emmett-Teller (BET), Freundlich, Langmuir, and Dubinin-Radushkevich equations. A commonly used isotherm describing the adsorption of single-component VOCs from gas streams, the Freundlich isotherm, is utilized in this activity:

$$q_i = K \, C_i^{1/n_i} ,$$ (13)

where C_i is the VOC concentration in the gas, and K and n_i are the Freundlich parameters. A method of correcting this single-component loading for the presence of other components is presented by Counce et al. (*25*). Basically, the correction factor is the ratio of carbon loadings for a single component to that for multiple components. Estimates of carbon loading are based extensively on work by J. C. Crittenden and others (26-28).

The adsorption of VOCs from the vapor phase can be affected by the relative humidity. This effect can be explained by noting the adsorption mechanism of water from the gas stream. At low relative humidity, the removal of water from the vapor phase is due mainly to adsorption onto the surface of the adsorbent. Due to a relatively low number of hydrophilic sites available on the activated carbon surface, the adsorption of water is fairly low. As the concentration of the water in the vapor phase increases, capillary condensation begins to occur in the pores of the activated carbon and the sites available for direct adsorption of the organic molecules from the vapor phase are reduced. The capillary condensation of water increases the resistance to mass transfer of other components in the pores because the compound must first dissolve in the water and then travel through the water to the adsorption site. The adsorption of VOCs from the vapor phase decreases rapidly as the relative humidity rises above 45 to 50% (29). Thus, the lifetime of the adsorption bed can be extended by lowering the relative humidity of the gas stream, which can be accomplished by raising the gas stream temperature. Increasing the temperature too much, however, can also reduce the adsorption capacity. The temperature and relative humidity of the gas stream need to be adjusted to obtain the optimum adsorption capacity.

In many cases, the amount of activated carbon used for vapor-phase VOC removal is small enough so that disposal of the spent carbon is a viable economic option. However, if the spent carbon is a hazardous waste, other disposal options need to be considered. One such option may be the regeneration of the carbon by a commercial vendor, which may be less expensive than on-site regeneration.

Steam stripping of activated carbon is a common method of reactivating carbon when the adsorption involves concentrated solvent vapors. Decisions on steam stripping as a method of regeneration are dependent on whether the condensate from this stripping operation will separate into two phases for easy removal of the organic phase. For adsorption of dilute VOC compounds onto activated carbon, the condensate from steam regeneration may not contain sufficient VOCs to form a separate phase and may require treatment of a secondary aqueous phase contaminated with dilute VOCs. To compound these problems, Crittenden et al. (29) report very high steam requirements for regeneration of activated carbon used in VOC adsorption service. Some estimated on-site costs for large-scale regeneration by multihearth, fluidized-bed, and infrared technologies are reported by Adams et al. (30).

Catalytic Destruction.   Another type of technology that has been used in industry to control the emission of VOCs from gas streams is catalytic destruction. In this process, a catalyst is used to promote the oxidation of organic compounds at lower temperature than required for thermal destruction. The catalyst increases the rate of the reaction by bringing the reactants together or by lowering the activation energy of the reaction. Approximately 500 to 2000 catalytic incinerators were in use in 1984 to control the emission of VOCs in various industries (31).

The performance of a catalytic destruction device depends upon temperature, type and concentration of compounds, space velocity (residence time), and type of catalyst. A typical catalytic destruction system is usually composed of four basic parts. A preheater is used to increase the temperature of the incoming gas stream. After the preheater, a mixing chamber is used to promote uniform temperature of the gas. The catalytic system then follows, which may be a fixed-bed or fluidized system. The catalyst is usually composed of either metal oxides or finely divided precious metal on either a metal or ceramic support structure. The final part of the system is a heat recovery system, which may be optional.

Spivey et al. (*32*) conducted a literature review on heterogeneous catalytic destruction of potentially environmentally hazardous compounds. They presented a review on the mechanism of catalytic oxidation reactions and a comparison of metal oxide and precious-metal catalysts. Listed below are some findings reported in their survey report:

1. Oxides of copper, manganese, cobalt, chromium, and nickel are the most active single-metal oxide catalysts.

2. Mixed-metal oxide catalysts generally have higher activity than single-metal oxide ones.

3. Metal oxide catalysts are less active than precious-metal catalysts, but the metal oxide catalysts are more resistant to certain poisons, such as halogens, arsenic, lead, and phosphorus.

Palazzolo et al. (*33*) studied the destruction of the mixtures of halogenated hydrocarbons using a fluidized-bed metal oxide catalyst and reported that the overall destruction efficiency varied from 97 to 99% for all the mixtures tested. They also noted that the catalyst inlet temperature had the largest effect on the destruction efficiency, while other variables, such as mixture composition, air-to-fuel ratio, space velocity, and inlet concentration, had only a marginal effect on the destruction efficiency. The destruction efficiency of tetrachloroethylene was the lowest of all the compounds tested. The destruction efficiency across the preheater ranged from 15 to 55%.

In the study by Wilson et al. (*1*), the noble metal catalyst was poisoned almost immediately in field tests.

Equipment Design and Sizing

The design of air strippers for removal of VOCs from groundwater in this activity is based on the removal of benzene. Benzene was chosen because it is an important contaminant and one of the less volatile VOCs. All VOCs of higher volatility (higher Henry's coefficient) can be assumed to be removed in the stripper with removal efficiencies at least that of benzene. It is more difficult to generalize stripper removal efficiencies for compounds of lesser volatility. A reasonable approach is to neglect the removal of these compounds for calculation of stripper performance and account for their presence in any prediction of carbon bed loading.

The cost of traditional packed-bed stripper towers is based on their physical size. The computer program developed in the current activity determines the size of the stripper from specification of stripping factor, fraction of gas velocity at flooding, and component removal efficiency. The selection of the stripping factor determines the ratio of gas to liquid flow rates for this contacting operation. The selection of the appropriate fraction of flooding (ratio of gas velocity to that at flooding, both at the same ratio of gas to liquid flow rates) determines the tower diameter. The internal computations of tower diameter and pressure drop across the column are based on the Eckert flooding and pressure drop diagram as presented by Treybal (6).

The height of the overall liquid transfer unit is based on calculations of individual gas and liquid transfer units. Computation of the individual liquid and gas transfer units, H_L and H_G, is based on correlations provided by Bolles and Fair (19). The number of transfer units is calculated based on expected inlet and outlet mole fractions of benzene.

The algorithm for estimating the costs of centrifugal air strippers does not involve detailed design of the unit. Design of these units may be accomplished based on information presented earlier. Costing of these units in the enclosed spreadsheet does not, however, require detailed design.

The design of fixed-bed carbon adsorbers is based on that for single-component adsorption and a correction factor used to account for decreased single-component bed loading based on the presence of other species. The humidity of this stream is assumed to have been reduced to $\leq 30\%$ by heating prior to adsorption. Values of the Freundlich isotherm parameters K and l/n are difficult to establish for gas-phase service but may be estimated using an approach presented by J. C. Crittenden et al. (35). The correction factor accounts for the presence of multiple components used to estimate the size of carbon beds for multicomponent systems.

The algorithm for calculation of the cost of catalytic incinerators is a function of gas throughput alone. The cost of fuel expense is a function of the operating temperature of the catalytic incinerator. For 90% destruction of benzene, an operating temperature of ~ 800°F is adequate (36). Information on the destruction of various gaseous mixtures of VOCs is presented by Tichenor and Palazzola (37). Proper selection of operating temperature is critical to the achievement of desired removal efficiencies.

For pumps and blowers, the operating expenses are related to the power requirements. The total power requirements for all pumps supplying groundwater are based on the estimated head at the pump discharge. This discharge head is largely due to the hydrostatic head, Z_T, and the head loss due to friction, which may be estimated as

$$Z_P' = (g/g_c)\, Z_T + fV^2 Z_L / 2g_c D$$

The velocity of water through a well-designed piping system is assumed to be

5 ft/s (*38*). A typical valve value of the Fanning friction factor, f, is 0.008. The required total power consumption of the pumps may be estimated as

$$P_B = \frac{L' \, Z_P'}{3957 \, \eta_P} \, , \tag{15}$$

where the efficiency of centrifugal pumps may be estimated as

$$\eta_P = 0.4 \, . \tag{16}$$

The cost of the blower is also related to the power consumption, which is estimated by

$$P_B = G' \, Z_B'' / \, 6356 \, \eta_B \, , \tag{17}$$

where the overall efficiency may be estimated (*39*) as

$$\eta_B = 0.70 \, . \tag{18}$$

The head requirements are primarily those due to flow through the relevant major equipment items.

The estimation of gas-phase pressure loss for the traditional packed tower was mentioned earlier. The pressure drop for the centrifugal air stripper is from a correlation presented in Section 2. The pressure drops across the catalytic incinerator and the activated carbon bed were assumed to be 5 and 4 in. of water, respectively.

Estimation of Costs for Removal of VOCs from Groundwater

This section focuses upon the estimation of the unit processing cost for the removal of VOCs from groundwater. The cost presented here is in 1990 U.S. dollars/1000 gal of water processed; the year for this cost estimation is an input quantity for the cost-estimating spreadsheets. The estimation procedure involves the following sequence, all of which is handled on spreadsheets:

1. Engineering design characterization, making use of design algorithms as discussed in the preceding section.

2. Estimation of the fixed capital costs for capital equipment. All costs presented for this study are adjusted to 1990 dollars. They may be accurate to ±30% overall.

3. Use of inputs from these two steps to estimate the cleanup costs per 1000 gal of groundwater processed.

The spreadsheet program allows for many of the engineering and financial parameters to be treated as input variables, which, in turn, allows for single-variable sensitivity analyses. Because of this, it is necessary to establish a set of "base-case" conditions, and these are summarized in Table II.

There are two important observations related to groundwater cleanup costs: (1) for the entire range of sensitivity analyses evaluated, the cleanup costs are dominated by annual expense rather than by fixed capital costs; and (2) the cost of carbon adsorption is sensitive to the concentration of contaminants in the groundwater, whereas the cost of catalytic incineration is not.

Estimation of the Fixed Capital Cost

The fixed capital costs for grass roots systems are estimated using correlation equations from the engineering literature, as summarized in the following subsections. Those values are then escalated by use of an equipment installation factor (or Lang factor) to account for additional costs such as installation, instrumentation and controls, yard improvements, piping, electrical, service facilities, etc. The adjusted amount then is the fixed capital cost used for subsequent cost analyses.

Equipment Installation Factors. Equipment installation factors in the range of about 1 to 6 are found in miscellaneous sources (*38* and *40*). Explicit values for selected pollution control equipment, however, seem to lie in the range of about 1.6 to 2.5.

The base-case values used in this study were 2.0 for either of the two strippers, and 1.6 for either of the two air treatment systems. Either or both may, however, be used as input variables for the spreadsheet simulation.

Fixed Capital Investment for Air Stripper and Associated Equipment. **(a) Stripper Air Fan and Motor.** Using the correlation from Walas (*41*), the cost of fans is estimated based on the use of backward-curved fan blades for low-pressure service (<1 psig). These fans are more efficient than others and are used for dust-free applications. The material of construction is stainless steel. The fan cost, C_t, in 1985 dollars is:

$$C_t = 1000 \ F_M \ \exp[a + b \ lnG' + c(lnG')^2] \ , \tag{19}$$

where F_M = 5.5 (stainless steel),
    a = 0.0400,
    b = 0.1821,
    c = 0.0786,
    $2000 < G' < 900,000$ scfm.

(b) Water Pump and Motor. Corripio et al. (*42*) use multiple equation sets for cost estimation purposes, which are somewhat awkward to install into a spreadsheet.

A few cost values were manually calculated from their sets of equations. A log-log scaling plot was then evaluated, and the following cost estimation equation was generated for a submersible centrifugal pump with a stainless steel head:

$$C_t = (2) (5984) (L'/1430)^{0.7} , \tag{20}$$

where L' is the water flow rate in gal/min, and C_t is the cost of pump and motor in 1977 dollars.

The scaling factor of 0.7 is consistent with median values for many equipment systems (*38*). Absent any data for the scaling factor, a common approach is to estimate on the basis of 0.6.

Stripper Shell. The weight in pounds, W_s, of the total stripper shell and structural accessories was estimated on the assumption of 0.25-in. steel construction.

The C_t (*43*) in 1977 dollars was then estimated as:

$$C_t = 1.1 \, C_B F_M , \tag{21}$$

where

F_M = 1.0 for carbon steel,
C_B = $\exp\{6.329 + 0.18255(\ln W_s) + 0.02297(\ln W_s)^2\}$,
W_s = $\pi \, D_i(Z_p + 0.8116 \, D_i) \, T_s \, \rho_{cs}$ (lb).

Plastic Flexirings. The cost of plastic Flexirings in 1990 dollars is:

$$C_P = \$11.00/ft^3 \quad (1\text{-}in. \; plastic \; Flexirings) \tag{22}$$

$$C_P = \$5.50/ft^3 \quad (2\text{-}in. \; plastic \; Flexirings) \tag{23}$$

$$C_P = \$3.50/ft^3 \quad (3\text{-}in. \; plastic \; Flexirings) \tag{24}$$

Fixed Cost of Emission Control Equipment Utilizing Activated Carbon Adsorption. The size chosen for the carbon adsorption units affects the initial capital costs, the recycle intervals, and the operating expenses. In order to narrow the range of sizes, the number of input days in a cycle was set up as an input variable. Thus, a carbon unit size may be calculated based on the other technical specifications that were input to the spreadsheet. One additional unit for standby and to minimize downtime during recycle is specified. The initial purchase requirements for carbon are then determined by the size and number of carbon adsorption units.

The cost of the carbon units, which includes everything except the initial carbon load, W_c, itself (*44*), is:

Table II. Design Input Table (Base-Case Values)
Independent Design Variable Inputs

GENERAL INPUTS	VALUE
Operating Temperature, °F	60
Inlet Gas Pressure, atm	1.00
GROUNDWATER STREAM	
Water Feed Rate, gal/min	500
Contaminant Concentration, ppb IN	10000
Contaminant Cleanup Factor	0.990
STRIPPER	
Stripping Factor (greater than unity)	4.0
Packing Factor (Treybal, 3rd Ed.)	
For Size: mm(in.)...METAL PALL RINGS ONLY	

Size-->	16(5/8)	25(1)	38(1 1/2)	50(2)
Select Cf-->	70	48	28	20

Packing Size, mm (in.)	50(2)
Packing Factor	20
Correlation Factor Term	1.502
Flooding Factor	0.4
Height of Each Packed Bed, ft	10
Height Design Safety Factor	1.7
ELECTRICAL EFFICIENCIES	
Air Heaters	0.7
Air Blowers	0.6
Water Pumps	0.7
STRIPPER-RELATED PARAMETERS	
Groundwater Depth, ft	30
Fanning Friction Factor	0.008
Horizontal Pipe Length, ft	100
Consolidated Friction Loss Coefficient for Valves, Elbows, etc.	15

Table II. Design Input Table (Base-Case Values)
Independent Design Variable Inputs (continued)

GENERAL INPUTS	VALUE
CARBON AIR STREAM CLEANUP	
Carbon Recycle Interval, d	7
Carbon Use Safety Factor	1.5
Carbon Bulk Density, lb/ft³	30
Freundlich Adjustment Factor (For BENZENE ONLY)	
	0.27566
CATALYTIC INCIN. AIR STREAM CLEANUP	
Catalytic Unit Size Safety Factor	1.5
OVERALL OPERATING CYCLE LOAD FACTOR	0.85
(365 d/year = 1.00)	
COST PARAMETERS	
Equip. Inst. Factor (Stripping)	2.20
EIF (Carbon and Catalytic)	1.60
Materials (SS) Factor, STRIPPER	1.70
Fuel Oil, 1990 $/MMBTu	6.15
Electricity Cost, 1990 $/kWh	0.060
Operations & Maintenance Factor	0.150
Overhead Rate (2) on Expense	100
Av. Annual Inflation: 1977-90	0.0484

$$C_t = 3.32 \ W_c + 36{,}400 \ , \tag{25}$$

where
   C_t is the cost in thousands of 1977 dollars,
   $10{,}000 \leq W_c \leq 200{,}000$ (lb).

Fixed Costs of Emission Control Equipment Utilizing a Catalytic Incinerator.
Vatavuk and Neveril (*36*) provide a nonlinear correlation equation for catalytic incineration costs. In the size range of interest to this study, however, a good linear fit can be made to the relevant portion of their graphical information. Thus, for "packaged units," the resultant equation for cost is:

$$C_t = 2.25 \ G' + 24{,}000 \ , \tag{26}$$

where
   C_t is in thousands of 1979 dollars.

Fixed Capital Investment Test for Rotary Stripper. Actual purchase prices for two units were available. These were for 50- and for 1000-gal/min capacities. Log-log scaling results in the following:

$$C_t = 180 \ (Q_L/2.23)^{0.31} \ , \tag{27}$$

where
   C_t is in thousands of 1990 dollars,
   $0.11 \leq Q_L \leq 2.23$.

The cost figures supplied were for complete systems, including packing.

Estimation of Operating Expense Cost

General Assumptions for Inflation and Interest Rates. The average annual inflation rate for the 1977-88 period in the chemicals industries has been 4.84% annually. This is a compound average of the CE and the M&S annual inflation indices (provided monthly in *Chemical Engineering* journal). In all cases, that annual rate has been used to adjust older price bases forward to 1990 dollars. Annual interest generally is related to annual inflation. Twenty or more years ago, applied interest rates tended to be ~3% higher than the inflation rate, or the "real interest" rate was averaging around 3% annually.

During the late 1970s, the real interest rate rose substantially. Over the past decade, it has been moving generally back down to the vicinity of 5% (*45*). That 5% annual real interest rate has been used throughout this study, although it is treated as an input variable and other values can also be examined.

General Assumptions for Overhead Rates. This category of cost estimating may well generate more diverse opinion than any other. Both a single-rate/single-base (*45-48*) and a multiple-rate/multiple-base method were evaluated. The maximum spread between the two methods is about ±10%, although the overall uncertainty of the total analysis may well be ±25% (*38*). Thus, it was felt that the more complex multiple-rate method was not warranted, and the bulk of the evaluation was done with the single-rate/single-base method.

In general, an overhead rate of 100%, shown as an input variable of Table II, was used for both the packed-column stripper and the rotary stripper. This was done because, in all cases, all of the direct labor is loaded into the expense base for the stripper unit.

A lower rate of 25% was applied to the two air treatment systems.

Fuel and Electricity Costs. The annual fuel and electricity requirements are provided by the design portion of the spreadsheet.

The fuel costs involve the assumption that the air stream will be preheated to 39°C using No. 2 oil as the energy source. This preheating is to reduce the relative humidity of the stream and thereby improve the performance of the carbon adsorption system.

Both fuel and electrical power costs vary around the country. The base-case values used here are typical of current costs in the southeastern United States in 1990. They are provided as input variables to the spreadsheet simulation, however.

Charcoal Recycle/Regeneration. The recycle and regeneration of charcoal is a major component of the annual expense base for the carbon adsorption units. This expense may be estimated by

$$E_R = 0.77 - 1.94 \ W_c' + 2.81 \ (W_c')^2 - 1.38(W_c')^3 \quad , \tag{28}$$

where C_c = \$1.00/lb (Midyear 1988).

This correlation expression was generated by curve-fitting to the multihearth method of graphical information of Adams et al. (*30*). The authors indicated that the regeneration cost estimates allow for 25% excess capacity, about 12% carbon replacement during each regeneration, and for all other related costs.

Maintenance. Annual maintenance costs in the range 5 to 15% of the fixed capital investment are commonly quoted (*38* and *40*). This wide range is simply indicative of the actual wide range of true maintenance costs.

Because of such uncertainties, the maintenance cost is handled as an input variable in the spreadsheet simulation. The base-case value was 15%, as shown in Table II.

Pall Ring "Maintenance". Total replacement on a 4-year cycle was assumed, which leads to an annual replacement cost equal to 25% of the initial investment cost.

Labor. One full-time equivalent (FTE) engineering technician level at $24,000/year direct salary in 1990 dollars was assumed. This is somewhat higher than suggested by Neveril (*40*), but the higher work load was used here as a conservative approach. No labor is loaded onto the two air treatment systems; all labor is loaded onto the stripper, as common to all three technical options.

Operating Lifetime Analysis

Cost studies sometimes involve only equipment and process design and capital and annual expense cost estimating, as described above. These two categories alone are not adequate for the economic evaluation of a proposed project. In the case of manufacturing operations, the next important step is to evaluate the unit manufacturing cost, e.g., the cost per unit of polyethylene produced by a manufacturing facility. However, no manufacturing is involved in the air stripping process. Thus, the "standard" selected for complete cost evaluation was the current year dollar cost per 1000 gal of groundwater processed. An operating lifetime of 20 years was arbitrarily chosen for the cost analysis presented here. The operating expense for each year of the operating lifetime was calculated using the expense estimate for the base year, determined as described above, and the assumed inflation rate. The capital costs were depreciated over the operating lifetime, with no salvage value. The current year dollar cost per 1000 gal of groundwater processed was then the sum of the capital costs and all of the annual operating expenses divided by the total gallons of groundwater processed over the operating lifetime.

With the above approach to the operating lifetime analysis, the unit processing cost for groundwater cleanup was dominated by the annual expense for all cases considered in this study. In fact, this is likely to be true for all similar purpose cases in which the depreciation period for the capital equipment was on the order of the time assumed here and the operating lifetime was equal to or less than the depreciation period (the assumption being that the equipment could be used for other projects). If the operating lifetime and the depreciation period were always equal (which assumes a one-time use of the equipment), the capital costs would become more of a factor as shorter times were examined. Different scenarios such as these may be readily analyzed using the values for capital cost and operating expense that were calculated for the base year.

Technical Combinations of Equipment

Six technical configurations were considered in this study.

Packed column stripper   ALONE

PLUS carbon adsorption treatment

PLUS catalytic incineration

Rotary stripper   ALONE

PLUS carbon adsorption treatment

PLUS catalytic incineration

Both fixed capital and expense components of the processing costs were presumed to be linearly additive for any combination of the applicable unit operations. More information on capital and operating costs for these options is found in Wilson et al. (*1*) and in Counce et al. (*25*).

Economics of Groundwater Remediation

Jet Fuel Spill

The situation considered here is that of a jet fuel spill which contaminated groundwater with about 10 ppmw of VOCs, the benzene content being 0.1 ppmw.   This is similar to the conditions described in the study by Wilson et al. (*1*). For the economic analysis, the carbon bed loadings were calculated based on the experimental data from this study. The results of the operating lifetime analysis are presented in Figure 6 for the six different cases discussed previously. Variables not being intentionally varied are as base-case values presented in Table II. In the figure, the cost per 1000 gal of groundwater processed is plotted as a function of the throughput rate of groundwater.

Figure 6 shows that the packed tower stripper and the rotary stripper produce comparable cleanup costs. This is because, as pointed out previously, the operating expenses were the dominant costs in the lifetime operating analyses. For a stripper with no emissions control, the cost per 1000 gal of groundwater ranged from about $3.5 to $0.5 with increasing groundwater flow rate. In the case of a stripper with catalytic incineration for emissions control, the cost range was about $5 to $2. With activated carbon, the cleanup cost was intermediate to these two cases. Since the carbon bed size is dependent upon the amount of VOCs adsorbed, the cost for activated carbon will approach and exceed the cost for catalytic incineration as the concentration of VOCs in the groundwater increases. Conversely, the cost will approach that for the case of a stripper with no emissions control as the VOC concentration decreases. The cost for catalytic incineration is not affected by VOC concentration since its cost is assumed to be dependent only upon the air flow rate. The stripper cost is dependent upon the fractional removal efficiency for the VOCs and not upon their concentration. In general, these costs compare reasonably well with those of Fang and Khor (*49*). When overhead charges are treated similarly, Fang and Khor considered large-scale air stripping of various VOCs from groundwater without emission control and with control by activated carbon adsorption.

Trichloroethylene Spill

Operating lifetime analyses were performed for the situation in which groundwater was contaminated with 10 ppmw of TCE. The rotary stripper cases were not included. Figure 7, which is similar to Figure 6, shows that the costs for both emissions control options were about the same. Again, as the TCE concentration in the groundwater varies, the cost for activated carbon will change. The difference in stripper costs between the jet fuel and the TCE spill situations was due to the difference in Henry's law constants.

Figures 8 through 13 present the results of sensitivity analyses performed for several independent parameters. In each figure, the base-case value of the parameter is indicated.

In Figure 8, the cost is relatively insensitive to stripping factor. However, the cost for the emissions control cases increases significantly with increasing stripping factor. Since the Henry's law constant is that for TCE and the groundwater flow rate is fixed at the base-case value, the stripping factor is a function of stripping gas flow rate. Thus, the cost for the activated carbon case increases with stripping factor because the carbon bed loading decreases as the TCE becomes more dilute in the air stream. The capital cost and operating expense of the catalytic incinerator increases with air flow rate.

Figure 9 shows that the stripper cost is affected by the flooding factor. As the flooding factor increases, the diameter, and therefore the cost, of the packed tower decreases. This effect is seen for all three cases in Figure 9.

As expected, the equipment installation factor and the overhead rate both have a direct effect upon the processing cost, as shown in Figures 10 and 11, respectively. Over a range of about 1 to 5, the installation factor causes the processing cost for the stripper-only case to increase by a factor of about 2. The costs for the emissions control cases increase by a factor of about 1.5. The effect of the overhead rate is somewhat less.

In Figure 12, the processing costs increase somewhat with the cleanup fraction (stripper removal efficiency). This increase is caused by the greater height required for the packed tower stripper. Figure 12 also shows that the TCE concentration in the groundwater affects the processing cost for the activated carbon case only, as discussed previously.

The carbon recycle interval causes an increase in processing cost for the activated carbon case, as shown in Figure 13. This is due to the greater size, and consequently high cost, of the carbon bed required at longer recycle intervals.

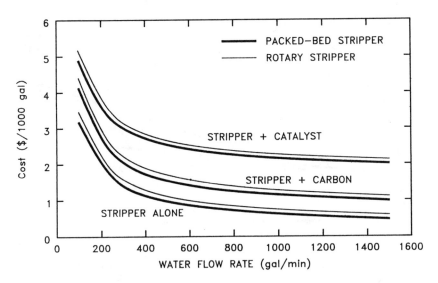

Figure 6. Comparison of lifetime processing costs for an air stripping system for benzene removal from groundwater featuring a traditional packed tower vs a similar purpose system featuring a rotary air stripper (Method A-1).

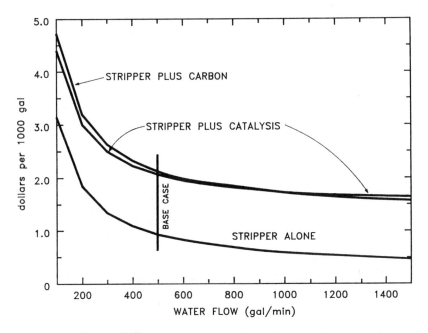

Figure 7. Operating lifetime processing costs as a function of capacity for an air stripping system utilizing a traditional packed tower for removing TCE from groundwater (Method A-1).

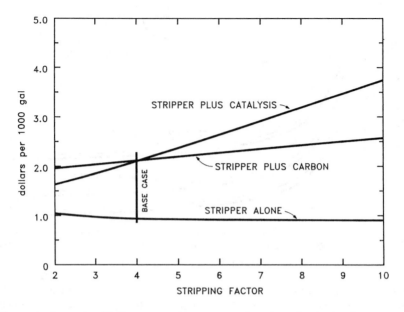

Figure 8. Operating lifetime processing costs as a function of stripping factor for an air stripping system featuring a traditional packed tower for removing TCE from groundwater (Method A-1).

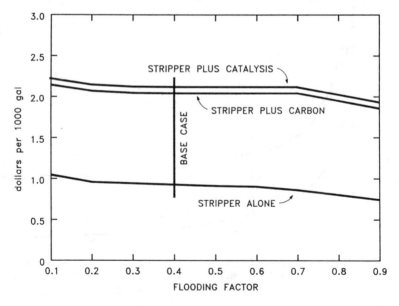

Figure 9. Operating lifetime processing costs as a function of the flooding factor (fraction of flooding) for an air stripping system featuring a traditional packed tower for removing TCE from groundwater (Method A-1).

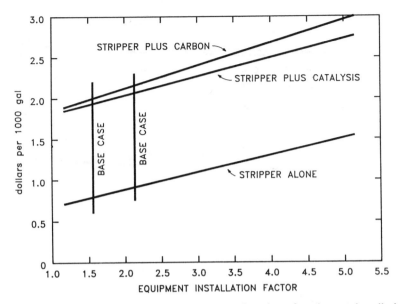

Figure 10. Operating lifetime processing costs as a function of equipment installation factor for an air stripper system featuring a traditional packed tower for TCE removal from groundwater (Method A-1). The base-case equipment installation factor is 2.20 for the stripper and 1.80 for the carbon and the catalytic units. For this graph, the abscissa value is applied to all three units.

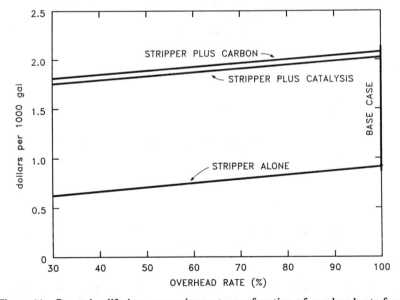

Figure 11. Operating lifetime processing costs as a function of overhead rate for an air stripping system featuring a traditional packed tower for TCE removal from groundwater (Method A-1).

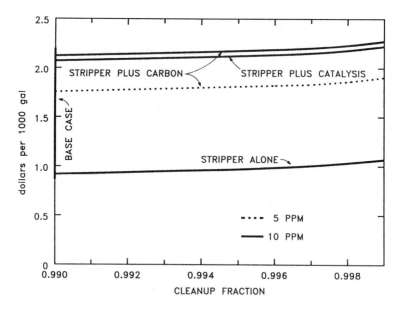

Figure 12. Operating lifetime processing costs as a function of the cleanup fraction of TCE from groundwater in an air stripper system featuring a traditional packed tower (Method A-1).

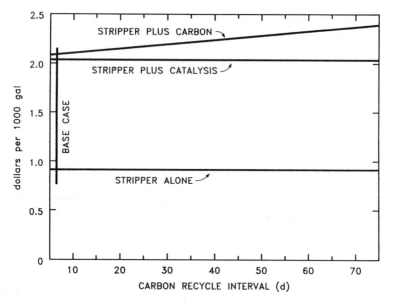

Figure 13. Operating lifetime processing costs as a function of the carbon recycle interval for an air stripping system featuring a traditional packed tower for TCE removal from groundwater (Method A-1).

Conclusions

A spreadsheet-based computer simulation has been developed to assist with the economic evaluation of technology for use in remediation of VOC-contaminated groundwater. This simulation utilizes referenced techniques for estimating the major components of the unit processing cost (U.S. dollars/1000 gal). The use of commercial spreadsheets for this simulation offers a simple and user friendly computer tool for this analysis.

Analysis of the lifetime operating costs for the application of air stripping technology for the remediation of contaminated groundwater indicates the following findings. (1) Little cost difference is observable for the use of a traditional packed tower compared to a rotary air stripper; as more experience is accumulated with rotary air strippers, better estimates may allow discrimination between the operating costs of these two technologies. (2) The use of emission control devices on systems for air stripping VOCs from groundwater considerably increases the costs of such operations; the costs of emission control vs activated carbon adsorption for catalytic incineration are strongly related to the concentration of contaminants in the groundwater, because the carbon requirements will vary proportionally to this contaminant concentration. (3) Cost considerations favor the use of lower values of the stripping factor. (4) Under the assumptions of this study, the costs of the remediation of contaminated groundwater with and without emissions control are largely controlled by annual operating costs.

Legend of Symbols

a_c	Acceleration, ft^2/s
a_p	Specific area of packing per unit of packed volume, ft^2/ft^3
B	Microporosity constant, mol^2/cal^2
C	Gas concentration of a contaminant, $mol/10^6$ L
C_B	Cost of traditional packed tower shell and skirt
C_C	Cost of activated carbon, dollars/lb
C_P	Cost of tower packing, dollars/ft^3
$C_p{'}_{,MG}$	Mean heat capacity of air in catalytic unit, BTU/mol F
C_T	Total delivered cost of equipment (U.S. dollars)
C_i	Fluid-phase concentration of component i, parts per million ($\mu g/g$)
D	Diameter of total water piping system based on a velocity of 5 ft/s
D_i	Diameter of packed tower, ft
E_R	Carbon reactivation expense, dollars/lb
F_m	Cost factor for materials of construction
f	Fanning friction factor
$G{'}$	Air flow rate, ft^3/min
G	Superficial gas molar velocity, lb mol/ft^2-s
G^*	Mass velocity of air, lb/ft^2-h
g	Acceleration of gravity, ft/s^2
g_c	Conversion factor, ft-lb/s^2-lb_f
H	Henry's constant, ft^3 atm/lb-mol

Legend of Symbols (continued)

H_G	Height of individual gas-phase transfer unit, ft
H_L	Height of individual liquid-phase transfer unit, ft
H_{toL}	Height of overall liquid-phase transfer unit, ft
K	Freundlich value
L	Superficial liquid molar velocity, lb-mol/ft^2-s
L'	Water flow rate, gal/min
L^*	Mass velocity of liquid, lb/ft^2-h
M	Molecular weight, lb/lb-mol
M_L	Average liquid molecular weight, M molecular weight, lb/lb-mol
M_s	Solute mass flow rate, lb/h
m	Equilibrium distribution ratio, y/x
N_{toL}	Number of overall liquid-phase transfer units, dimensionless
n	Refractive index
P	Power, hp
P_B	Brake power, hp
P_T	Total pressure, atm
P_i	Partial pressure of component i, atm
P_s	Saturation pressure, atm
P'	Power, kW
Q_L	Volumetric liquid rate, ft^3/s
q	Carbon capacity, solute/10^6 g C or lb solute/10^6 lb C
q_i	Loading of component i on activated carbon, μlb/lb (μg/g)
q_w	Working capacity of activated carbon, lb solute/lb C
R	Ideal gas law constant
r	Radius
S	Stripping factor, dimensionless
T	Temperature, Kelvin
TC	Treatment capacity, g C/L H$_2$O
T_s	Traditional packed-tower shell thickness, in.
U_G	Gas velocity, ft/h
V_G	Superficial gas velocity, ft/s
V_L	Superficial liquid velocity, m/s
Vw_k	Wave front velocity of component k, m/s
W	Adsorption space occupied by the adsorbate, cm/g
W_c	Mass of activated carbon required, lb
W_o	Maximum space available on carbon, cm^3/g
W_s	Weight of tower shell, lb
W^1	Carbon usage rate, 10^6 lb/year
x_1, x_2	Liquid-phase mole fraction exiting and entering the stripper, respectively
x	Mole fraction in the liquid phase, lb-mol/lb-mol
y	Mole fraction in the gas phase, lb-mol/lb-mol
y_1, y_2	Gas-phase mole fraction entering and leaving the stripper, respectively
Z	Depth of packing, ft

Legend of Symbols (continued)

Z_B''	Total static head at blower, in. H_2O
Z_L	Total length of water supply piping, ft
Z_T	Total vertical head requirements for pump, ft
Z_A''	Static pressure loss for equipment components, in. H_2O
Z_p	Depth of packing including safety factor, ft
Z_p'	Head requirement for water supply, ft_r-lb_f/lb_M
1/n	Freundlich parameter

Greek symbols

β	Affinity coefficient
μ	Polarizability of the adsorbate
μ_L	lb/ft-h
μ^*	Polarizability of the reference molecule
n_B	Overall blower efficiency
n_m	Moles efficiency
n_p	Efficiencies of pump head
ρ	Density, lb/ft^3
ρ_G	Gas density, lb/ft^3
ρ_L	Liquid density, lb/ft^3
ρ_c	Bulk density of activated carbon, lb/ft^3
ρ_{cs}	Density of carbon steel, lb/ft^3
θ_c	Adsorption time, h
ϵ	Void fraction
ω	Rad/s

Subscripts

G	Gas
L	Liquid
M	Logarithmic average

Literature Cited

(1) Wilson, J. H.; Counce, R. M.; Lucero, A. J.; Jennings, H. L.; Singh, S. P. *Remediation of Contaminated Groundwater by Air Stripping with Emission Control Project*, Final Report, Oak Ridge National Laboratory, Oak Ridge, TN (in progress).

(2) Ashworth, R. A.; Howe, G. B.; Mullins, M. E.; and Rogers, T. N. *J. Haz. Mater.* 1988, 25.

(3) Mackay, D.; Shiu, W. Y. *J. Phys. Chem. Ref. Data* 1981, *10*(4), 1175-1199.

(4) Singh, S. P.; Counce, R. M. *Removal of Volatile Organic Compounds from Groundwater: A Survey of the Technologies*, ORNL/TM-10724; Oak Ridge National Laboratory, Oak Ridge, TN 1989.

(5) Treybal, R. *Mass Transfer Operations*; McGraw-Hill Book Company: New York, NY, 1980.

(6) Chopey, N. P.; Hicks, T. G. *Handbook of Chemical Engineering Calculations*; McGraw-Hill Book Company: New York, NY, 1984.

(7) Fair, J. R.; Steinmeyer, D. E.; Penny, W. R.; Croker, B. B. In *Chemical Engineer's Handbook, Sixth Edition*, Section 18; Perry, R. H.; Green, D. W.; Maloney, J. O., Eds McGraw-Hill Book Company: New York, NY, 1984.

(8) Eckert, J. S. *Chem. Eng. Prog.* 1961, *54*, 57-59.

(9) Strigle, R. E., Jr. *Random Packings and Packed Towers, Design and Applications*, Gulf Publishing Co.: Houston, TX, 1987.

(10) Bravo, J. L. et al. *Hydrocarbon Process* 1985, *65*(3), 91-95.

(11) Selby, G. W.; Counce, R. M. *Ind. Eng. Chem. Res.* 1988, *27*(10), 1917-1922.

(12) Lucero, A. J.; Wilson, J. H.; Counce, R. M.; Singh, S. P.; Ashworth, R. A.; Elliott, M. G. "Air Stripping of VOC from Groundwater Using Towers Filled with Structured Packing;" paper presented at AIChE 1990 Spring National Meeting, Orlando, Florida, 1990.

(13) Perry, R. H.; Chilton, C. H. *Chemical Engineer's Handbook*, 5th Edition; McGraw-Hill Book Company: New York, NY, 1984.

(14) Kunesh, J. G.; Lahm, L.; Yanagi, T. *Ind. Eng. Chem. Res.* 1987, *26*, 1845-1850.

(15) Kunesh, J. G. *Chem. Engr.* 1987, *94*(18), 101-105.

(16) Fair, J. R. In *Handbook of Separation Process Technology*; Rousseau, R. W., Ed.; Wiley and Sons, New York, NY, 1987.

(17) Norton Company. *Design Information for Packed Towers*, Bulletin DC-11; Norton Company, Akron, OH, 1971.

(18) Bolles, W. L.; Fair, J. R. *Chem. Eng.* 1982, *78*, 109-116.

(19) Hill, J. H. "Packings for VOC Stripper Columns;" paper presented at 1987 AIChE Annual Meeting, New York, NY, 1987.

(20) Singh, S. P. *Air Stripping of Volatile Organic Compounds from Groundwater: An Evaluation of a Centrifugal Vapor-Liquid Contactor*; Ph.D. Dissertation, University of Tennessee, Knoxville, TN, 1989.

(21) Leonard, R. A. *Prediction of Hydraulic Performance in Annular Centrifugal Contactors*, Argonne National Laboratory, Argonne, IL, ANL-80-57, 1980.

(22) Metcalf & Eddy, Inc. *Wastewater Engineering: Treatment, Disposal, Reuse*; McGraw-Hill, Inc.: New York, NY, 1979.

(23) Foster, M. L. "Evaluation of Parameters Affecting Activated Carbon Adsorption of a Solvent-Laden Air Stream;" paper presented at the 78th Annual Meeting of the Air Pollution Control Association, Detroit, MI, June 1985.

(24) Counce, R. M.; Wilson, J. H.; Thomas, C. O. "Manual for Estimating Costs of VOC Removal from Groundwater Contaminated with Jet Fuel," (in preparation).

(25) Crittenden, J. C.; Speth, T. F.; Hand, D. W.; Luft, P. J.; Lykins, B. *J. Environ. Eng.* 1987, *113*(6), 1363.

(26) Crittenden, J. C.; Luft, P.; Hand, D. W.; Oravitz, J. L.; Loner, S. W.; Arl, M. *Environ. Sci. & Technol.* 1985, *19*, 1037.

(27) Speth, T. F. *Predicting Equilibria for Single and Multicomponent Aqueous Phase Adsorption Onto Activated Carbon*, Master's Thesis, Michigan Technol. Univ., Houghton, MI, 1986.

(28) Crittenden, J. C.; Cortright, R. C.; Rick, B.; Tang, S.-R.; Perram, D. *J. Am. Water Works Assoc.* 1988, 73.

(29) Adams, J. Q.; Clark, R. M.; Miltner, R. J. *J. Am. Water Works Assoc.* 1989, 132-140.

(30) Jennings, M. S.; Krohn, N. E.; Berry, R. S. *Control of Industrial VOC Emissions by Catalytic Incineration: Volume 1, Assessment of Catalytic Incineration and Competing Controls*, EPA-600/2-84-118a; Washington, DC, 1984.

(31) Spivey, J. J.; Tichener, B. A.; Ashworth, R. A. *Catalytic Oxidation of VOCs: A Literature Review*; Research Triangle Institute: Research Triangle Park, NC, 1987.

(32) Palazzolo, M. A.; Jamgochain, C. L.; Steinmetz, J. I.; Lewis, D. L. *Destruction of Chlorinated Hydrocarbons by Catalytic Oxidation*, EPA-600/2-86-079; Washington, DC, 1986.

(33) Crittenden, J. C.; Regg, T. J.; Perram, D. L.; Tang, S. R. *J. Environ. Eng.* (in press).

(34) Crittenden, J. C.; Speth, T. F.; Hand, D. W.; Luft, P. S.; Lykins, B. "Evaluation of Multicomponent Competitive Adsorption in Fixed-Beds," *J. Environ. Chem.* Eng. 1987, 113(6), 363, 1987.

(35) Vatavuk, W. M.; Neveril, R. B. *Chem. Eng.* 1982, *89*(14), 129-132.

(36) Tichenor, B. A.; Palazzolo, M. A. *Environ. Prog.* 1987, *6*(3), 172.

(37) Peters, M. S.; Timmerhaus, K. D. *Plant Design and Economics for Chemical Engineers*, 3rd Ed.; McGraw Hill Book Co.: New York, NY, 1980.

(38) Clarke, L.; Davidson, R. L. *Manual for Process Engineering Calculations*, 2nd Ed; McGraw Hill Book Co.: New York, NY, 1986.

(39) Neveril, R. B. *Capital and Operating Cost of Selected Air Pollution Control Systems*, EPA 450/5-80-002; Research Triangle Institute: Research Triangle Park, NC, 1978.

(40) Walas, S. M. *Chemical Process Equipment Selection and Design*; Butterworth Publishing: New York, NY, 1988.

(41) Corripio, A. B.; Chrien, K. S.; Evans, L. B. *Chem. Eng.* 1982, *82*(4), 115-118.

(42) Mulet, A.; Corripio, A. B.; Evans, L. B. *Chem. Eng.* 1981, *88*(26), 77-82.

(43) Vatavuk, W. M.; Neveril, R. B. *Chem. Eng.* 1983, *90*(2), 131-132.
(44) Thomas, C. O. *Methanol as a Carrier for Alaskan Natural Gas*; prepared for the U.S. General Accounting Office, Contract No. 3130090, Washington, DC, November 1982.
(45) Thomas, C. O. "Alaskan Methanol Concept," A policy analysis study for the Federal Energy Administration; Institute for Energy Analysis, Oak Ridge Associated Universities: Oak Ridge, TN, 1975.
(46) Thomas, C. O. "Alaskan Methanol Concept," Testimony for the Senate Commerce Committee and the Senate Interior and Insular Affairs Committee; Institute for Energy Analysis, Oak Ridge Associated Universities: Oak Ridge, TN, 1976.
(47) Thomas, C. O. In *Methanol Technology and Applications in Motor Fuels*; Paul, D. K., Ed.; Noyes Data Corporation: Park Ridge, NJ, 1978.
(48) Fang, C. S.; Khor, S.-L. *Environ. Prog.* 1989, 8(4), 270-278.

RECEIVED April 5, 1991

BIOLOGICAL TREATMENT

Chapter 11

Biodegradation of Munition Waste, TNT (2,4,6-Trinitrotoluene), and RDX (Hexahydro-1,3,5-Trinitro-1,3,5-Triazine) by *Phanerochaete chrysosporium*

Tudor Fernando and Steven D. Aust

Biotechnology Center, Utah State University, Logan, UT 84322–4430

Rapid biodegradation of TNT (2,4,6-trinitro-toluene) and RDX (hexahydro-1,3,5-trinitro-1,3,5-triazine) by the white rot fungus *Phanerochaete* chrysosporium BKM F-1767 was observed. At an initial concentration of 1.25 nmoles/10 ml cultures of P. chrysosporium, 50.8 ± 3.2% of the [14C] TNT was degraded to $^{14}CO_2$ in 30 days and only 2.8% of the initial TNT could be recovered. In RDX contaminated liquid cultures, (1.25 nmoles/10 ml cultures), 66.6 ± 4.1% of the [14C] RDX was converted to $^{14}CO_2$ over a 30 day period and only 4% of the initial RDX was recovered. In soil, slow but sustained degradation, 6.3 ± 0.6% of the [14C] TNT initially present, i.e., 57.9 nmoles/10 g soil, was mineralized over a 30 day incubation. In contrast, 76.0 ± 3.9% of the [14C] RDX added, (1.25 nmoles/10 g soil), was degraded to $^{14}CO_2$ in 30 days. Mass balance analyses of liquid and soil cultures contaminated with TNT revealed that polar metabolites were formed, and HPLC analysis indicated that not more than 5% of the radioactivity remained as undegraded TNT. However, the mass balance analyses of liquid (10 ml) and soil (10 g) cultures contaminated with 1.25 nmoles of [14C] RDX showed no metabolites. When the concentration of TNT in cultures (both liquid and soil) was adjusted to contamination levels in the environment, i.e., 10,000 mg/kg in soil and 100 mg/liter in water, 18.4 ± 2.9% and 19.6 ± 3.5% of the initial TNT was converted to $^{14}CO_2$ in 90 days in soil and liquid cultures, respectively. In both systems, about 85% of the initial TNT

0097–6156/91/0468–0214$06.00/0

disappeared over a 90 day incubation period.
These results suggest that P. chrysosporium may
be useful for the reclamation of sites
contaminated with TNT and RDX.

One of the major environmental problems associated with
U.S. military is the presence of soil, sediment and
water contaminated with toxic explosive residues at and
around many major military facilities. In these
facility sites the soil, sediment and the water were
contaminated over several years by improper disposal of
waste and waste water generated during the production
of munitions, burning or detonation of off-
specification material, and demilitarization of out-of-
date munitions. TNT (2,4,6-trinitrotoluene) and RDX
(hexahydro-1,3,5-trinitro-1,3,5-triazine) are the
predominant conventional explosives used by military
forces (10,19,26). The major waste water problems that
originate from TNT and RDX are associated with the
finishing processes, shell washout practices and
equipment and building washoff operations (13,16,22).
TNT and RDX are only slightly soluble in water and
therefore large volumes of water have been used to wash
off residues (26,32). For example, a single munition
manufacturing plant can generate and dispose as much as
500,000 gallons of waste water per day (22,31). As a
result waste water containing these toxic contaminants
leach through the soil contaminating the soil and
eventually the ground water (16,22).
    The toxic and mutagenic effects of these compounds
in different kinds of organisms including humans have
been well documented (4,11,12,33). Exposure to TNT is
known to cause pancytopenia, a disorder of the blood-
forming tissues characterized by a pronounced decrease
in the number of leukocytes, erythrocytes and
reticulocytes in humans and other mammals (9). Toxic
effects including liver damage and anemia have been
reported in workers engaged in large-scale
manufacturing and handling practices (6,8,25,29).
Also, TNT is toxic to fathead minnow (Pimephales
promelas) and blue gills (Lepomis macrochirus) at
concentrations of 2 to 3 μg/ml (20,27). It is also
toxic to certain green algae (Selenastrum
capricornutum, Microcystis aeruginosa, Chlamydomonas
reinhartii), tidepool copepods, (Tigriopus
californicus), and oyster (Crassostrea gigus) larvae
(27,33). Furthermore, TNT is a mutagen as assayed by
the Ames test (33). On the other hand, RDX has been
found to be a potent convulsant in the rat with an
intraperitoneal minimum lethal dose of 25 mg/kg (30).
It's acute toxicity has been investigated in mice,
cats, pigs and dogs (26,28,30). RDX related toxic
effects to humans have been reported from all over the

world. For example, occupationally related cases of
human RDX intoxication have been reported in many
countries including the United States (11). As in
animal studies, the main symptoms of RDX intoxication
in humans were related to the central nervous system
(11,1).

The white rot fungus, Phanerochaete chrysosporium
has shown to degrade a wide variety of environmentally
persistent organopollutants, including a number of
organohalides and a variety of toxic chemicals (2,5).
Also, this fungus is one of the relatively few
microorganisms known to be able to degrade lignin, a
naturally occurring and recalcitrant biopolymer, to
carbon dioxide (14). Recent studies in a number of
laboratories (2,5,14) have shown that the ability to
degrade such a diverse group of compounds is dependent
on the nonspecific and nonstereoselective lignin
degrading system which is expressed by the fungus under
nutrient (nitrogen, carbon, or sulfur) limiting
conditions (14,15). The lignin degrading system
consists, in part, of a family of lignin peroxidases
(commonly known as ligninases) which are able to
catalyze the initial oxidative depolymerization of
lignin (14). It has recently been shown that
ligninases are also able to catalyze the initial
oxidation of a number of environmentally persistent
toxic chemicals (24).

The purpose of this investigation was to determine
whether P. chrysosporium would grow in the presence of
TNT and RDX and degrade these compounds as it did other
xenobiotics. Because P. chrysosporium has the ability
to degrade a wide variety of environmentally persistent
organopollutants to carbon dioxide, we have suggested
that this organism may be useful in certain hazardous
waste treatment systems (2). In this study, the
ability of P. chrysosporium to degrade TNT contaminated
water and soil was investigated by adjusting the TNT
concentration equivalent to some of the highest
contamination levels found in the environment. The
results of this study indicate that biological
treatment systems that utilize P. chrysosporium might
be used as effective, and economical methods for
remediation of TNT and RDX contaminated water and soil.

Materials and Methods

Organism

P. chrysosporium BKM-F-1767 was obtained from the
Forest Products Laboratory, U.S. Department of
Agriculture, Madison, WI. The organism was maintained
at room temperature on 2% (w/v) malt agar slants.
Subcultures were routinely made every 30-60 days.

Chemicals

Radiolabeled [^{14}C]TNT (ring labeled, specific activity 21.58 mCi/mmol) and [^{14}C]RDX (2,4,6-^{14}C, specific activity 7.75 mCi/mmol) were purchased from Chemsyn Science Laboratories, Lenexa, KS and the non-radioactive TNT was obtained from the U.S. Army Toxic and Hazardous Materials Agency, Aberdeen, MO and from Chem Service, Inc. (West Chester, PA). The purity of the chemicals was determined by thin layer chromatography (TLC) and high performance liquid chromatography (HPLC) and was found to be greater than 98% in all cases.

Soil

The soil used in the culture experiments was classified as a silt loam consisting of 19% sand, 54% silt and 27% clay. The organic matter content in the soil was 3.62%, the organic carbon content was 2.10% and total nitrogen was 0.19%. The soil pH was 6.4 and the cation exchange capacity was 23.6 meq/100 g. The concentrations of selected extractable cations were as follows: Na, 0.1; K, 0.99; Ca, 16.85; and Mg, 3.59 meq/100 gm. The concentrations of selected divalent cations extractable with diethylenetriamine-pentaacetic acid were as follows: Zn, 2.3; Fe, 35.1; Cu, 1.2; Mn, 29.4; Cd, 1.4; Pb, 0.86; Ni, 0.47; and Cr, 0.16 mg/kg.

Liquid Culture Conditions

The liquid culture medium consisted of 56 mM glucose (w/v), 1.2 mM ammonium tartrate (nitrogen-limited), trace metals (14), and thiamine (1 mg/ml) in 20 mM 2,2'-dimethyl succinate (sodium) buffer, pH 4.2 (15). The culture medium was sterilized by filtration through a cellulose acetate membrane filter (0.22 μm). Culture bottles were sterilized by autoclaving at 121°C and 15 psi for 20 min. Aliquots of the culture medium (9 ml) were dispensed into each of several 250 ml Wheaton bottles equipped with a gas exchange manifold with a Teflon seal. Cultures were inoculated with a spore suspension of P. chrysosporium BKM F-1767 (1 ml; optical density = 0.5 absorbance units at 650 nm; ~ 2.5 x 10^6 spores) and grown at 39°C. Control incubations contained everything except 1 ml of P. chrysosporium spore suspension. The cultures were grown under ambient atmosphere for 6 days. On day 6, 1.25 nmoles of [^{14}C] TNT or [^{14}C] RDX, were added to separate cultures and at 3 day intervals thereafter the headspaces were flushed with oxygen (99.9%) and the liberated CO_2 was passed through a volatile organic trap consisting of 10 ml scintillation cocktail (Safety Solve, Research Products International Corp., Mt.

Prospect, IL) prior to a vial containing 10 ml of CO_2 trap. The CO_2 trap was a mixture of ethanolamine in methanol and scintillation cocktail (1:4:5, v/v/v). The volatile organic trap was used to ensure that the radiolabeled material trapped in the CO_2 trap was not contaminated with volatile organics as a result of air stripping during flushing. The amount of radioactivity in each trap was determined by liquid scintillation spectrometry.

Mass Balance Analyses

At the end of the incubation period, mass balance analyses were performed on liquid cultures of P. chrysosporium that contained [^{14}C] TNT or [^{14}C] RDX. The contents of each 250 ml Wheaton bottle were extracted three times with 30 ml of methylene chloride and water (1:1, v/v). The methylene chloride extracts were then pooled and concentrated by evaporation under a gentle stream of nitrogen. Following these extractions, particulate material, i.e., fungal mat, was separated from the aqueous fraction by filtration through glass wool. The amount of [^{14}C] which was bound to the fungal mat and not extractable by organic solvent was determined by combustion in a Harvey biological oxidizer (R.J. Harvey Instrument Corp., Hillsdale, NJ). Measurement of radioactivity in the liberated CO_2 was determined by trapping CO_2 as described above. The amount of $^{14}CO_2$ in the trap was determined by liquid scintillation spectrometry. The aqueous fraction was also assayed for radioactivity by liquid scintillation spectrometry (Beckman LS 5801 liquid scintillation counter, Beckman Instruments, Irvine, CA).

HPLC Analyses

Analyses for [^{14}C] TNT and [^{14}C] RDX metabolites were performed using an HPLC system equipped with a Spectra-Physics model SP 8810 pump (Spectra Physics, San Jose, CA), a Rheodyne injector (Rheodyne Inc., Cotati, CA), a reverse phase column (5 μ, 4.6 by 250 mm; Rsil C-18; Beckman Instrument Inc., San Ramon, CA), and a Spectra-Physics model SP 8450 (Spectra Physics, San Jose, CA) variable wavelength absorbance detector. Isocratic elution was performed in a solvent system composed of methanol:water (50:50, v/v). The flow rate was 1.0 ml/min. In a typical injection, the methylene chloride extract (obtained during mass balance analysis) was evaporated under gentle stream of nitrogen and the residue was redissolved in 0.5 - 2.0 ml of methylene chloride. Twenty μl aliquots of this material were used for injection into the HPLC. The retention time for TNT and RDX was determined by monitoring elution at

254 nm . Fractions (1 ml) were collected in
scintillation vials. Safety Solve (9 ml) was then
added to each fraction and radioactivity was determined
by liquid scintillation spectrometry as described
previously.

Soil Cultures

The ability of P. chrysosporium to degrade [^{14}C] TNT
and [^{14}C] RDX in soil was also determined. An
agricultural silt loam soil was used in this study.
Soil samples (10 gm) were placed in 250 ml Wheaton
bottles and then contaminated separately, with 57.9
nmoles and 1.25 nmoles of [^{14}C] TNT and [^{14}C] RDX,
respectively, dissolved in 100-160 μl of acetone. The
acetone was allowed to evaporate and the soil was then
mixed with 6.7 gm of ground corn cobs (Mt. Polaski,
Inc., Mt. Polaski, IL) that had been previously (10
days) inoculated with P. chrysosporium. Preinoculated
corn cobs were made by autoclaving the ground corn cobs
(121°C and 15 psi) for 30 min and aseptically
inoculating with a spore suspension of P. chrysosporium
as described above. The moisture content of the soil
cultures was brought up to the optimal moisture content
required for the organism, 40% w/w (7), by adding
unsterilized water. Unlike the liquid cultures, the
soil was not buffered or supplemented with trace metals
or other nutrients or glucose as a carbon source. Most
importantly, sterile conditions were not observed to
simulate environmental conditions. Cultures were
incubated at 39°C for 30 days or longer. Control
cultures were identical except they contained 6.7 gm of
uninoculated corn cobs. During the first 3 days of
incubation, cultures were grown under ambient air.
Every three days thereafter, head spaces of the culture
bottles were flushed with oxygen for 25 min, and
liberated CO_2 was trapped and the amount of
radioactivity in the CO_2 and in the volatile organic
trap was determined by liquid scintillation
spectrometry as described above.

Mass Balance Analyses

Mass balance analysis were performed on soil cultures
using extraction procedures previously described by
Jenkins and Walsh (10). At the end of the incubation
period, the contents of each 250 ml Wheaton bottle was
extracted three times with 30 ml acetonitrile. The
soil, corn cob and acetonitrile mixtures were dispersed
by using a vortex mixer for 10 min followed by
sonication in a Branson ultrasonic bath (Branson Inc.,
Shelton, CT) for 18 hr. The acetonitrile extracts were
pooled and concentrated by evaporation under a gentle
stream of nitrogen. The concentrated extract was then

centrifuged for 5 min at 1500 rpm and the clear
supernatant was removed using a volumetric pipet and
mixed with an equal volume of water in a glass
scintillation vial (10). The contents of the vial were
thoroughly mixed and allowed to stand for 15 min and
filtered through a Gelman Sciences 0.45 μm ARCO LS-25
disposable filter assembly. Twenty μl aliquots of this
material were used for HPLC analyses as described
above. Fractions (1 ml) were collected in
scintillation vials and the radioactivity was
determined as described previously. Radioactivity
which was retained in the soil/corn cob/fungal matrix
and not recovered by organic solvent extraction was
determined by combustion to CO_2 in a Harvey biological
oxidizer followed by measurement of $^{14}CO_2$ by liquid
scintillation spectrometry.

Degradation of TNT at Levels Found in the Environment

In a separate experiment, P. chrysosporium was tested
for its ability to mineralize the [^{14}C] TNT in both
liquid and soil cultures at levels encountered in the
TNT contaminated sites, i.e., 100 mg/l of water and
10,000 mg/kg in soil (personal communication, U.S.
Naval Civil Engineering Laboratory, Port Hueneme, CA).
Culture conditions were as described above except for
the concentration of TNT and the period of incubation.
Unless otherwise stated, $^{14}CO_2$ evolution studies were
performed in quadruplicate. Rates of mineralization
were obtained and mass balance analyses were performed
as described above after 30, 60, and 90 days for the
liquid and the soil cultures. One culture bottle was
extracted at the end of each incubation period except
for day 90 when two cultures were used.

Results

Phanerochaete chrysosporium mineralized 51% of the
[^{14}C] TNT over a period of 30 days incubation (Figure
1). A mass balance analysis of the cultures revealed
that 50.8 $\pm$ 3.2% of the total radioactivity was evolved
as $^{14}CO_2$, 21.9% was present as water soluble
metabolites, 10.7% was found in the methylene chloride
fraction, and 9.3% was associated with the mycelial
fraction. A total mass recovery of 92.7% was achieved.
HPLC analysis (Fig. 2) of the methylene chloride
extract revealed that only about 2.8% of the [^{14}C] TNT
initially present be identified as undegraded TNT. The
remaining 9% represented unidentified metabolites
formed during the 30 day incubation period. Almost all
of the unidentified metabolites formed in the
incubation period were more polar than TNT. In control
cultures incubated under the same culture conditions
but not inoculated with P. chrysosporium, 97-98% of the

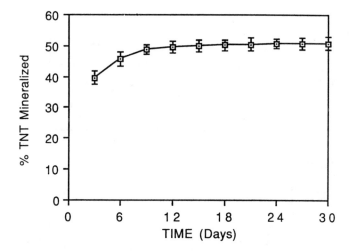

Figure 1.   Mineralization of [^{14}C] TNT in nutrient nitrogen limited cultures of P. chrysosporium. Each culture contained 1.25 nmoles of [^{14}C] TNT. Data points represent the mean ± standard deviation (N = 4).

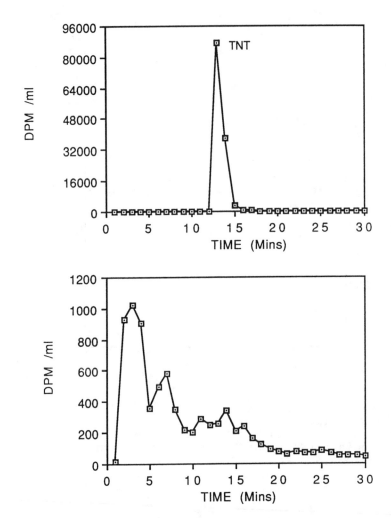

Figure 2. HPLC elution profile of a methylene chloride extract of a nutrient nitrogen limited culture of P. chrysosporium that was incubated with [¹⁴C] TNT (1.25 nmoles) for 30 days. Authentic [¹⁴C] TNT (top) eluted from the column about 12—15 minutes. DPM = disintegrations per minute. (Reprinted with permission from ref. 34. Copyright 1990 American Society for Microbiology.)

radioactivity added was found in the methylene chloride fraction and was unmetabolized [^{14}C] TNT. In [^{14}C] RDX contaminated liquid cultures, P. chrysosporium mineralized 67% of the initial [^{14}C] RDX over a period of 30 days incubation (Fig. 3). Mass balance was done at the end of the 30 day incubation period and a total of 66.6 ± 4.1% of the initial radioactivity added to the cultures was liberated as $^{14}CO_2$, 20.2% was recovered as water soluble metabolites, 4.8% was present in the methylene chloride fraction and 2.1% was associated with the mycelial fragments. A total of 93.7% radioactivity added to the cultures were recovered. HPLC analysis of the methylene chloride fraction demonstrated that only about 4% of the [^{14}C] RDX added to the cultures was undegraded. No metabolites were found in the methylene chloride fraction. In control cultures, where the P. chrysosporium was not added but identical culture parameters were included, 97-98% of the radioactivity initially added was recovered in the methylene chloride fraction and was identified as undegraded [^{14}C] RDX.

Biodegradation was also examined in a system in which [^{14}C] TNT and [^{14}C] RDX was separately adsorbed onto soil that was amended with corn cobs previously inoculated with P. chrysosporium (10 days earlier). In this soil-corn cob mixture, 6.3 ± 0.6% of the [^{14}C] TNT initially present was degraded to $^{14}CO_2$ over a 30 day incubation period (Fig. 4). Mass balance analysis of cultures of P. chrysosporium incubated with [^{14}C] TNT in a soil-corn cob matrix for 30 days revealed that 6.3 ± 0.6% of the recovered radioactivity was liberated as $^{14}CO_2$, 63.6% was found in the acetonitrile fraction, and 25.2% was not extractable by the organic solvents and was found in the soil-corn cob mixture. The material could not be identified as it could not be extracted from the matrix. A total mass recovery of 95.1% was observed. HPLC analysis of the radiolabeled material in the acetonitrile extract revealed that only about 2.2% of the [^{14}C] TNT initially added to the cultures was identified as undegraded TNT.

Mass balance analysis of cultures of P. chrysosporium incubated with [^{14}C] RDX in a soil/corn cob matrix for 30 days revealed that 76.0 ± 3.9% of the recovered radioactivity was evolved as $^{14}CO_2$, 4.5% was present in the acetonitrile extract and 9.7% was unextractable and present in the soil/corn cob matrix (Fig. 3). A total mass recovery of 90.2% was achieved. Analysis of the radiolabeled material in the acetonitrile extract revealed that only 4% of the [^{14}C] RDX initially present was identified as undegraded RDX.

In control uninoculated cultures, incubated under the same non-sterile conditions, greater than 99% of the radioactivity added was found in the acetonitrile fraction and was unmetabolized [^{14}C] RDX. The assay

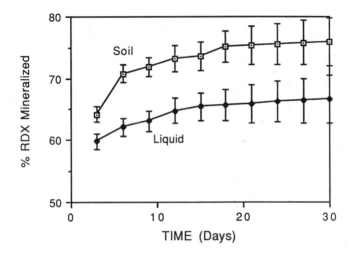

Figure 3. Mineralization of [^{14}C] RDX in liquid and in soil/corn cob cultures of <u>P. chrysosporium</u>. Each culture contained 1.25 nmoles of [^{14}C] RDX. Data points represent the mean ± standard deviation (N = 4).

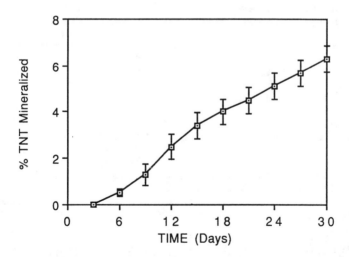

Figure 4. Mineralization of [^{14}C] TNT in soil/corn cob cultures of P. chrysosporium. Each culture contained 57.9 nmoles of [^{14}C] TNT in 10 g of soil and 6.7 g corn cobs. Data points represent the mean ± standard deviation (N=4). (Reprinted with permission from ref. 34. Copyright 1990 American Society for Microbiology.)

for radioactivity in the volatile organic trap revealed that less than 0.5% of the [^{14}C] RDX was volatilized or air stripped during the flushing of cultures with oxygen.

At the end of 30 and 60, or 90 days, liquid and soil cultures contaminated with 100 mg/l and 10,000 mg/kg TNT, respectively, were extracted and mass balance analyses performed as described above. The results of mass balance analysis of 100 mg/l TNT contaminated liquid cultures showed that 19.0 $\pm$ 3.0% of the recovered radioactivity was evolved as $^{14}CO_2$, 19.5% was found in the methylene chloride extract, 51.6% was present as water soluble compounds, and 5.1% was bound to the fungal mat after a period of 60 days incubation. A total mass recovery of 95.2% was achieved. When the methylene chloride fraction was analyzed by HPLC (Fig. 5), the amount of unmetabolized [^{14}C] TNT remaining in liquid cultures was 22.1% and 14.9%, respectively, over a period of 30 and 60 days of incubation. In control cultures, which were incubated under the same conditions but not inoculated with P. chrysosporium, greater than 99% of the radioactivity was found in the methylene chloride extract and identified as TNT by HPLC.

The results of mass balance analysis of soil cultures contaminated with 10,000 mg/kg TNT showed that 18.4 $\pm$ 2.9% had been evolved as $^{14}CO_2$, 62.6% was found in the acetonitrile extract, and 11.5% was bound to soil/corn cob/fungal matrix after 90 days. The total mass recovery was 92.5% after a period of 90 days incubation. When the acetonitrile extracts of day 30, 60, and 90 cultures were analyzed by HPLC, they showed that the amount of residual ^{14}C-TNT that was not degraded to $^{14}CO_2$ or intermediates was approximately 50, 30, and 15%, respectively (Fig. 6). A major metabolite eluted from the HPLC column at about 3 minutes in all cases. The amount of this metabolite increased and then decreased with time (Fig. 6). Also, the HPLC elution profile of both liquid and soil cultures revealed that most of the metabolites formed during the incubation periods were more polar than TNT itself. In control cultures, incubated under the same non-sterile conditions but not inoculated with P. chrysosporium, greater than 99% of the radioactivity added was found in the acetonitrile fraction and was unmetabolized [^{14}C] TNT. The assay for radioactivity in the volatile organic trap revealed that less than 0.5% of the ^{14}C-TNT was volatilized or air stripped during the flushing of cultures with oxygen. In some control experiments in which CO_2 was trapped in Ba(OH)$_2$ rather than the ethanolamine based scintillation cocktail, it was shown that the $^{14}CO_2$ was quantitatively precipitated as BaCO$_3$ (data not shown).

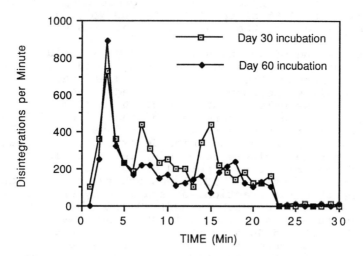

Figure 5. HPLC elution profiles of methylene chloride extracts of nutrient nitrogen limited liquid cultures of P. chrysosporium that were incubated with [^{14}C] TNT (100 mg/l) for 30 and 60 days. TNT eluted at approximately 15 minutes.

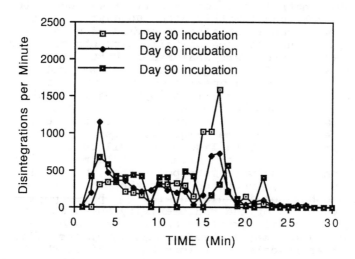

Figure 6. HPLC elution profiles of acetonitrile extracts of soil-corn cob cultures incubated with P. chrysosporium and 10,000 mg/kg TNT for 30, 60 and 90 days. TNT eluted at approximately 15 minutes.

Discussion

Previous studies of TNT biodegradation showed that this compound is relatively resistant to biodegradation by microorganisms in the environment (3,13,17). However, some microorganisms, such as bacteria, biotransform TNT into hydroxyl amino compounds by stepwise reduction of nitrogroups and azoxy compounds by conjugation of amino compounds, but further biodegradation of these biotransformed products has not been observed (13,17,33). According to Carpenter et al. (1978), solubility properties and infrared spectra confirm that these TNT biotransformation products react with microbial lipids, fatty acids and protein constituents to form polyamide type macromolecular structures. Also, Carpenter et al. (1978), further confirmed this hypothesis by correlating the infrared spectrum of the lipid component with that of a model compound synthesized from TNT transformation products and lipid precursors. The recalcitrance nature of these polymers to further microbial degradation is paralleled with the documented resistance of some alkyl and aryl polyamides containing similar linkages (23). Also, most of the previous investigations regarding the biodegradation of TNT did not provide any evidence to indicate that the TNT aromatic nucleus was cleaved and ultimately degraded (3,13,17,18,21,32,33). Instead in most studies it has been shown that the TNT was degraded by a reduction mechanism (3,13,22,27). The reduction mechanism appeared to involve a stepwise reduction of the nitro groups, through the nitroso and hydroxy amino to the amino group (3,13,22,27,33). However, in these reductive degradations, NO_2 groups that are reduced to hydroxyl amine form dimers such as 2,2',6,6'-tetranitro-4,4'-azoxytoluene, 4,4',6,6'-tetranitro-2,2'-azoxytoluene, and 2',4,6,6,-tetranitro-2,4'-azoxytoluene (13,17,33). Thus, the reduction mechanism represent only a modification of the molecule and not decomposition of the aromatic nucleus (3,13,17,33). Furthermore, the TNT reduction products may also pose environmental hazards since they have been reported to be toxic and mutagenic (4,12).

However, in contrast to TNT biodegradation, McCormick et al. (19) observed the biodegradation of RDX under anaerobic conditions yielding a number of metabolites, e.g., hexahydro-1-nitroso-3,5-dinitro-1,3,5-triazine, 1-hydroxylamino-3,5-dionitroso-1,3,5-triazine, N-hydroxymethylmethyl-enedianitramine, formaldehyde, methanol, etc. The biodegradation of RDX occurred via successive reduction of the NO_2 groups to a point where destabilization and fragmentation of the ring occurred. According to McCormick et al., the non cyclic degradation products arise via subsequent reduction and rearrangement of the fragments. Although the ring was cleaved and non-cyclic degradation products were formed, according to McCormick et al., several of the products were mutagenic or carcinogenic or both. However, this study shows that the wood rotting (white rot) fungus P. chrysosporium BKM F-1767 is able to degrade [^{14}C] TNT and [^{14}C] RDX in a reasonably short period of time. Degradation was demonstrated by mineralization of [^{14}C] TNT and [^{14}C] RDX, metabolite formation in the case of [^{14}C] TNT, and mass balance analyses. Biodegradation of [^{14}C] TNT and [^{14}C] RDX was also shown to occur in a soil and corn cob matrix inoculated with P. chrysosporium. Compared with liquid cultures substantially less [^{14}C] TNT was converted to $^{14}CO_2$ in soil cultures. However, it is worth noting that in liquid cultures, mineralization of [^{14}C] TNT virtually ceased after 15 days whereas in soil cultures, it continued throughout the 30 day incubation period at a nearly continuous albeit relatively slow rate. Most importantly, even in soil-corn cob cultures, although the mineralization was slow, the amount of TNT degraded or converted to other metabolites was very high. Also, this experiment suggests that the extent of mineralization in a soil-corn cob matrix could be extended by increasing the incubation period.

In contrast to other microbial degradation systems, in which degradation occurs by a stepwise reduction of NO_2 groups, P. chrysosporium appears to degrade TNT by an oxidative mechanism. The family of enzymes involved in these oxidations could be lignin peroxidases, which are produced under nutrient limiting (carbon, nitrogen and sulfur) conditions. Contrary to other reports (17,27,33) P. chrysosporium was able to cleave the ring as evidenced by the liberation of $^{14}CO_2$ from the ^{14}C-ring labeled TNT and RDX. The oxidation of the ring was extensive in that 51% and 67% of [^{14}C] TNT and [^{14}C] RDX, respectively, was mineralized in liquid cultures within 30 days of incubation. Mineralization of [^{14}C] RDX in soil was higher than [^{14}C] TNT. Furthermore, as a result of the degradation of TNT and RDX by an oxidative mechanism, the possibility of conjugation of metabolites for the formation of azoxycompounds, i.e, dimers as observed in bacterial degradation, should be reconsidered in P. chrysosporium mediated system. The possibility of formation of toxic and mutagenic metabolites by synthesis of reduced nitro compounds and conjugated azoxy compounds during the degradation of TNT by P. chrysosporium may need to be reevaluated.

A major objective of our research is the development of bioremediation systems using P. chrysosporium to treat water, soils, sediments and other materials that are contaminated with toxic and

persistent organopollutants. In this study we have
shown that P. chrysosporium is able to degrade [^{14}C]
TNT or [^{14}C] RDX extensively. A particular interest
was the fact that substantial amounts of the chemicals
were converted to $^{14}CO_2$ by this fungus. The
concentration of [^{14}C] TNT used in the liquid culture
experiments were of the same magnitude as is found in
waters contaminated by TNT in the environment (20,22).
The concentration of TNT in effluents from
manufacturing processes are, on average, about 20 mg/L
(20) and the concentration of TNT in contaminated soil
may be as high as 10,000 mg/kg. High concentrations of
TNT were not lethal to the fungus in these experiments
and considerable quantities of TNT were degraded. We
have suggested that P. chrysosporium may be useful in
the biodegradation of hazardous wastes in waste
treatment systems (2). When compared to costly and
tedious physical decontamination processes, this study
suggests that a versatile organism such as P.
chrysosporium may provide a more economical, biological
treatment system that could be applied to in situ
decontamination processes when the conditions are
suitable for the growth of the fungus.

Acknowledgments

This research was supported by the U.S. Naval Civil
Engineering Laboratory, Port Hueneme, CA and by grant
No. ES04922 from the National Institute of
Environmental Health Sciences. We thank Shelly
Rasmussen for technical assistance and Terri Maughan
for her expert secretarial assistance during the
preparation of this manuscript.

Literature Cited

1. Barsotti, M.; Crotti, G.; "Epileptic Attacks as
    Manifestations of Industrial Intoxication caused
    by Trimethylene Trinitroamine (T_4)"; Med. Lavoro.
    1949, 40, 107-112.
2. Bumpus, J.A.; Tien, M.; Wright, D.; Aust, S.D.;
    "Oxidation of Persistent Environmental Pollutants
    by a White Rot Fungus"; Science 1985, 228, 1434-
    1436.
3. Carpenter, D.F.; McCormick, N.G.; Cornell, J.H.;
    Kaplan, A.M.; "Microbial Transformation of ^{14}C-
    labeled 2,4,6-Trinitrotoluene in an ACtivated-
    sludge System"; Appl. Environ. Microbiol. 1978,
    35, 949-954.
4. Dilley, J.V.; Tyson, C.A.; Newel, G.W. Mammalian
    toxicological evaluation of TNT waste waters.
    VIII.   Acute and subacute mammalian toxicity of
    condensate water. SRI International: Menlo Park,
    CA, 1979; Vol. 111, pp 1-115.

5. Eaton, D.C.; "Mineralization of Polychlorinated
    Biphenyls by Phanerochaete chrysosporium: A
    Ligninolytic Fungus"; Enzyme Microb. Technol.
    1985, 7, 194-196.
6. Eddy, J.H. Jr.; "Methionine in the Treatment of
    Toxic Hepatitis"; Am. J. Med. Sci. 1945, 210, 374-
    380.
7. Fernando, T.; Aust, S.D.; Bumpus, J.A.; "Effects
    of Culture Parameters on DDT [1,1,1-trichloro 2,2-
    bis(4-chlorophenyl)ethane Biodegradation by
    Phanerochaete chrysosporium"; Chemosphere 1989,
    19, 1387-1398.
8. Hamilton, A.; "Trinitrotoluene as an Industrial
    Poison"; J. Ind. Hyg. 1921, 3, 102-116.
9. Harris, J.W.; Killermeyer, R.W. The red cell:
    production, metabolism, destruction, Normal and
    abnormal. Rev. ed. Harvard University Press:
    Cambridge, MA, 1970; pp 1-71.
10. Jenkins, T.F.; Walsh, M.E. Development of an
    analytical method for explosive residues in soil.
    Cold Regions Research and Engineering Laboratory:
    Springfield, VA, 1987; Report 87-7, pp 1-51.
11. Kaplan, A.S.; Berghout, C.F.; Peczenik, A.; "Human
    Intoxication from RDX"; Arch. Environ. Health
    1965, 10, 877-883.
12. Kaplan, D.L.; Kaplan, A.M.; "Mutagenicity of TNT
    surfactant Complexes"; Bull. Environ. Contam.
    Toxicol. 1982, 28, 33-38.
13. Kaplan, D.L.; Kaplan, A.M.; "Thermophilic
    Biotransformations of 2,4,6-trinitrotoluene Under
    Simulated Composting Conditions"; Appl. Environ.
    Microbiol. 1982, 44, 757-760.
14. Kirk, T.K.; Chang, H.m.; "Decomposition of Lignin
    by White Rot Fungi. II. Characterization of
    Heavily Degraded Lignins from Decayed Spruce";
    Holzforschung 1975, 29, 56-64.
15. Kirk, T.K.; Schultz, E.; Connors, W.J.; Lorenz,
    L.F.; Zeikus, J.G.; "Influence of Culture
    Parameters on Lignin Metabolism by Phanerochaete
    chrysosporium"; Arch. Microbiol. 1978, 117, 277-
    285.
16. Klausmeier, R.E.; Osmon, J.L.; Walls, D.R.; "The
    Effect of Trinitrotoluene on Microorganisms"; Dev.
    Ind. Microbiol. 1973, 15, 309-317.
17. McCormick, N.G.; Feeherry, F.E.; Levinson, H.S.;
    "Microbial Transformation of 2,4,6-trinitrotoluene
    and Other Nitroaromatic Compounds"; Appl. Environ.
    Microbiol. 1976, 31, 949-958.
18. McCormick, N.G.; Cornell, J.H.; Kaplan, A.M.;
    "Identification of Biotransformation Products from
    2,4,-dinitrotoluene"; Appl. Environ. Microbiol.
    1978, 35, 945-948.
19. McCormick, N.G.; Cornell, J.H.; Kaplan, A.M.;
    "Biodegradation of Hexahydro-1,3,5-trinitro-1,3,5-

triazine"; Appl. Environ. Microbiol. 1981, 42,
817-823.
20. Nay, Jr., M.W.; Randall, C.W.; King, P.H.;
"Biological Treatability of Trinitrotoluene
Manufacturing Waste Water"; J. Wat. Pollut.
Control Fed. 1974, 46, 485-497.
21. Osmon, J.L.; Klausmeier, R.E.; "The Microbial
Degradation of Explosives"; Dev. Ind. Microbiol.
1972, 14, 247-252.
22. Pereira, W.E.; Short, D.L.; Manigold, D.B.; Ross,
P.K.; "Isolation and Characterization of TNT and
its Metabolites in Groundwater by Gas
Chromatograph-Mass Spectrometer-Computer
Techniques"; Bull. Environ. Contam. Toxicol. 1979,
21, 554-562.
23. Rogers, M.R.; Kaplan, A.M.; "Effects of
Penicillium janthinellum on Parachute Nylon - Is
There Microbial Deterioration"; Int. Biodeterior.
Bull. 1971, 7, 15-24.
24. Sanglard, D.; Leisola, M.S.A.; Fiechter, A.; "Role
of Extracellular Ligninases in Biodegradation of
Benzo(a)pyrene by Phanerochaete chrysosporium";
Enzyme. Microb. Technol. 1986, 8, 209-212.
25. Sax, N.I. Dangerous properties of industrial
materials. 2nd ed. Reinhold Publishing Corp.:
New York, NY, 1963; pp 10-95.
26. Schneider, N.R.; Sharon, L.B.; Andersen, M.E.;
"Toxicology of Cyclotrimethylenetrinitramine:
Distribution and Metabolism in the Rat and the
Miniature Swine"; Toxicol. Appl. Pharmacol. 1977,
39, 531-541.
27. Smock, L.A.; Stoneburner, D.L.; Clark, J.R.; "The
Toxic Effects of Trinitrotoluene (TNT) and its
Primary Degradation Products on Two Species of
Algae and the Fathead Minnow"; Water Res. 1976,
10, 537-543.
28. Sklyanskaya, R.M.; Poznariskii, F.I.; "Toxicity of
Hexogen"; Farmakol. i. Toksikol. 1945, 7, 43-47.
29. Voegtlin, C.; Hooper, C.W.; Johnson, J.M.;
"Trinitrotoluene Poisoning"; U.S. Public Health
Res. 1919, 34, 1307-1313.
30. Von Oettingen, W.F.; Donahue, D.D; Yagoda, H.;
Monaco, A.R.; Harris, M.R.; "Toxicity and
Potential Dangers of Cyclotrimethylene
Trinitramine (RDX)"; J. Ind. Hyg. Toxicol. 1949.
31, 21-31.
31. Walsh, J.T.; Chalk, R.C.; Merritt, C. Jr.;
"Application of Liquid Chromatography to Pollution
Abatement Studies of Munition Wastes"; Anal. Chem.
1973, 45, 1215-1220.
32. Won, W.D.; Heckley, R.J.; Glover, D.J.;
Hoffsommer,J.C.; "Metabolic Disposition of 2,4,6-
trinitrotoluene"; Appl. Microbiol. 1974, 27, 513-
516.

33. Won, W.D.; DiSalo, L.H.; Ng, J.; "Toxicity and Mutagenicity of 2,4,6-trinitrotoluene and its Microbial Metabolites"; Appl. Environ. Microbiol. **1976**, <u>31</u>, 576-580.

34. Fernando et al.; Appl. Environ. Microbiol. **1990**, <u>56</u>, 1666-1671.

RECEIVED April 5, 1991

Chapter 12

Biodegradation of Volatile Organic Chemicals in a Biofilter

V. Utgikar[1], R. Govind[1], Y. Shan[1], S. Safferman[2], and R. C. Brenner[2]

[1]Department of Chemical Engineering, University of Cincinnati,
Cincinnati, OH 45221
[2]Risk Reduction Engineering Laboratory, U.S. Environmental Protection
Agency, Cincinnati, OH 45268

Emission of the volatile organic compounds (VOCs) has recently received increased attention due to its environmental impact. It is possible to treat these compounds by biodegradation. A mathematical model has been developed in this paper to describe the biodegradation of the VOCs in a biofilter. Numerical solutions of a mathematical model describing the steady state biodegradation of VOCs in the biofilter have been presented in the paper. The use of the model in design of the biofilter is demonstrated for a given load of leachates. Preliminary experimental data on the removal of toluene and methylene chloride have been presented. Calculations have been made for removal of the most common constituents of the leachate. It was found that for an inlet gas flow rate of 0.175 m^3/s(370 ft^3/min), a biofilter 3 m in diameter and 5.3 m in height is required for 90% removal of the contaminants.

Landfill is a traditional method for the disposal of solid wastes. Nearly 200 million tons of solid waste are disposed off in this manner in more than 9000 landfill sites in the U.S.A. [Varian Associates (*1*)]. This waste consists mostly of the common discards such as household refuse, plastic containers, aerosol cans, paints etc. This method of solid waste disposal is now known to contribute to a dangerous and difficult problem of landfill leachate pollution.

The various harmful chemicals present in the solid waste are leached into runoff water, thereby creating water pollution which also poses potential health hazard to the people. Table 1 shows the volatile organic constituents of a typical leachate from a landfill site [Bramlett et al (*2*)]. It can be seen that the leachate contains a number of volatile organic chemicals(VOCs) designated as primary pollutants by the EPA.

It has been shown that VOCs in landfill leachates discharged to conventional activated sludge systems will volatilize in addition to being biodegraded. For some VOCs with high Henry's law constant, significant removal due to stripping and volatilization can occur [Lai and Govind (*3*)].

0097–6156/91/0468–0233$08.00/0
© 1991 American Chemical Society

Table 1 : Composition of Landfill Leachates- Volatile Organics

No.	Compounds	Concentration (µg/L)	
		Range	Mean
1	Benzene and alkylated Benzenes	1100-2600	1850
2	Toluene	4400-12300	8350
3	Acetone	14000-32000	23000
4	Higher Ketones (Methylethyl,Methyliso- butyl and Methylbutyl)	11000-27000	19000
5	Chloroethylenes(di-,tri- and tetra-)	1300-4300	2800
6	Chloroform	1000-3100	2050
7	Methylene Chloride	8000-22000	15000
8	Chloroalkanes (mostly trichloroethane)	10-3850	1930
9	Chlorobenzene	190-770	480

As seen from the Table 1, the VOCs are present at as high a concentration level as 23000 µg/L. The atmospheric pollution from the emission of the VOCs from the leachate can lead to severe environmental and safety problems. Hence, it is necessary to develop a treatment scheme for controlling the VOC pollution.

Biofilter Treatment Scheme

The common treatment schemes for removal of gaseous VOCs pollutants are :
  1. Physical : Adsorption on a solid and absorption in a liquid
  2. Chemical : Incineration, catalytic conversion, etc.
Compared to these methods, biodegradation offers the following advantages:
  1. It is cost effective
  2. It requires less energy
  3. It has minimum adverse impact on the environment
  Biodegradation can be carried out in a biofilter. Biofilter is a packed column containing biologically active mass attached to a suitable support medium.Biofilters for treatment of air or gaseous streams have not been extensively investigated in literature. Pomeroy (4) and Carlson and Leiser (5) presented a similar approach for the removal of sewage related odors. However, the main mechanism for the removal of odor compounds was adsorption, and there was no clear evidence of biodegradation. Smith et al (6) determined that soil beds are effective in removal of sulfur containing gases and can serve as sinks for hydrocarbons. Hartenstein (7) presented a range of operating conditions for a biofilter. He found that an important operating parameter governing efficiency was the moisture content in the filter bed. Eitner (8) determined that the most active microbes in a biofilter are the heterotrophic and chemoorganotrophic groups. He also noted that Actinomycete spp. are able to exploit a wide variety of organics and are reported to kill pathogens under certain conditions.
  The distribution of microbes in the biofilter has been described by Ottengraf and van den Oever(9), Eitner (8) and Kampbell et al (10). The population density of

the microbes is highest at the gas entrance at the bottom of the biofilter and these microbes preferentially metabolize the more readily degradable influent compounds. The less degradable compounds are assimilated in the upper portion of the biofilter. Eitner (*8*) presented data which indicated that significant reduction of hydrocarbon concentration was achieved in approximately one week and maximum removal rates were attained within one month of operation. Ottengraf and van den Oever (*9*) investigated the survivability of the microbial flora in the biofilters which were not loaded and showed that biofilter operation can be suspended for 14 days with only minimal loss in microbial activity.

High removal efficiencies in biofilters (>98%) have been reported by Eitner and Gethke (*11*) for odors from sewage plants and by Prokop and Bohn (*12*) for odors from a rendering plant. Kampbell et al (*10*) have reported removal efficiencies between 95% and 99% for propane, isobutane, n-butane and trichloroethylene. Ottengraf and van den Oever (*9*) have also reported high removal rates for toluene, butanol,ethyl acetate and butyl acetate. Don and Feenstra (*13*) presented data comparing several alternating technologies for treatment of waste hydrocarbon gas streams and showed that biofilter is the most cost effective treatment method.

It is proposed to treat the VOCs by biodegradation using the biofilter treatment scheme shown in Figure 1. The landfill leachate is first fed to a stripper where the VOCs are stripped off from the leachate by air. The liquid effluent from the stripper proceeds to the wastewater treatment plant. The gases coming out of the stripper are fed to a biofilter where the VOCs are microbially degraded. The air exiting the biofilter is substantially free of the contaminants and is discharged to the atmosphere. Activated carbon will be used as support medium for the biomass in the treatment of the VOCs. The biofilter will be operated as a trickle bed with the supply of various nutrients necessary for biomass growth.

Model for Biodegradation on Activated Carbon

The biofilter forms a gas-liquid-solid system and the mathematical models of gas-liquid-solid system are presented in this section. The physical concept of the biofilm on the carbon support is shown in Figure 2. The model describing the steady state biodegradation in a biofilter have been presented earlier [Utgikar et al(*14*), Utgikar and Govind(*15*), Utgikar et al (*16*)]. In this section, a dynamic model has been presented and from which a steady state model is derived.

The basic assumptions involved in the formulation of the model are as follows :

1. Carbon support is homogeneous and has a flat geometry
2. No biodegradation takes place inside the carbon and all biodegradation takes place inside the biofilm.
3. The gas-liquid and biofilm-carbon interfaces are at equilibrium
4. The biofilm is homogeneous and the biofilm density is constant.
5. The biodegradation reaction is first order with respect to the substrate concentration. This assumption is justified at low substrate concentrations, i.e., $S \ll Ks$ (biodegradation half saturation constant), where Monod kinetics gets reduced to the first order kinetics.
6. Any growth in the biomass results in increase in biofilm thickness.
7. The adsorption of substrate on carbon follows the Freundlich Isotherm.

The assumption of homogeneity of biofilm implies that the substrate concentration does not vary laterally in the carbon film and the diffusion is unidirectional. The biofilm is dynamic in nature. The constant density assumption implies that growth

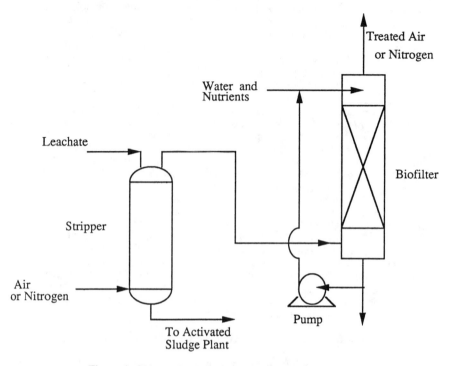

Figure 1: Schematic of the Biofilter Treatment Scheme

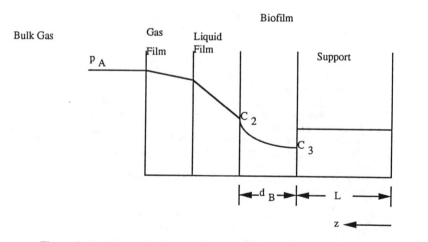

Figure 2: Biofilm on Activated Carbon: Substrate Concentration Profile

of biofilm takes up space given by the decay of part of biofilm in such a manner that the biofilm density remains constant throughout the biofilm.

The mass balance for the substrate over a differential element in the support material yields the following equation:

For $0 \le z \le L$;

$$\frac{\partial \overline{C}}{\partial t} = D_s \frac{\partial^2 \overline{C}}{\partial z^2} \qquad \text{.. (1)}$$

where,
D_s = substrate diffusivity in the support material,
$\overline{C}$ = substrate concentration in the support material,
z = dimension co-ordinate,
t = time, and
L = support dimension.

The initial and boundary conditions are given by equations 2, 3 and 4 respectively:

$$\overline{C} = 0, t \le 0, \text{ all } z \qquad \text{.. (2)}$$

$$\frac{\partial \overline{C}}{\partial z} = 0, \ z = 0, t \ge 0 \qquad \text{.. (3)}$$

$$D_s \frac{\partial \overline{C}}{\partial z} (\text{at } z = L) = \frac{\partial}{\partial t} \left[\int_0^L \overline{C} \, dz \right] \qquad \text{.. (4)}$$

The first boundary condition implies that there is no flux at $z = 0$, while the second boundary condition is the overall material balance for the substrate in the support material.

The material balance for the substrate over a differential element in the biofilm yields the following equation:

For $L \le z \le L + d_B$;

$$\frac{\partial C}{\partial t} = D_e \frac{\partial^2 C}{\partial z^2} - k_1 C \qquad \text{.. (5)}$$

where,
C = concentration of the substrate in the biofilm,
D_e = effective diffusivity of the substrate in the biofilm,
k_1 = first order biodegradation rate constant, and
d_B = biofilm thickness.

The initial and boundary conditions are given by equations 6, 7 and 8 respectively:

$$C = 0, t < 0, \text{ all } z \qquad \text{.. (6)}$$

$$D_e \frac{\partial C}{\partial z} = \rho_P D_s \frac{\partial \overline{C}}{\partial z}, \quad \text{at } z = L \qquad \text{.. (7)}$$

$$D_e \frac{\partial C}{\partial z} = R_A \text{ , at } z = L + d_B \qquad \qquad \text{.. (8)}$$

Equations 7 and 8 describe the continuity of the fluxes at the biofilm-support and biofilm-liquid film respectively. R_A, the effective substrate removal rate is also given by equation 9:

$$R_A = K_L (p_A/H - C_2) \qquad \qquad \text{.. (9)}$$

where,
K_L = overall mass transfer coefficient,
H = Henry's constant,
p_A = partial pressure of substrate A in the gas, and
C_2 = substrate concentration at the liquid-biofilm interface.

In addition to the above equations, one more equation is obtained for biofilm growth. From assumption 6, biofilm growth results in increase in biofilm thickness; which yields the following equation:

$$X \frac{d}{dt} (d_B) = \int_L^{L + d_B} (y\, k_1\, C - b\, X)\, dz \qquad \qquad \text{.. (10)}$$

where,
y = bacterial yield,
b = bacterial decay coefficient, and
X = biomass concentration.

The first term in the bracket under the integral sign gives the biomass yield due to the biodegradation of the substrate, and the second term describes the natural decay of the biomass. The difference between the two, integrated over the entire biofilm thickness gives the overall growth of biomass and hence the biofilm thickness.

This system of equations is solved to obtain the substrate concentration profile in the biofilm, which is given by equation 11.

$$C = \frac{C_3 \sinh [\, \phi\, (1+ d_B^* - z^*)] + C_2 \sinh [\, \phi\, (z^*-1)]}{\sinh (\phi\, d_B^*)}$$

$$+ \frac{2\pi D_e}{(d_B)^2} \sum_{n=1}^{\infty} \exp\{[-D_e\, (n\pi/d_B)^2 + k_1]t\} \frac{(C_2 \cos (n\pi) - C_3)\, n \sin (n\pi(z^*-1)/d_B^*)}{D_e(n\pi/d_B)^2 + k_1} \qquad \text{.. (11)}$$

where,
$z^* = z/L$,
$dB = d_B/L$, and
C_3 = substrate concentration at the biofilm-support interface.

The substrate concentrations C_2 and C_3, and biofilm thickness d_B are given by equations 12, 13 and 14:

$$C_2 = \frac{P_A}{H} - [\frac{D_e\phi}{L\,K_L} \frac{C_2 \cosh[\phi\,d_B^*] - C_3}{\sinh(\phi\,d_B^*)}$$

$$+ \frac{2\pi^2 D_e^3}{K_L\,(d_B)^3} \sum_{n=1}^{\infty} \exp\{[-D_e\,(n\pi/d_B)^2 + k_1]t\} \frac{(C_2 - (-1)^n C_3)\,n^2}{D_e(n\pi/d_B)^2 + k_1}] \qquad .. (12)$$

$$\frac{2\,m_1\,D_s\,\rho_P\,C_3^{1/m_2}}{L} \sum_{n=1}^{\infty} \exp\{-D_s\,[(2n-1)\frac{\pi}{2L}]^2 t\} = \frac{D_e\phi}{L} \frac{C_2 - \cosh[\phi\,d_B^*]\,C_3}{\sinh(\phi\,d_B^*)}$$

$$+ \frac{2\pi^2 D_e^2}{(d_B)^3} \sum_{n=1}^{\infty} \exp\{[-D_s\,(n\pi/d_B)^2 + k_1]t\} \frac{((-1)^n C_2 - C_3)\,n^2}{D_e(n\pi/d_B)^2 + k_1} \qquad .. (13)$$

$$X \frac{d}{dt}(d_B) = \frac{2\,y\,L\,k_1\,C_3}{\phi}[(\cosh(\phi\,d_B^*))^2 - 1]$$

$$+ \frac{4y\,k_1\,D_e}{d_B} \sum_{n=1,3,5,..}^{\infty} \exp\{[-D_e\,(n\pi/d_B)^2 + k_1]t\} \frac{C_2 + C_3}{D_e(n\pi/d_B)^2 + k_1} - b\,X\,d_B \qquad .. (14)$$

where, m_1 and m_2 are Freundlich adsorption isotherm parameters.

At steady state, the above equations can be reduced to three dimensionless equations - 15, 16 and 17. Steady state implies that the concentration profiles in the biofilm are independent of time. There is no net growth of biofilm and the thickness of the biofilm reaches a constant value. The biodegradation rate attains a final value that does not change with time. The rate of transport of the substrate through the biofilm equals the rate of biodegradation. This results in a constant, non-zero concentration of the substrate in the biofilm-support material.

$$d_B^* = \frac{r}{\phi} C_3^* ((\cosh(\phi d_B^*))^2 - 1) \qquad .. (15)$$

$$C_2^* = C_3^* \cosh(\phi d_B^*) \qquad .. (16)$$

$$C_3^* = 1 - \frac{\phi}{\phi_L} \frac{C_2^* \cosh(\phi d_B^*) - C_3^*}{\sinh(\phi d_B^*)} \qquad .. (17)$$

where,

$$r = \frac{y\,k_a\,p_A\,/\,H}{b\,X} \qquad .. (18)$$

$$\phi = L\sqrt{\frac{k_1}{D_e}} \qquad .. (19)$$

$$\phi_L = L\frac{K_L}{D_e} \qquad \text{.. (20)}$$

r is the biomass growth to decay ratio, ϕ is the Thiele type modulus for biodegradation reaction and ϕ_L is the Thiele type modulus for mass transfer. The dimensionless concentrations are obtained by the following equation:

$$C^* = C / (p_A/H) \qquad \text{.. (21)}$$

Also, the dimensionless reaction rate is defined by the following equation:

$$R_A^* = R_A / (K_L \ p_A /H) = 1 - C_2^* \qquad \text{.. (22)}$$

Solution of the Kinetic Model

The kinetic model for the biodegradation on the particle is solved as a function of the following parameters :
1. Biomass growth/decay ratio (r)
2. Thiele type modulus for reaction (ϕ)
3. Thiele type modulus for mass transfer (ϕ_L)

The range of the values for the parameters is as shown below :
1. Biomass growth/decay ratio : 0.01-10

2. ϕ :   0.5 - 62.5

3. ϕ_L :   250 - 1250

The ranges of values are chosen to account for the variations in the kinetic parameters and the physical properties of the system; e.g., the range of r covers the growth to decay ratio for cases from high growth and low decay to low growth and high decay. The variations in ϕ are essentially variations in k_1, since the diffusivities are of the same order of magnitude. The hundredfold variation in ϕ corresponds to ten-thousandfold variation in k_1 expected between easily degradable and difficult to degrade substrates.

The results are plotted in the form of dimensionless graphs of the concentrations at the liquid-biofilm and the biofilm-support interface, dimensionless thickness and the dimensionless reaction rate (defined as $R_A/(K_L$ $P_A/H)$ as a function of ϕ and these are shown in Figures 3a,3b,4,5,6a and 6b.

Discussion

The above results have been obtained for steady state conditions. Steady state implies that the concentration profiles in the biofilm are independent of time. There is no net growth of biofilm and the thickness of biofilm reaches a constant value. The biodegradation rate attains a final value that does not change with time. The rate of transport of the substrate equals the rate of biodegradation. This results in a constant non-zero concentration of substrate in the biofilm-support material.

The results indicate that:
1. Steady state does not exist at the lower values of biomass growth/decay ratio(below 0.1). The biofilm thickness depends upon two factors - biofilm

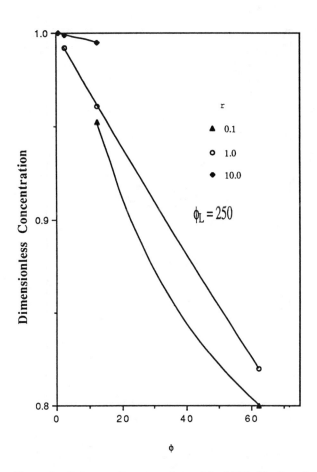

Figure 3a: Substrate Concentration at Liquid-Biofilm Interface

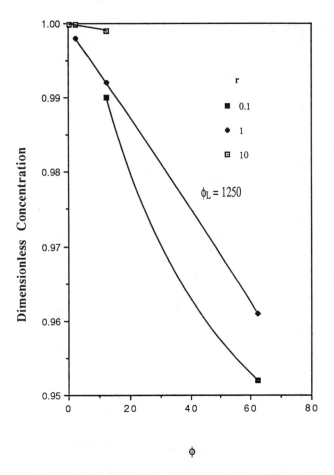

Figure 3b: Substrate Concentration at Liquid-Biofilm Interface

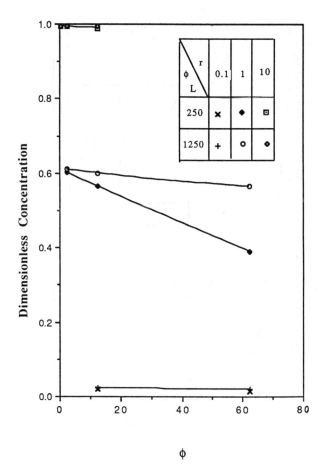

Figure 4: Substrate Concentration at Biofilm-Support Interface

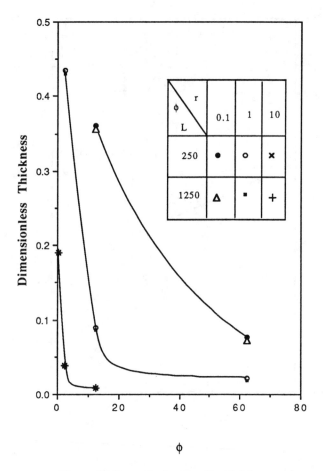

Figure 5: Dimensionless Biofilm Thickness

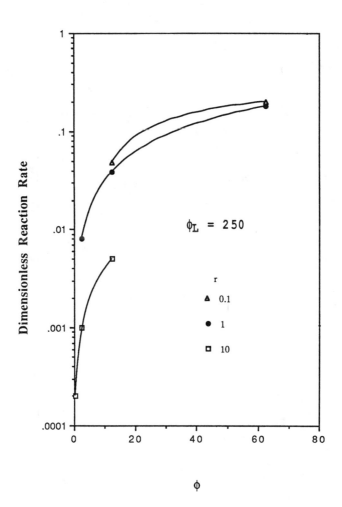

Figure 6a: Dimensionless Reaction Rate

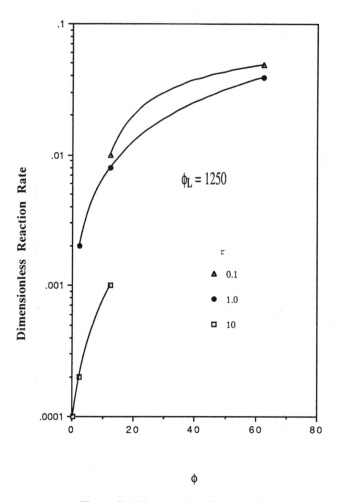

Figure 6b: Dimensionless Reaction Rate

growth due to substrate utilization and biofilm decay due to bacterial metabolic activity. These two factors are opposing in their effects on biofilm. Low values of this parameter imply low yield of biomass compared to biomass decay which may be due to either one or combination of the following factors - low substrate concentration, low rate constant(k_1), low bacterial yield(y) or high value of bacterial decay constant(b). Therefore, at low values of the parameter biomass growth is insignificant compared to the decay and the biofilm thickness decreases and tends to reach zero. This results in a trivial solution of zero biofilm thickness, absence of concentration profile and no biodegradation. In other words, the steady state is nonexistent.

2. C_2^*(fig.3), C_3^*(fig.4) and d_B^* (fig.5) decrease with increase in ϕ. This can be attributed to the increased reaction rate which causes faster depletion of the substrate. Lower substrate concentrations result in lower biofilm thickness.

3. The reaction rate increases with ϕ, as expected. However, it also shows an increase with decrease in ϕ_L. This behavior can be explained on the basis of increased diffusivity of the substrate through the biofilm.

4. The liquid film thickness affects the reaction rate through ϕ_L. A decrease in liquid film thickness results in increase in K_L and hence an increase in ϕ_L. As discussed above, this results in lower value of dimensionless reaction rate. This apparent discrepancy between high mass transfer coefficient and a low reaction rate can be explained in the following way: The reaction rate is made dimensionless by dividing the actual rate by $K_L H/p_A$. The dimensionless reaction rate is lower for higher K_L simply because the actual reaction rate is divided by a higher value. The actual reaction rate is higher for higher K_L values, i.e., for lower liquid film thickness. The quantitative analysis of liquid film thickness and partial wetting of support will be presented in a subsequent publication.

5. It can be seen from figures 6a and 6b that the reaction rate is lower for higher value of r. This apparent discrepancy is similar to one discussed above. For any particular substrate, r depends upon p_A/X. Lower value of r can be due to 1)higher X or 2) lower p_A. At same p_A value, a lower r means high X, i.e., high biomass concentration. This would result in higher rate as obtained. On the other hand, if lower value of r is due to lower p_A, then a high value of reaction rate is merely due to the division of actual reaction rate by a lower quantity (p_A). It can be seen that for order of magnitude increase in r, the decrease in dimensionless reaction rate is less than one order of magnitude. This means that the actual reaction rate has increased with r, as expected.

6. The equations developed have been solved for the case of biodegradation of a single substrate A. In a multicomponent mixture, the biodegradation rate of A will be affected by the presence of other components. The steady state behavior of the multicomponent system can be modelled along similar lines as single component systems. For a system with n components, there are 2n+1 equations - n for C_2^*, n for C_3^* and 1 for d_B^*. The equations for C_2^* and C_3^* are identical to equations 16 and 17 and the equation for d_B^* is as follows:

$$d_B^* = \frac{\displaystyle\sum_{i=1}^{n} b_i(r_i/\phi_i)C_{3,i}^*((\cosh(\phi_i d_B^*))^2-1)}{\displaystyle\sum_{i=1}^{n} b_i} \quad ..(23)$$

Biofilter Design

The kinetic model for the biodegradation on individual support particle can be utilized in sizing up of the biofilter. The mass balance on the differential element of the biofilter, as shown in the Figure 7, yields:

$$-G \, dy_A = R_A a \, A \, dZ \qquad \qquad .. (24)$$

Substituting for R_A from equation (2) and using $p_A = p_t \, y_A$

$$dZ = - \frac{H \, G \, dy_A}{K_L a \, A \, p_t \, (1-C_2^*) \, y_A} \qquad \qquad .. (25)$$

or,

$$Z = \frac{H \, G}{K_L a \, A \, p_t} \int_{y_{Aout}}^{y_{Ain}} \frac{dy_A}{(1-C_2^*) y_A} \qquad \qquad .. (26)$$

Experimental Verification

Experiments were carried out on a bench scale biofilter to verify the model predictions. Biomass from activated sludge plant was charged into a bioreactor and the suspension was aerated. The biomass was acclimated to three substrates - toluene, trichloroethylene and methylene chloride - in the bioreactor. The acclimation was carried out by dosing the bioreactor periodically with the three substrates along with the nutrients necessary of the growth of the biomass. The composition of nutrient solution is shown in Table 2. The acclimated biomass was transferred to the biofilter by recirculating the suspension through the biofilter. The biomass was effectively transferred to the carbon support. After the transfer, the recirculation loop was cut off and the biofilter was operated independently. The biofilter was operated as a trickle bed column with the substrate containing air and the nutrient solution constituting the gas and liquid phases respectively. Figure 8 shows the schematic of the experimental apparatus. The operating parameters and system properties are listed below:

support dimension	: 3 mm
column diameter	: 25 mm
packed height	: 600 mm
bed voidage	: 0.4
gas flow rate	: 160 ml/min
(superficial velocity)	: 0.53 cm/s
liquid flow rate	: 10 ml/min
(superficial velocity)	: 0.033 cm/s
Temperature	: 25 ^{0}C.

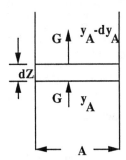

Figure 7: Mass Balance in the Biofilter

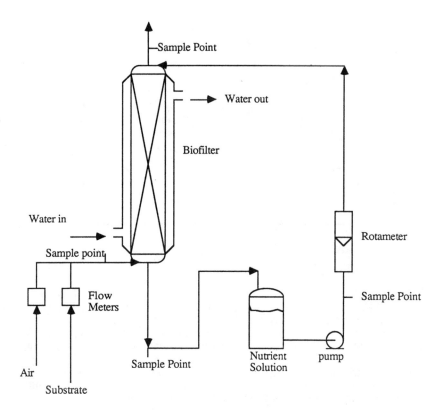

Figure 8: Schematic of Experimental Biofilter

The gas and liquid phase samples from the inlets and the outlets of the biofilter were analyzed for all VOCs. It was found that the changes in the concentrations of the VOCs in the liquid phase were negligible. The mass balance for the substrates showed that less than 1% of the substrates removed from the gas phase is accounted for in the liquid phase. The removal efficiency of the biofilter was calculated for each substrate. Figures 9 and 10 show the removal efficiencies as a function of inlet gas phase concentrations for toluene and methylene chloride respectively. The removal efficiency is defined as the difference between the inlet and outlet VOC concentrations expressed as a percentage of the inlet VOC concentration.

Table 2: Composition of Nutrient Solution

Salt	Concentration (mg/lit)
KH_2PO_4	85
K_2HPO_4	217.5
Na_2HPO_4	266.4
NH_4Cl	25
$MgSO_4.7H_2O$	22.5
$CaCl_2$	27.5
$FeCl_2.6H_2O$	0.25
Trace elements -	
$MnSO_4.4H_2O$	0.0399
H_3BO_3	0.0572
$ZnSO_4.7H_2O$	0.0428
$(NH_4)_6Mo_7O_{24}$	0.0347
$FeCl_3.EDTA$	0.10
Yeast Extract	0.15

Stripper-Biofilter Design

Stripper Design. The air requirement for reducing the VOCs to 10% of their initial level is calculated by assuming that the outlet concentration in gas is the equilibrium concentration corresponding to the inlet concentration in the liquid. This is calculated by using the Henry's law constants for each component. The results of these calculations are shown in Table 3. The maximum flow requirement is 8 m^3 air (at STP) per m^3 of leachate for methylene chloride and this value is selected for stripping of the organics. Acetone and other ketones are not stripped from the liquid phase and are excluded from these calculations. The resulting effluent concentrations in the aqueous and the gaseous phases are shown in Table 4.

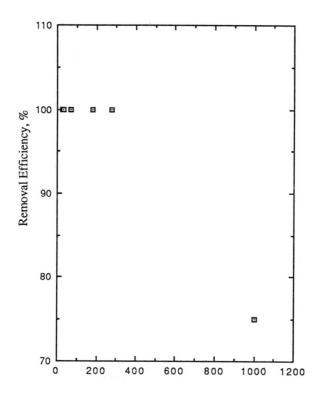

Inlet Gas Phase Concentration, ppm

Figure 9: Removal Efficiency of the Biofilter for Toluene (Measured Experimentally). [Removal Efficiency = (1-outlet concentration) X 100.]

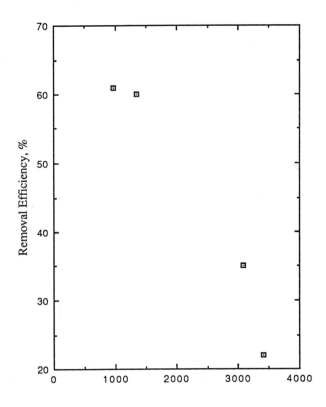

Inlet Gas Phase Concentration, ppm

Figure 10: Removal Efficiency of the Biofilter for Methylene Chloride (Measured Experimentally). [Removal Efficiency = (1-outlet concentration) X 100.]

Table 3: Air Requirements for 90% Removal in the Stripper

No.	Compound	Air Requirement (m^3 at STP/m^3 liquid)
1	Benzenes and Alkyl Benzenes	2.9
2	Toluene	3.05
3	Chloroethylenes	2.2
4	Chloroform	5.3
5	Methylene Chloride	7.8
6	Chloroalkanes	0.7
7	Chlorobenzene	6.5

Table 4 : Effluent Concentrations after the Stripper - Volatile Organics

No.	Compound	Effluent Concentrations (ppm)			
		Gas		Liquid	
		Range	Mean	Range	Mean
1	Benzene and Alkyl-benzenes	29-70	50	~0	~0
2	Toluene	157-440	300	~0	~0
3	Acetone	1.6-3.7	3	13.9-22.5	18.2
4	Higher Ketones	9-21	15	10.6-18.3	14.4
5	Chloroethylenes	28-90	60	~0	~0
6	Chloroform	23-70	48	~0	~0
7	Methylene Chloride	242-650	415	0.65-1.1	0.87
8	Chloroalkanes	0.2-81	40	~0	~0
9	Chlorobenzene	5-19	12	~0	~0

Biofilter Design. The design basis is 0.5 Mgd of the leachate containing the contaminants. The air requirement is 0.175 m^3/s or 7.8 mol/s.

The main problem encountered in the design is the unavailability of the kinetic data for the compounds. Most of these compounds are degraded with great difficulty and it is difficult to obtain reliable estimates of k_1, b and y. Therefore the sizing calculations for any of the compounds is quite difficult. C_2^* is a function of parameters ϕ, ϕ_L and r. The parameter r involves the ratio of p_A and X. It is assumed that this ratio remains independent of p_A. This implies that as the partial pressure of substrate A in air decreases, the biomass concentration that it can support also decreases proportionately. In other words, the value of r is independent of y_A and is constant in the biofilter. With this simplifying assumption, the term $(1-C_2^*)$ can be pulled out of the integral and integration of equation 26 becomes straightforward.

The values of r and ϕ for different substrates are shown in Table 5.

Table 5: Parameter Values for Substrates

No.	Substrate	ϕ	r
1	Benzene and Alkyl Benzenes	30	1
2	Toluene	30	1
3	Acetone	50	10
4	Higher Ketones	50	10
5	Methylene Chloride	10	0.1
6	Chlorobenzene	10	0.1
7	Chloroalkanes	10	0.1
8	Chloroform	10	0.1
9	Chloroethylenes	10	0.1

Using typical values [Satterfield(17)] of superficial gas and liquid velocities (0.02 kg/m^{2}s and 6 kg/m^{2}s respectively), the column sizing is carried out using equation (26). The value of $K_L a$ used in the calculation is 0.2 s^{-1} and is obtained from Goenaga et al(18). The cross sectional area of the column works out to be ~7 m^2 and hence the diameter of the column 3 m. The height of the column is obtained by integrating equation (25). For 90% reduction, the height required is ~5.3 m. The concentration profiles of the different substrates are calculated using equation (26) and these are shown in figures 11 and 12 respectively. Similar calculations have been carried out for the multicomponent mixture by solving 2n+1 simultaneous equations. The resulting concentration profiles for the substrates are shown in figures 13 and 14.

Conclusions

1. A kinetic model has been developed to describe the steady state biodegradation of VOCs on activated carbon in a biofilter.
2. The model is solved for a range of values of the parameters involved.
3. The kinetic model for the individual support particle is used in the design of biofilter.
4. Experimental data are obtained for the removal of toluene and methylene chloride in the biofilter.

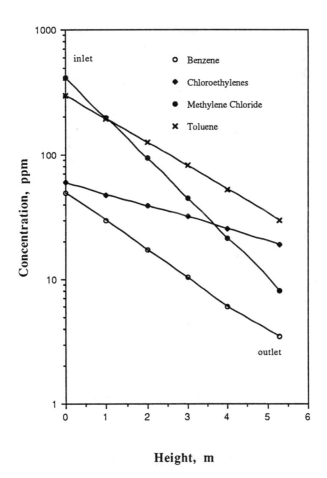

Height, m

Figure 11: Concentration Profiles in Biofilter

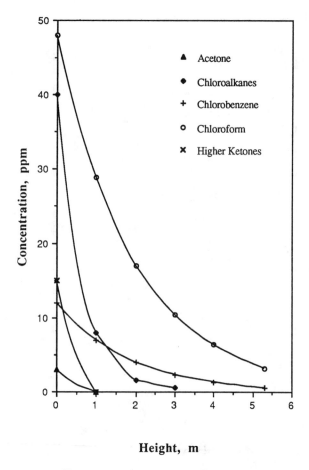

Height, m

Figure 12: Concentration Profiles in Biofilter

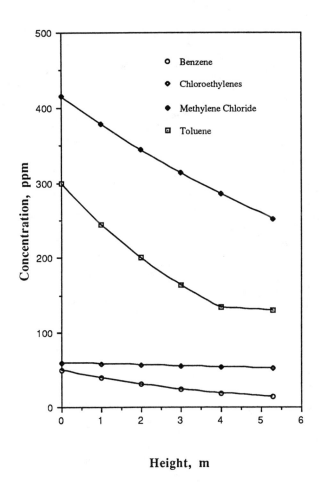

Height, m

Figure 13: Concentration Profiles in Biofilter - Multicomponent Mixture

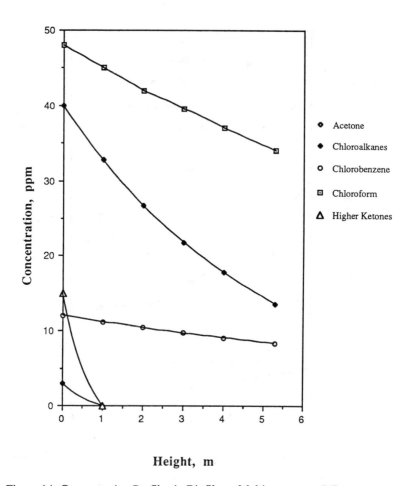

Figure 14: Concentration Profiles in Biofilter - Multicomponent Mixture

Notation

A	Cross sectional area of the biofilter, cm^2
b	Bacterial decay, s^{-1}
C	Substrate concentration in the biofilm, mol/cm^3
$\overline{C}$	Substrate concentration in the support, mol substrate/gm support
C_2	Substrate concentration at liquid-biofilm interface, mol/cm^3
C_3	Substrate concentration at biofilm-support interface, mol/cm^3
d_B	Biofilm thickness, cm
D_e	Effective diffusivity of the substrate in biofilm, cm^2/s
D_s	Effective diffusivity of the substrate in the support, cm^2/s
G	Gas flow rate, mol/s
H	Henry's law constant, $kPa\ m^3/mol$
k_1	First order biodegradation rate constant, s^{-1}
K_L	Overall gas-liquid mass transfer coefficient, cm/s
$K_L a$	Overall volumetric mass transfer coefficient, s^{-1}
L	Characteristic dimension of the support, cm
m_1	Freundlich adsorption isotherm parameter, $mol^{(1-1/m_2)}\ cm^{(3/m_2)}/gm$
m_2	Freundlich adsorption isotherm parameter, -
n	Number of components in the mixture, -
P_t	Total gas pressure, kPa
p_A	Partial pressure of substrate A, kPa
r	Biomass growth/decay Ratio, -
R_A	Biodegradation rate of A, $mol/cm^2\ s$
$R_A a$	Volumetric degradation rate of A, $mol/cm^3\ s$
S	Substrate concentration (p_A/H), mol/cm^3
t	time, s
X	Biomass concentration, mol/cm^3
y	Bacterial yield, gm biomass/mol substrate
y_A	Mol fraction of A in the gas, -
z	Dimension coordinate, cm
Z	Height of the biofilter, cm
ϕ	Thiele modulus for reaction, -
ϕ_L	Thiele modulus for mass transfer, -
ϕ_L	Support density, gm/cm^3

superscript

*	Dimensionless quantity

subscript

i	ith component of the mixture
in	at the inlet of the biofilter
out	at the outlet of the biofilter

Acknowledgment

The authors would like to thank the U.S. Environmental Protection Agency for supporting the research and Professor M. Suidan, Department of Civil and Environmental Engineering, University of Cincinnati for his helpful suggestions.

Literature Cited

1. Varian Associates, 'Landfill Site Air Pollution: A Problem at our Feet', *Western Waste*, **March 1989**, p.32.
2. Bramlett, J.; Furman, C.; Johnson, A.; Nelson, N.; Ellis, W. and Vick, W., 'Composition of Leachates from Actual Hazardous Waste Sites', U.S. Environmental Protection Agency Report No. EPA/600/2-87/043, **June 1987**.
3. Lai L. and R. Govind, **1990**, submitted for publication to *Environ.Prog.*
4. Pomeroy, R.D., 'Biological treatment of odorous air', *J. Water Pollut. Contr. Fed.*, **1982**, *54*, p. 1236.
5. Carlson, D.A. and Leiser, C.P., 'Soil beds for the control of sewage odors', *J.Water Pollut. Contr. Fed.*, **1966**, *38*, pp.829-840.
6. Smith, K.A.; Bremmer, J.M. and Tatabai, M.A., 'Sorption of gaseous atmospheric pollutants by soils', *Soil Science*, **1973**, pp.313-319.
7. Hartenstein, H., 'Assessment and redesign of an existing biofiltration system', M.S. Thesis, University of Florida, **1987**.
8. Eitner, D., 'Untersuchnungen uber Einsatz und Leittungsfahigkeit von Konpostfilteranlagen zur biologischen Abluftreiningun im Bereich van Klaranlagen unter besonderer Berucksichtigung der Standzeit', Investigations of the use and ability of compost filters for the biological waste gas purification with special emphasis on the operation time aspects, **1984**, GWA, Band 71, TWTH Aachen, West Germany.
9. Ottengraf, S.P.P. and van den Oever, H.C., Kinetics of Organic Compound Removal from Waste Gas by Biological Filter, *Biotech. Bioeng.*, **1983**, *25*, pp.3089-3102.
10. Kampbell, D.H.; Wilson, J.T.; Reed, H.W. and Stocksdale, T.S., Removal of Volatile Aliphatic Hydrocarbons in a Soil Bioreactor, *JAPCA*, **1987**, *37*, pp. 1236-40.3.
11. Eitner, D. and Gethke, H.G., 'Design, construction and operation of bio-filters for odor control in sewage treatment plants', **1987**, paper presented at the 80th meeting of the air pollution control association, New York.
12. Prokop, W.H. and Bohn, H.L., Soil Bed System for Control of Rendering Plant Odors, *J.Air Pollut. Contr. Assoc.*, **1985**, 35, pp. 1332-8.
13 Don, J.A. and Feenstra, L., 'Odor abatement through biofiltration', **1984**, paper presented at Symposium Louvain-La-Neuve, Belgium
14. Utgikar, V.; Govind, R. and Brenner, R.C., 'Biological Pretreatment followed by POTW Treatment of CERCLA Leachates: Biodegradation of Stripped Toxics in a Biofilter', Presentation at Biosystems Symposium of U.S. EPA, Arlington, VA, February **1990**.
15. Utgikar, V. and Govind, R., 'Steady State Analysis of Biodegradation of Toxics in a Biofilter', presentation at I & EC Division Symposium on Emerging Technologies for Hazardous Waste Treatment, Atlantic City, New Jersey, June **1990**.
16. Utgikar, V.; Govind, R.; Shan, Y.; Safferman, S. and Brenner, R.C., 'Biodegradation of Toxics in a Biofilter', Paper submitted to 17th annual EPA Hazardous Waste Research Symposium to be held in Cincinnati, OH, April 1991.
17.Satterfield, C.N., 'Trickle Bed Reactors', *AIChE J.*, **1975**, *21*, pp. 209-28.
18. Goenaga, A.; Smith, J.M.and McCoy, B.J., 'Study of gas to Liquid Mass Transfer by Dynamic methods in Trickle beds', *AIChE J.*, **1989**, *35*, pp.159-63.

RECEIVED April 5, 1991

Chapter 13

Detoxification of Organophosphorus Pesticide Solutions

Immobilized Enzyme System

P. L. Havens[1] and H. F. Rase

Department of Chemical Engineering, The University of Texas at Austin, Austin, TX 21218

Increased regulation of pesticidal chemicals has made the disposal of wastes generated by pesticide users an increasing concern from both an environmental and an economic standpoint. This work utilizes an immobilized enzyme system to degrade pesticides in the organophosphate (OP) class which, while extensively used worldwide on many crops, are often highly toxic to non-target organisms. An enzyme derived from an overproducing strain of *Pseudomonas diminuta*, called parathion hydrolase (PH), carries out a hydrolysis of the phosphate ester bond in the OP molecule, resulting in an as much as 100-fold reduction in toxicity. Partially purified PH was covalently immobilized upon several rigid supports and retained a large degree of its activity upon immobilization. Three activated controlled pore glasses and a beaded polymer were found to be promising supports for application of the enzyme to the detoxification of contaminated waste waters, offering high mechanical stability and high enzyme capacities. A comparison was made of the four supports on the basis of specificity of binding, amount of protein loading and stability over time. The immobilized enzyme was capable of reducing the concentration of several OP pesticides to sub-ppm levels in both solutions of purified OP's and commercial formulations of commonly used compounds. A field-scale system was designed and its economics were evaluated.

There has been considerable progress in the past fifteen years toward understanding, isolating, and genetically engineering bacterial strains for production of enzyme systems active in the detoxification of organophosphate pesticides. It is timely, indeed, to place greater emphasis on developing and testing efficient means for applying this knowledge in organophosphate waste remediation. The research herein described was designed to study the effect of support chemistry and configuration on

[1]Current address: DowElanco, Environmental Chemistry Laboratory, 9001 Building, Midland, MI 48641–1706

0097–6156/91/0468–0261$06.25/0

the performance of immobilized enzyme systems for organophosphate detoxification. Additionally, a viable scaled-up portable system for detoxification of commercial pesticide waste streams is proposed and analyzed.

With the banning of chlorinated pesticides such as DDT for environmental reasons, the organophosphates (OP's) have undergone a resurgence in usage since 1970 (*1*), with U.S. production reaching 157 million pounds in 1985 (*2*). Nearly 100 compounds are in either commercial or experimental use (*3*). The OP's offer the advantage of quick knockdown of pests, combined with limited lifetimes in the environment. However, their low persistence may require more frequent spraying to obtain the same pest control effect of chlorinated compounds.

The OP's are highly toxic to pests due to their inhibition of the vital enzyme acetylcholinesterase, whose action carries out the normal degradation of the neurotransmitter acetylcholine in the central nervous system. If not broken down after causing a neuron to fire, acetylcholine will produce continuous firing of the nerve or contraction of muscle fibers which leads to paralysis of the muscles, convulsions and respiratory arrest. Warm-blooded animals as well as insect pests utilize cholinesterase pathways, and thus OP's are highly toxic to mammalian species, including humans. This is a serious safety concern for formulators, applicators and agricultural workers. Contaminated equipment and wastewater can be poisonous even at low concentrations of active ingredients. Although improvements in application technology as well as in the development of compounds less toxic to mammal have eased some of these concerns, increasingly stringent hazardous waste regulations have made detoxification of pesticide wastes a vital issue.

Regulation and Sources of Hazardous Wastes

The regulation of waste materials took on a more proactive character with the passage by Congress of the Resource Conservation and Recovery Act of 1975 (RCRA). RCRA directed the U. S. Environmental Protection Agency (EPA) to develop and implement regulations to protect human health and the environment from mismanagement of hazardous wastes. RCRA regulates the management of all solid waste, defined as any material (liquid, wet or dry solids) that is to be discarded or has served its useful purpose. To be classified as "hazardous waste," the waste material must exhibit one of several characteristics outlined in Title 40 of the Code of Federal Regulations (40 CFR).

There are two classifications of hazardous commercial products which appear in the RCRA listings: the "E" list, termed "Acutely Hazardous," and the "F" list, termed "Toxic." The E list (40 CFR §261.33(e)) contains many pesticidal active and inert ingredients, including several organophosphate pesticides. The regulation of these wastes was first focused on large companies, who generate the majority of waste volume. The 1980 amendments to RCRA specifically exempted from regulation generators of less than 1000 kilograms of wastes per month (defined as "small quantity generators"). However, in 1984, the Hazardous and Solid Waste Amendments to RCRA were enacted. Small quantity generators of between 100 and 1000 kg of hazardous wastes per month then came under the auspices of the regulatory system. Final regulations were issued by EPA in March of 1986, with requirements becoming effective September 22, 1986.

There are several possible sources of wastes associated with using pesticides. The main sources are the mixtures left over from spraying operations and aqueous rinsates (*4*). Farmers may dispose of excess materials without a permit by spraying on the target area or on the fringes of the target area. Containers which have been completely emptied by normal means are not regulated; however, if they contained acute hazardous materials (RCRA E list), the containers must be triple rinsed (40 CFR §261.7) or they themselves are regulated as hazardous solid waste. The rinsates from containers and wash water from equipment and tank washdown may be added to spray mix as diluent. However, this may not always be possible, for it may not be legal to apply a compound present as carry-over diluent to certain crops or contaminants (e.g., dirt or insects) may be present in the wash water. In these cases, the wastewaters and left over spray mixes, if generated over the 100 kg (or the 1 kg acute hazardous) limit, become hazardous wastes and fall under regulation. Commercial operators are not allowed the exemptions from permitting available to individual applicators and must carefully manage pesticides and wastes generated from pesticides to avoid the extreme expense of large-scale regulation (5). The amount of wastewater generated during pesticide operations can be substantial. For example, it has been estimated that a single piece of application equipment generates 20,000 liters of wash water per year (44). The volume of these wastes and the rinsates generated during container rinsing measure in the hundreds of millions of gallons per year (6). In the current and future regulatory climate, cost-effective technologies will be required to detoxify and dispose of these wastes.

Pesticide Disposal Technology. There has been much activity in recent years in the development of technology for the disposal of formulated pesticides and their associated wastes. Conventional treatment schemes, such as activated sludge treatment and incineration, have been used successfully by pesticide manufacturers, but adaptation to the small-scale use typical of agricultural activity is not feasible. Activated sludge treatment requires intensive management to avoid killing the sludge organisms, high capital investment, and high operating and maintenance costs. Small-scale incinerators are not feasible for similar reasons of high cost and the need for special operator skills to ensure complete combustion and no air pollution.

Several different studies have attempted to develop technology for use by smaller-scale operators in the field for the treatment of left over formulations and wastewaters. Most of these techniques rely on the natural degradative capacity of native soil organisms (7,8). Adsorption by activated carbon has also been explored and piloted for the cleanup of pesticide wastewaters (9,10). Ultraviolet irradiation-ozonation treatment has also been studied (11,12).

As can be seen above, many different techniques are available for on-site treatment of pesticide wastes. A contract report to the EPA (6) gives rankings for different disposal technologies based on comparisons of environmental safety, effectiveness, versatility, availability and applicator factors. For reasons of cost and degree of operational skill required, both activated sludge and incineration treatments are impractical for the field user. Containment and soil degradation systems received high ranking, although many of the systems are experimental and have not been tried in a range of climatic conditions. Adsorption has the possibility of being highly successful; being a closed system, there is no danger of release of volatile residues. In addition, the technology is well known and widely applicable on almost any scale. However, the wastes must be pretreated to remove particulate matter and the spent carbon must be disposed of properly, adding to expense. Chemical and other physical treatment methods also hold promise, although much research remains to be done. The ideal waste treatment system would be a closed system of low cost which would be able to completely remove or break down a wide range of pesticidal ingredients without generating additional waste materials.

From a regulatory standpoint, an operator who generates more than the RCRA limits for conditionally exempt generators (100 kilograms F list; 1 kilograms E list) must apply for a permit under 40 CFR 260.10 to treat wastes on-site. This can be a time-consuming and expensive process (13,45). However, changes in RCRA, passed by Congress in 1989, will require streamlining of the permitting procedures, hopefully reducing the time and cost involved. These amendments will also encourage waste recycling and reduction efforts and make on-site treatment more practicable for the pesticide end user.

Biodegradation of Organophosphates by Biological Systems. The study of the biodegradation of organophosphates has been developed over about the last fifteen years. A concerted effort has been directed at isolating the organisms responsible for the degradation, identifying the specific enzymes carrying out the actual chemical change, understanding this enzyme system at the genetic level and utilizing these enzymes for organophosphate waste remediation. A short review of organophosphate degradation appears in a CRC handbook (*14*).

In 1972, Hsieh and Munnecke (*15*) related the use of an enrichment technique to select parathion degrading organisms from a mixed soil culture. They used a chemostat to adapt a mixed bacterial culture to utilize parathion and *p*-nitrophenol as its sole carbon source. In a later paper (*16*), seven bacterial isolates, mostly pseudomonads, were identified in the mixed culture. The authors found their adapted mixed culture could hydrolyze 50 mg of parathion per liter per hour in a chemostat. *Para*-nitrophenol, a metabolite of parathion hydrolysis, was shown to inhibit the initial growth in the chemostat.

This pioneering study was followed by a series of advancements by Munnecke and his associates in both the understanding and utilization of the bacterial system and the active enzyme agent. Parathion metabolism by the mixed bacterial culture on technical grade parathion and emulsifiable concentrate (parathion/xylene) was found to be favored by an alkaline medium (*17*), and the metabolic pathways were studied (*18*). They found the active enzyme to be localized in the cells and stable up to 55°C for 10 minutes. Munnecke (*19*) compared the rates of hydrolysis of twelve commonly using organophosphates by crude enzyme extracted from the bacterial cells with alkaline hydrolysis rates. For eight of the twelve compounds, enzymatic rates were higher (e.g., 2450 times greater for parathion). Next Munnecke immobilized the crude enzyme extract on controlled pore glass (CPG) and silica (*20,21*) and obtained 99% removal of parathion in wastewater with no loss of activity over a period of 70 days. He also demonstrated detoxification of commercial formulations of parathion and container residues (*22*) using a non-immobilized lyophilized enzyme mixture. Barik and Munnecke (*23*) studied the degradation of concentrated diazinon in soil using a crude extract from mixed culture. Similar experiments were carried out by Honeycutt and his co-workers (*24*).

Work on the genetics and enzymology of the parathion hydrolase (PH) enzymes was initiated by Serdar et al. (*25*) by isolating an organism capable of hydrolyzing parathion. The organism was identified as *Pseudomonas diminuta*. By curing the cells with mitomycin C to remove plasmids, the authors isolated hydrolase-negative colonies. These colonies lacked a plasmid which was present in the native strain. The plasmid, named pCMS1, was isolated by agarose electrophoresis and found to have a molecular weight of about 44×10^6 daltons.

Once the plasmid containing the PH gene had been isolated, Serdar and Gibson (*26*) carried out a study to clone and express the gene. The strain of *Ps. diminuta* designated as MG pCMS55 exhibited the highest activity of all the strains developed. The pCMS55 plasmid was developed from the native pCMS1 plasmid and its high activity was postulated by the authors to be the result of the presence of two PH codons.

In a further genetic study, McDaniel, et al. (*27*) performed additional cloning studies with the pCMS1 parathion hydrolase gene. The native plasmid was digested with restriction enzymes and subcloned into pBR322. This vector was used to transform *E. coli* and the transformant was found to have hydrolytic activity. Researchers at Amgen Incorporated (*28*) performed further genetic work on the PH gene from *Pseudomonas diminuta* MG. They expressed the gene in *E. coli* under the control of a heat-inducible promoter and obtained high levels of hydrolase activity. A complete sequence of 365 amino acids was deduced, including a 29 amino acid leader sequence which was removed after translation. The authors stated that this leader sequence was a "typical" signal sequence, i.e., suited for membrane transport, but they did not state if the enzyme was periplasmic, cytoplasmic or excreted by their *E. coli* strain.

The activity of PH was applied *in situ* in a study carried out by the U.S.D.A. (*29*). A species of *Flavobacterium* exhibiting parathion hydrolase activity was introduced into cattle dip vats containing coumaphos. A 650 gallon vat was inoculated and as the bacteria grew, coumaphos was hydrolyzed into products amenable to UV/ozonation degradation. Essentially complete hydrolysis was observed after 48 hours of growth with vigorous agitation. To promote the growth of the desired bacteria, the authors had to add xylose as a carbon source, as well as buffering salts and a nitrogen source.

Mulbry and Karns (*30*), working at the U.S. EPA Pesticide Degradation Laboratory, have purified and characterized three different parathion hydrolase enzymes from three microbial sources. All three organisms were gram-negative strains isolated from enrichment cultures.

Materials and Methods

Cell Growth. Culture plates (1% tryptone, 0.5% yeast extract, 0.5% NaCl, 1.5% agar, 50 μg/ml kanamycin) were inoculated with cells from frozen glycerol suspensions of *Ps. diminuta* MG pCMS55. After overnight incubation at 32°C, single colonies were transferred to 250 ml shake flasks containing 50 ml of TYE medium (1% tryptone, 0.5% yeast extract, 0.5% NaCl, 50 μg/ml kanamycin). Following overnight incubation at 32°C, these flasks were used to inoculate a 1 or 1.5 liter fermenter, which was maintained at 32°C, pH 7. Fermentation was continued until the cells were at the upper log growth phase.

Cells were harvested by centrifugation (8000 g, 20 minutes, 4°C) and resuspended in 1 ml of phosphate buffer (50 mM, pH 8.5) per gram of cells. The cells were then disrupted by sonication and centrifuged again (60000 g) yielding the crude cell extract as the supernatant. Partial purification of the extract was obtained by ammonium sulfate precipitation (20-40% cut) and the salt was removed by dialysis against 50 mM phosphate buffer, pH 8.5.

Enzyme Assays. Standard assays were carried out by pipeting 20 μl of protein solution into a quartz spectrophotometer cuvette and adding 400 μl of parathion solution which had been prepared from reference standard material (US EPA Pesticide and Industrial Chemicals Repository). The production of a hydrolysis product of parathion, *p*-nitrophenol, was followed by tracking the increase in absorbance at the 410 nm wavelength. Since the hydrolysis stoichiometry is one-to-one, the appearance of the phenol corresponds to the disappearance of parathion. The concentration of *p*-nitrophenol was determined by comparison with a standard absorbance curve. Regression upon the initial linear portion of the concentration vs. time curve yielded an initial rate in nmol of parathion per μl per minute. Protein assays were carried out using premixed Coomassie Blue dye reagent according to the manufacturer's instructions (*36*). The rate was divided by the amount of protein to yield a specific activity, defined as nmol of parathion consumed per minute per milligram of protein

Analysis of chlorpyrifos was accomplished using reversed-phase HPLC. A linear gradient of 50:50 methanol:water (+ 1% acetic acid) to 100% methanol over 30 minutes on a Waters μBondapack C-18 column at 1 ml per minute flow eluted chlorpyrifos at 28.3 minutes and trichloropyridinol at 17.3 minutes. A calibration curve was prepared by injecting known concentrations of chlorpyrifos.

Immobilizations. The PH enzyme was immobilized upon four different supports for use in continuous-flow operation. Three are derivatized controlled-pore glasses (CPG's): Aminoaryl CPG (AA), Carbonyl diimidazole CPG (CDI) and N-Hydroxysuccinimide CPG (NHS). The fourth, VA-Epoxy Biosynth™ (VA) is a beaded epoxy polymer. The supports utilize different chemistries to covalently link to free amino groups on the enzyme (N-termini or amino side chains); all are microporous resulting in high surface areas per unit weight. The beads are supplied pre-activated and ready for protein linking (except aminoaryl glass which requires an activating reaction). The physical characteristics of each support are summarized in Table I and coupling chemistries are described below.

Table I. Packed-Bed Supports: Physical Properties

Support	Particle size, μm^a	# of groups, $\mu mol/g^a$	Arm length, $\mathring{A}^b$	Pore size, $\mathring{A}^a$
Aminoaryl CPG (AA)	125-177	6.8	9.4	500
Carbonyl-diimidazole CPG (CDI)	74-125	50	4	200
N-Hydroxy-succinamide CPG (NHS)	125-177	50	4	500
VA-Epoxy Biosynth (VA)	50-200	300	7	300

[a]Manufacturers' data.
[b]From estimated bond lengths taken from CRC Handbook of Chemistry and Physics.

The conditions under which the immobilizations were performed were varied somewhat from run to run. In general, however, the immobilizations were carried out with agitation by shaking either on ice or at room temperature, followed by a non-shaken period at 4 °C. The protein solution in contact with the beads (termed the supernatant) was periodically assayed for protein and activity. The supernatant which remained following the immobilization reaction was removed from the beads by filtration and assayed also for activity and total protein. The bead were then washed with buffer until the wash buffer was free of protein. Prepared beads were stored under refrigeration.

AA contains aminoaryl groups linked to the glass surface. These groups must be activated by conversion to a diazo group (21,31). This is done by reacting the beads with nitrous acid under vacuum. The activated beads are then mixed with protein solution and the diazo groups react with primary amino groups on the proteins, coupling the proteins to the support.

CDI is treated with carbonyl diimidazole (a bifunctional reagent), linking an imidazole group to the surface. This linking, however, is not directly to the glass surface; the surface is coated with a hydrophilic, non-ionic, "glycophase" layer (32).

The glycophase prevents possible interaction of the proteins with the highly charged glass surface. Protein binding is accomplished by simply mixing a protein solution with the beads (*33*).

NHS is synthesized by reacting glycophase-coated glass with carbonyl di-N-hydroxysuccinimide, resulting in a highly reactive N-hydroxysuccinimide group attached to the support surface. Coupling is initiated by mixing the beads with protein solution (*34*).

VA is a beaded copolymer of vinyl acetate and divinylethylene-urea whose surface is modified by hydrolysis of acetate groups with oxirane groups. The three-membered oxirane ring is readily reactive toward primary amines and protein coupling is accomplished by simply mixing the protein solution with the beads (*35*).

Mini-Column System. A micro-column system (Figure 1) was constructed to study the detoxifying activity of the enzyme immobilized on the beaded supports. The substrate reservoir is a Erlenmeyer flask, stirred magnetically and connected by Teflon chromatography tubing to a flow-through spectrophotometer cell. The outlet of the cell was connected to a peristaltic pump and then to the immobilized enzyme column. A nylon coupler connects the top of the column to additional tubing which leads back to the substrate reservoir.

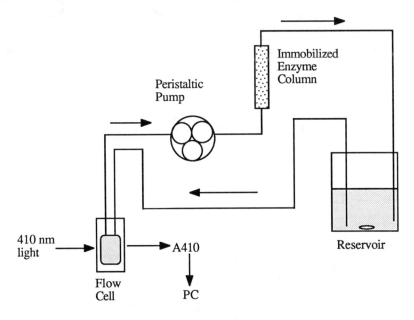

Figure 1. Mini-column system

The controlled-pore glass beads were packed into 0.8x10 cm disposable polystyrene chromatography columns by pipetting a slurry of packing and buffer into the columns and allowing the bed to form by gravity settling (*37*). When the desired bed height was reached, a fritted disk was inserted and pressed it into place. Additional buffer was passed through the bed to equilibrate the beads.

The columns were run in upflow mode which allows good distribution of the liquid phase over the entire reactor cross-section and good contact of the liquid with the catalyst. Low bypassing and wall flow are assured because the liquid completely fills the void volume of the bed. The spectrophotometer was calibrated on buffer at 410 nm, and the parameters for collection of time course data were set using data capture software. For most runs, the software was set to collect data at 10 points per minute for one hour. A volume of degassed Tris buffer was measured into the substrate reservoir and a stirring bar added. A known volume of substrate stock solution in methanol was added with stirring. The pump was started and data collection began when the liquid front had returned to the reservoir. After the data had been collected, the system was flushed with fresh buffer and the column removed to storage. The absorbance at 410 nm versus time was stored, ported to spreadsheets and plotted. The initial rate in absorbance per time was determined by performing linear regression on the initial part of the reaction progress curve. This rate was converted to the change in concentration of parathion per time by using a p-nitrophenol calibration curve. For chlorpyrifos, periodic samples were analyzed by HPLC.

Results and Discussion

Comparison of the Supports. In order to compare directly the rates of reaction from column runs with the same operating conditions, the initial rates (from the linear portion of the time-course curve) were assumed to be first order. With this assumption, a pseudo-first order rate constant, k, could be directly compared by dividing the rate by the initial concentration of parathion. Since for batch operation the rate is also proportional to the amount of enzyme, k was further weighted by the milligrams of total protein immobilized (computed by the difference in protein between the initial extract and final supernatant), yielding the quantity k/mg. This number, analogous to the specific activity in solution, was used for comparisons of enzyme activity.

The calculation of rates described above are dependent on the assumption that the reactor system in operating in true batch-recycle mode, i.e., the conversion per reactor pass is small. Smith (*38*) derives a criterion to confirm differential operation of the reactor as follows:

For differential operation, by a mass balance:

$$C_{res} - C_e = -\frac{V_{bed}}{Q}\left(\frac{V_{res}}{V_{bed} + V_{res}}\right) r$$

Where C_{res} is the concentration of substrate in the reservoir, C_e is the concentration at the reactor exit, V_{bed} is the volume of the catalyst bed, V_{res} is the volume of the reservoir, Q is the flowrate, and **r** is the rate. If the quantity $C_{res} - C_e$ is small compared to the initial concentration of substrate, differential operation is indicated, i.e., the conversion per pass is small. A NHS column run (*39*) exhibited the highest rate of all the experimental runs so this calculation is illustrated using the data from this run. For the hydrolysis reaction, the rate is highest at the beginning of the run so $C_{res} = C_0$

$r_{max} = 2.71E\text{-}2$ nmol parathion consumed/($\mu l \cdot min$)
$C_0 = 4.645E\text{-}2$ nmol parathion/μl
$V_{bed} = 1.56$ ml
$V_{res} = 50$ ml
$Q = 20$ ml/min

Filling in the previous equation:

$$C_{res} - C_e = - \frac{1.56 \text{ ml}}{20 \text{ ml/min}} \left(\frac{50 \text{ ml}}{50 \text{ ml} + 1.56 \text{ ml}} \right) \left(-2.71\text{E-}2 \frac{\text{nmol}}{\mu\text{l}\cdot\text{min}} \right)$$

$$= 2.05\text{E-}3 \text{ nmol}/\mu\text{l}$$

This concentration difference is an order of magnitude lower than C_0, yielding a conversion per pass, Δx_{pass}, of:

$$\Delta x_{pass} = \frac{C_{res} - C_e}{C_{res}} = \frac{2.05\text{E-}3}{4.63\text{E-}2} \times 100 = 4.4\%$$

Differential or batch-recycle operation is thus confirmed by this small conversion per pass.

Rate Comparison. The sections to follow compare the suitability of the four packed-column supports for continuous operations. The data plotted below show the results from the immobilizations carried out with partially purified enzyme extracts. The conclusions arrived at by the relationships plotted below are primarily qualitative.

The rate constants of the fresh immobilized preparations are plotted in Figure 2. There is a definite difference between NHS and the other three supports. NHS exhibits a consistently higher rate, while CDI, AA and VA show very similar rates. A Student's *t* test was invoked to evaluate the validity of this conclusion. Based on this test it can be stated with 95% confidence that the NHS rates are significantly different from that exhibited by the other supports.

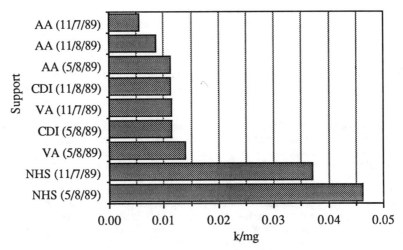

Figure 2. Rate constants for each support

Possible Reasons for Observed Differences in Supports. An attempt was made to correlate the appropriate physical and protein binding properties of the various supports with the degree of catalytic activity observed. One possibility is the pore diameter; Figure 3 shows this relationship.

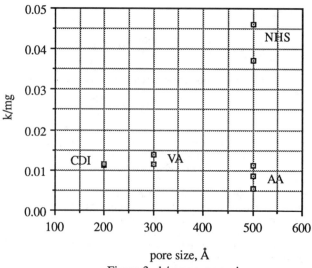

pore size, Å

Figure 3. k/mg vs. pore size

There does seem to be some relationship between the pore diameter and rate, especially if AA is neglected. Since AA and NHS, both with 500 Å pores, show widely different rate characteristics, it is apparent that substrate transport is not the controlling factor. As an additional confirmation from diffusion theory, the dimensions of the substrate, determined from approximate bond lengths (less than 20 Å for parathion), is much smaller than the pore diameter (hundreds of Å). The mean free path between collisions is so small (especially for liquids) that a substrate molecule does not "see" the walls of the pore and moves in essentially free diffusion (*40*). This points to the hypothesis that the rate is somehow a function of the protein binding. An immediate test is to examine the rate versus the amount of protein bound, shown in Figure 4.

It can be seen from this plot that there is not a similar relationship between the amount of protein immobilized and the level of activity of the catalyst. In the case of the NHS, intermediate amounts of protein are immobilized, but activity is significantly higher; AA, with the highest protein loadings, showed the lowest activities. These data suggest that the different proteins, some active and some inactive for the reaction, are not taken up equally, a conclusion further reinforced by the fact that different proteins have widely varying properties of charge, size, and availability of amino groups for covalent binding. This conclusion seems logical if we recall that the AA groups are attached directly to the glass surface while the other CPG's have a deactivating glycophase layer between the glasses and active groups (the VA polymer surface should also be relatively neutral). This property should have no effect on transport in solution since the substrate molecules are uncharged; however, the possibility of an effect on the specificity of protein binding and/or denaturation by the charged glass surface cannot be neglected (*32*).

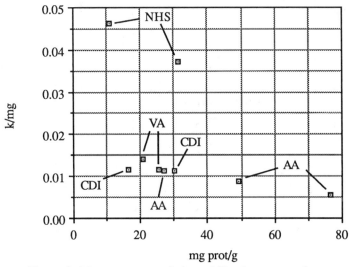

Figure 4. k/mg vs. mg protein immobilized per gram of support

This possibility can be confirmed by plotting the k/mg versus the specific activity of the final supernatant (i.e., the protein solution remaining after immobilization is complete). This relationship is a possible indication of the specificity of binding in that the specific activity is a measure of the units of enzymatic activity present per mg of total protein. A low specific activity for the final supernatant indicates that most of the units of activity have been bound to the support, while a high specific activity for the final supernatant indicates that other proteins (besides the desired enzyme) have been preferentially bound to the support. Figure 5 is the plot of these quantities.

Again, NHS differs greatly from the other supports; according to the supernatant specific activity hypothesis, it has taken up parathion hydrolase preferentially to the other proteins in solution. The other supports, on the other hand, have bound proportionally more of the other proteins present in the solution. Except for AA, this can perhaps be explained on the basis of pore size. McDaniel et al. (*27*) have suggested that PH is a dimeric protein, with a molecular weight of 65000 Daltons, and may, therefore, be larger than many of the other proteins present. If such is the case, the larger pores of NHS may make it possible for the parathion hydrolase molecule to more effectively penetrate into, and bind onto, the bead interior. The hydrophobic glass surface near the binding sites of AA seems to repulse parathion hydrolase or attract other proteins, for the AA support binds large amounts of protein, but shows low activity. The pore size hypothesis can perhaps be confirmed by combining the information from Figures 3 and 5 into Figure 6, showing an inverse relationship between the specific activity of the final supernatant and pore size.

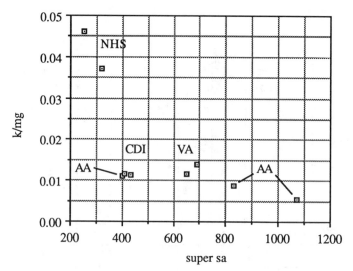

Figure 5. k/mg vs. supernatant specific activity (super sa)

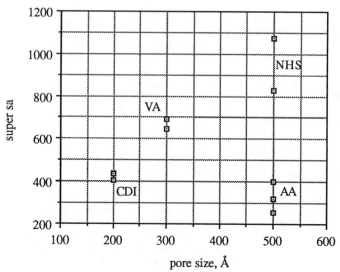

Figure 6. Final supernatant specific activity vs. pore size

The hypothesis seems to be confirmed by this plot, again if we neglect the AA supports for charge reasons. NHS exhibits a higher specificity of binding than the other supports.

Another physical property known is the number active groups available for protein binding. This varies from 6.8 to 300 μmoles of sites per gram of support. Figure 7 shows the relationship between the number of sites and the initial activity.

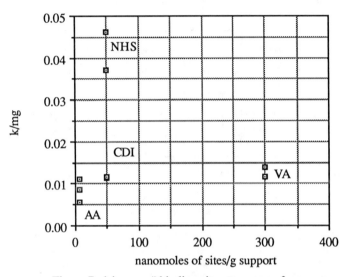

Figure 7. k/mg vs. # binding sites per gram of support

Again, the trend shown is not consistent with expectations. NHS seems to show a somehow more "efficient" use of its binding sites. VA, on the other hand, does not seem to have been able to utilize many of its much greater number of sites. This is perhaps an effect of the conditions used for binding in that the conditions of agitation, binding time, pH, etc. may not have been optimal for the particular chemistry of each support.

The other physical property known is what is termed the arm length. This is an approximate distance between the support surface and the protein which was calculated by approximate bond lengths. It is generally thought (*41*) that increasing this arm length should increase the activity of the catalyst by allowing the bound enzyme to move more freely, increasing the chances an enzyme active site will contact a substrate molecule, although for small substrate molecules the effect is not very important. This supposition was tested by plotting the initial k/mg versus this arm length in angstroms (Figure 8). As might be expected, the arm length does not affect the activity.

To summarize, the NHS glass support gives superior reaction rates, possibly by virtue of its ability to preferentially bind parathion hydrolase. This is postulated to be an effect of pore size, although this cannot be conclusively proven by the data shown here. Experiments would be required comparing glasses with different pore sizes, each with the same coupling chemistry.

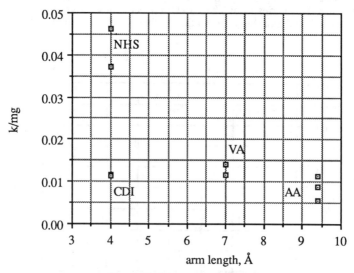

Figure 8. k/mg vs. arm length

Stability Comparison. The stability of each support was examined by carrying out column runs over time. A curve fit was then performed on each set of k/mg data to determine the time dependence of activity. The supports were stored in buffer under refrigeration between runs. An exponential decrease in activity over time was observed for all of the supports. It was thus possible to interpolate a mean half-life for each support (Table II).

Table II. Support Half-lives

Support	$t_{1/2}$ (mean), days	standard deviation
AA	23.2	12.7
CDI	66.9	12.2
NHS	17.3	1.6
VA	103	58.2

It is difficult to make any predictions based upon this data, but it can perhaps be postulated that for a "good" batch of catalyst, i.e., one that exhibits high initial activity, the half life for CDI could be as long 90 days and as long as 160 days for VA-Epoxy. NHS and AA exhibit much shorter half lives, an unfortunate result since the NHS formulation is the most active catalyst. The differences shown in stability between the various supports cannot be readily explained other than to say that there may be some difference in the strength of the bonds holding the enzyme to the support, causing differing degrees of eventual enzyme washout.

One can speculate also upon the actual causes of the loss of activity shown by all the supports besides washout. The most obvious possibility is ordinary time-dependent denaturation of the enzyme. This occurs quite readily in solution and could easily be happening to the immobilized formulation. Another possibility is that the substrate runs are in some way deactivating the enzyme. However, if this were the case, supports which have undergone more runs should deactivate more quickly. Comparing, for example, the deactivation coefficients for NHS 5/19/88 and NHS 4/25/89 shows this is not the case; the two coefficients are nearly identical. This implies that deactivation is only a time-dependent, and not a run-dependent phenomena. To test this hypothesis conclusively would require running columns continuously over a long period and comparing activity loss with columns run only intermittently.

Activity with Commercial Formulations. In order for the immobilized enzyme supports to be applicable to actual field use, the enzyme must be active toward commercial formulations of pesticides. Pesticide active ingredients are supplied in many forms, one of which is the emulsifiable concentrate (EC) form. Emulsifiable concentrates are made up of active ingredient in an inert mixture of solvents (generally xylenes), emulsifiers and detergents. The EC is mixed with water to form an emulsion containing concentrations of active ingredient appropriate for the final usage.

It is generally thought that organic solvents and detergents are an anathema to enzymatic activity, as both types of compounds tend to denature proteins, with concomitant loss of enzymatic activity. It was thought that the immobilizations carried out here would sufficiently stabilize PH, allowing it to remain active in the presence of these denaturant compounds. A preliminary experiment was carried out using an EC of chlorpyrifos (O,O-diethyl O-(3, 5, 6-trichloro-2-pyridyl) phosphorothioate). EC was diluted with water to a concentration of 0.25% active ingredient, a concentration widely used on a variety of fruit trees and ornamentals. The spray mix was pumped through a NHS-immobilized enzyme column. Samples were taken periodically and assayed by HPLC for the disappearance of chlorpyrifos. Figure 9 shows the result of this run.

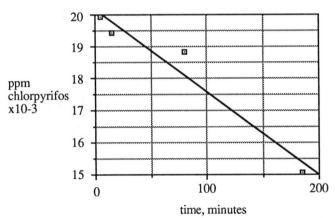

Figure 9. 0.25% Chlorpyrifos EC mix on NHS column

It can be seen from these results that immobilized PH is active in the presence of large amounts of EC inert ingredients. It is obvious that the complete degradation of all the chlorpyrifos in this solution would require a long period of time (assuming the same rate as above, about 13 hours to 1 ppm), but these concentrations are not likely to be encountered for the application envisioned for this immobilized enzyme system. Rinsate waters would have a much lower concentration of pesticide and could be treated in a practical amount of time (discussed in more detail in the next section). The important fact derived from this experiment is that the enzyme is not immediately denatured by commercial formulation ingredients.

Projection to Field Scale. A system is envisioned at the field scale which could successfully detoxify the aqueous waste materials generated during the course of ordinary spray operations. It has been estimated that a single washing of a typical spray plane generates about 100 gallons of waste water (42) containing widely varying concentrations of active ingredients. This wash water, often containing dirt, insect parts, etc., is often unusable as diluent for the next spray run and thus must be disposed of safely. This disposal can be a very expensive proposition; for example, the cost to dispose of liquid toxic waste material can cost more than $600 per 55 gallon drum (Vosburg, C., University of Texas, personal communication, 1989), depending on transportation costs. This cost is for disposal by burning in a cement calcining furnace which cannot accept many waste materials. The costs for more toxic materials, which must be disposed of in a hazardous waste incinerator, can be $1200 (or more) per drum.

A system has been designed to utilize the detoxifying power of the PH enzyme on these waste materials. It is envisioned that the system would be set up locally, either on individual farms and airstrips, or more centralized, for instance, by county, where operators would bring their waste to be treated. A design and cost analysis was carried out for a centralized system, capable of treating about 100 gallons of wastewater per batch. The system would operate in batch recycle mode, just as the laboratory system used in this study. A schematic appears in Figure 10.

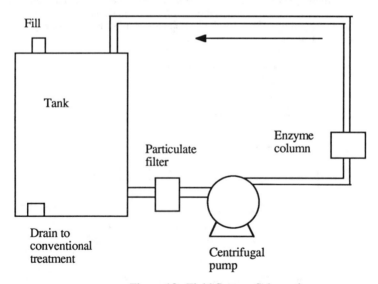

Figure 10. Field System Schematic

A major decision presents itself as to what packing material to use for the column and in what configuration to supply it. The CDI support offers reasonable rates and has one of the longer half-lives of the four supports studied, as much as 80 days. Thus, to be on the safe side, the catalyst should be changed every month. To make the changing easy, the bed would be supplied in a easy-to-handle cartridge configuration which could be readily loaded into a column holder.

Scaleup and Economic Evaluation of System. A rough scaleup calculation is as follows: A 1.5 ml (V_{bed}) column of CDI packing treated 50 ml (V_{res}) of solution at a flow rate (F) of 20 ml/min, reducing the concentration of parathion from 13.5 ppm to less than 1 ppm in two hours. Scaling to a V_{res} of 100 gallons yields a scaleup factor of 7570. Scaling the bed volume directly gives V_{bed} = 7570 x 1.5 ml = 11.4 liters. This is obviously not a practical size for the bed, given that the packing material costs $2400 per liter (laboratory cost-bulk would assuredly be less). If V_{bed} is reduced by a factor of 22.8 for a V_{bed} of 500 ml ($1200 packing cost), the time for treatment must be increased by the same factor, giving a treatment time of about 46 hours. This should not be a problem, since the system can run perfectly well without close attention. The flow rate must also be scaled, to keep the mean residence time, τ, constant for the large column. This assures a volume of liquid has the same contact time with catalyst.

$$\text{for the original bed:} \quad \tau = \frac{V_{bed}}{F} = \frac{1.5 \text{ ml}}{20 \text{ ml/min}} = 0.075 \text{ min}$$

$$\text{for the scaled bed:} \quad F = \frac{V_{bed}}{\tau} = \frac{500 \text{ ml}}{.075 \text{ min}} = 6.7\text{E}3 \text{ ml/min} = 1.8 \text{ gal/min}$$

The flow rate is easily attainable using a common centrifugal pump costing less than $400. The height to diameter ratio of the bed is set by pressure drop considerations. A bed of 8 cm diameter x 10 cm height would have an estimated pressure drop of 27 psia, well within the capacity of inexpensive pumps. Good flow distribution is assured by the use of fritted material as column endpieces.

Capital costs for the system would be minimal, probably less than $2000 including the tank, filter, pump, piping and column enclosure. The major cost of operation would be the enzyme column, which would be replaced periodically. In bulk amounts, the support material would cost about $600 (assuming an at least 50% savings from laboratory cost). In the minicolumns studied here, approximately 2 ml of dialysate solution containing about 20 mg of protein were used for the immobilizations. Scaling to a 500 ml column, about 667 ml of dialysate would be required, containing about 6700 mg of protein. A cost estimate for larger scale production of the enzyme was obtained from the Pennsylvania State University Biotechnology Processing Center (Burkhart, T., personal communication, 1989). Including post-fermentation processing (cell lysis, ammonium sulfate precipitation and dialysis) they estimate a cost of $4500 per 250 liter fermentation batch. This would yield approximately 5 liters of dialysate, sufficient material for about 7.5 columns. However, $4500 is a custom manufacturing cost, with high overhead expenses included. An estimate from Novo Industries, a manufacturer of industrial enzymes, stated that manufacturing at the industrial scale (40,000-300,000 liters) would results in a unit cost savings of about 50% (MacMullin, S., Novo Industries, personal communication, 1989). Thus, on a large scale, the unit protein cost would be about $300 per column and the total materials cost would be about $900. Total manufactured cost is estimated at about two times this figure (*43*), or $1800, the fixed operating cost per month. Assuming 15 batches of material could be treated per month, a breakeven treatment cost would be $1800/15 = $120 per batch of waste, or about $60 per drum (two drums treated per batch). This is an at least 10-fold savings over current treatment costs of >$600 per drum.

It must be stressed that the scaleup calculations above are based upon an experimental system kept at very rigidly controlled conditions, i.e., between runs, the minicolumns were stored under refrigeration in buffer. In addition, a system for monitoring activity on the field scale would have to be devised; for compounds with strong absorbance maxima, a colorometric system could be easily utilized. For the field-scale system to be economical, it would have to be running almost continuously in a non-controlled environment. Thus, the stability projections arrived at above cannot be projected to full field operation with complete confidence. A more detailed study of the temperature coefficient of denaturation of the enzyme would be required on a continuously operating pilot system to thoroughly quantify the scaleup. However, the work of Munnecke (21) offers some hope for the applicability of the system. The author, using an aminoaryl-linked CPG, observed no loss of enzymatic activity over a seventy day period of continuous operation with parathion production waste water.

The question also arises of whether the system as designed here would yield sufficient levels of degradation to meet federal and state standards for disposal. Standards enforced by the Texas Water Commission (Pedde, S., State of Texas Water Commission, personal communication, 1989) for release of pesticide solutions into surface waters are extremely stringent. The standard for parathion is 0.65 parts per billion (ppb), while the standard for chlorpyrifos is 0.83 ppb. Obviously, the system described here could not treat wastes for direct discharge into bodies of water (without large-scale dilution). However, the system could serve as a pretreatment system for wastes before they are discharged into conventional (activated sludge) treatment. The City of Austin, Texas, for example, will not accept pesticide wastes directly, but if they are pretreated, the city sewage facilities will accept toxics at the level of about 0.5 ppm (Gatlin, J., City of Austin, personal communication, 1989). These levels are attainable by the immobilized enzyme system in a practical time frame and the system would meet the guidelines for pretreatment.

Summary

The results of this study have indicated that an immobilized PH enzyme system holds promise for the detoxification at of organophosphates at the end-user level. The largest challenge will be to make the system available to applicators and demonstrate the economy of its use. The users of organophosphates will be of course be reluctant to change from current methods of waste disposal, but increasing pressure at both government and public levels will inevitably force more responsible handling of waste materials. This will motivate additional work to develop practical and cost-efficient waste treatment technology.

Literature Cited

1. World Health Organization; *Organophosphorus Insecticides: A General Introduction*, Environmental Health Criteria 63; Geneva, Switzerland, 1986.
2. United States International Trade Commission; *Synthetic Organic Chemicals, U.S. Productions and Sales, 1985*; USITC Pub. No. 1892; U.S. Government Printing Office: Washington, DC, 1986; p 229.
3. McEwen, F.L.; Stephenson, G.R.; *The Use and Significance of Pesticides in the Environment*; New York: Wiley-Interscience, 1979.
4. Krueger, R.F.; Seiber, J.N.; "Introduction"; In *Treatment and Disposal of Pesticide Wastes*; Krueger, R.F.; Seiber, J.N.; ACS Symposium Series 259; American Chemical Society: Washington, DC, 1984, p xi.

5. Krueger, R.F.; "Federal Regulation of Pesticide Disposal"; In *Pesticide Waste Disposal Technology*; Bridges, J.S.; Dempsey, C.R.; Pollution Technology Review No. 148; Noyes Data: Park Ridge, NJ, 1988, pp 23-34.
6. SCS Engineers; *Disposal of Dilute Pesticide Solutions*, EPA Contract SW-174c; Long Beach, CA; 1979.
7. Hall, C.V.; "Pesticide Waste Disposal in Agriculture"; In *Treatment and Disposal of Pesticide Wastes*; Krueger, R.F.; Seiber, J.N.; ACS Symposium Series No. 259; American Chemical Society: Washington, DC, 1984, pp 27-36.
8. Winterlin, W.L.; Schoen, S.R.; Mourer, C.R.; "Disposal of Pesticide Wastes in Lined Evaporation Beds"; In *Treatment and Disposal of Pesticide Wastes*; Krueger, R.F.; Seiber, J.N.; ACS Symposium Series No. 259; American Chemical Society: Washington, DC, 1984, pp 97-116.
9. Kobylinski, E.A.; Dennis, W.H.; Rosencrance, A.B.; "Treatment of Pesticide-Laden Wastewater by Recirculation Through Activated Carbon"; In *Treatment and Disposal of Pesticide Wastes*; Krueger, R.F.; Seiber, J.N.; ACS Symposium Series No. 259; American Chemical Society: Washington, DC, 1984, pp 125-151.
10. Dennis, W.H.; "A Practical System of Treat Pesticide-Laden Wastewater"; In *Pesticide Waste Disposal Technology*; Bridges, J.S.; Dempsey, C.R.; Pollution Technology Review No. 148; Noyes Data: Park Ridge, NJ, 1988, pp 50-54.
11. Kearney, P.C.; Zeng, Q.; Ruth, J.M.; "A Large Scale UV-Ozonation Degradation Unit: Field Trials on Soil Pesticide Waste Disposal"; In *Treatment and Disposal of Pesticide Wastes*; Krueger, R.F.; Seiber, J.N.; ACS Symposium Series No. 259; American Chemical Society: Washington, DC, 1984, pp 195-209.
12. Dillingham, S. *Insight* December 4, **1989**, 25.
13. Flechas, F.W.; "Resource Conservation and Recovery Act (RCRA) Permitting of On-Site Pesticide Waste Storage and Treatment"; In *Pesticide Waste Disposal Technology*; Bridges, J.S.; Dempsey, C.R.; Pollution Technology Review No. 148; Noyes Data: Park Ridge, NJ, 1988, pp 270-275.
14. Munnecke, D.M.; Johnson, L.M.; Talbo, H.W.; Barik, S.; "Microbial Metabolism and Enzymology of Selected Pesticides"; In *Biodegradation and Detoxification of Environmental Pollutants*; Chakrabarty, A.M.; CRC Press: Boca Raton, FL,1982, pp 1-32.
15. Hsieh, D.P.H.; Munnecke, D. "Accelerated Microbial Degradation of Concentrated Parathion"; In *Fermentation Technology Today*; Terui, G.; Society of Fermentation Technology: Tokyo, Japan, 1972, pp 551-554.
16. Munnecke, D.M.; Hsieh, D.P.H. "Microbial Decontamination of Parathion and *p*-Nitrophenol in Aqueous Media"; *App. Microbiol.* **1974**, *28*, 212.
17. Munnecke, D.M.; Hsieh, D.P.H.; "Microbial Metabolism of a Parathion-xylene Pesticide Formulation"; *App. Microbiol.* **1975**, *30*, 575.
18. Munnecke, D.M.; Hsieh, D.P.H.; "Pathways of Microbial Metabolism of Parathion"; *App. Environ. Microbiol.* **1976**, *31*, 63.
19. Munnecke, D.M.; "Enzymic Hydrolysis of Organophosphate Insecticides, a Possible Pesticide Disposal Method"; *App. Environ. Microbiol.* **1976**, *32*, 7.
20. Munnecke, D.M.; "Properties of an Immobilized Pesticide-Hydrolyzing Enzyme"; *App. Environ. Microbiol.* **1977**, *33*, 503.
21. Munnecke, D.M.; "Hydrolysis of Organophosphate Insecticides by an Immobilized Enzyme System"; *Biotechnol. Bioeng.* **1979**, *21*, 2247.
22. Munnecke, D.M.; "Enzymic Detoxification of Waste Organophosphate Pesticides"; *J. Agric. Food Chem.* **1980**, *28*, 105.
23. Barik, S.; Munnecke, D.M.; "Enzymatic Hydrolysis of Concentrated Diazinon in Soil"; *Bull. Environm. Contam. Toxicol.* **1982**, *29*, 235-239.

24. Honeycutt, R.; Ballantine, L.; LeBaron, H.; Paulson, D.; Seim,V.; "Degradation of High Concentrations of a Phosphorothioic Ester by Hydrolase"; In *Treatment and Disposal of Pesticide Wastes*; Krueger, R.F.; Seiber, J.N.; ACS Symposium Series No. 259; American Chemical Society: Washington, DC, 1984, pp 343-352.
25. Serdar, C.M.; Gibson, D.T.; Munnecke, D.M.; Lancaster, J.H.; "Plasmid Involvement in Parathion Hydrolysis by *Pseudomonas diminuta*"; *Appl. Environ. Microbiol.* **1982**, *44*, 246.
26. Serdar, C.M.; Gibson, D.T.; "Enzymic Hydrolysis of Organophosphates: Cloning and Expression of a Parathion Hydrolase Gene from *Pseudomonas diminuta*"; *Bio/Technology* **1985**, *3*, 567.
27. McDaniel, C.S; Harper, L.L.; Wild, J.R.; "Cloning and Sequencing of Plasmid-borne Gene (opd) Encoding a Phosphotriesterase"; *J. Bacteriol.* **1988**, *170*, 2306.
28. Serdar, C.M.; Murdock, D.C.; Rohde, M.F.; "Parathion Hydrolase Gene from *Pseudomonas diminuta* MG: Subcloning, Complete Nucleotide Sequence, and Expression of the Mature Portion of the Enzyme in *Escherichia coli*"; *Bio/Technology* **1989**, *7*, 1151.
29. Karns, J.S.; Muldoon, M.T.; Kearney, P.C.; "A Biological/Physical Process for the Elimination of Cattle-Dip Pesticide Wastes"; In *Pesticide Waste Disposal Technology*; Bridges, J.S.; Dempsey, C.R.; Pollution Technology Review No. 148; Noyes Data: Park Ridge, NJ, 1988, pp 215-220.
30. Mulbry, W.W.; Karns, J.S.; "Purification and Characterization of Three Parathion Hydrolases from Gram-Negative Bacterial Strains"; *App. Env. Microbiol.* **1989**, *55*, 289.
31. Mason, R.D.; Weetall, H.H.; "Invertase Covalently Coupled to Porous Glass. Preparation and Characterization"; *Biotechnol. Bioeng.* **1972**, *14*, 637.
32. Pierce Chemical Company; *1989 Handbook and General Catalog*; Rockford, IL, p. 58.
33. Hitchcock-DeGregory, S.T.; Yiengen, B.E.; "Ligand-Independent Binding of Serum Protein to Sepharose 4B Activated using Cyanogen Bromide and Blocked with Ethanolamine"; *Biotechniques* **1984**, Nov/Dec, 326-332.
34. Bio-Rad Laboratories; *Price List L*; Richmond, CA, 1986, pp. 52-55.
35. Crescent Chemical Company; *Instruction Manual for VA-Epoxy Biosynth*; Hauppenhauge, NY, 1988.
36. Pierce Chemical Company; *Instructions for Coomassie Protein Assay Reagent*; Pub. no. 23200; Rockford, IL; 1988.
37. Pierce Chemical Company; *Instructions for Disposable Polystyrene Columns*; Pub. no. 29920; Rockford, IL; 1988.
38. Smith, J.M.; *Chemical Engineering Kinetics*; 3rd ed.; McGraw-Hill: New York, 1981, pp. 199-202.
39. Havens, P.L.; *Detoxification of Organophosphorus Pesticide Solutions: An Immobilized Enzyme System*; Ph.D. dissertation; The University of Texas at Austin: Austin, Texas, **1990**.
40. Satterfield, C.N.; *Mass Transfer in Heterogeneous Catalysis*; MIT Press: Cambridge, MA, 1970, p 33.
41. Parikh, I.; March, S.; Cuatrecasas, P.; "Topics in the Methodology of Substitution Reactions with Agarose"; In *Methods in Enzymology*; Mosbach, K.; vol. 34; Academic Press: New York, 1974, pp 77-102.
42. Keane, W.T.; "Applicator Disposal Needs"; In *Pesticide Waste Disposal Technology*; Bridges, J.S.; Dempsey, C.R.; Pollution Technology Review No. 148; Noyes Data: Park Ridge, NJ, 1988, pp 13-15.
43. Peters, M.S.; Timmerhaus, K.D.; *Plant Design and Economics for Chemical Engineers*, 3rd ed.; McGraw-Hill: New York, 1980, pp 207-208.

44. Nye, J.C.; "Treating Pesticide-Contaminated Wastewater: Development and Evaluation of a System"; In *Treatment and Disposal of Pesticide Wastes*; Krueger, R.F.; Seiber, J.N.; ACS Symposium Series No. 259; American Chemical Society: Washington, DC, 1984, pp 153-160.
45. United States Environmental Protection A gency; *Understanding the Small Quantity Generator Hazardous Waste Rules: A Handbook for Small Business*; EPA/530-SW-86-019; U.S. Government Printing Office: Washington, DC, 1986.

RECEIVED April 5, 1991

Chapter 14

Sequencing Batch Reactor Design in a Denitrifying Application

Basil C. Baltzis, Gordon A. Lewandowski, and Sugata Sanyal[1]

Department of Chemical Engineering, Chemistry, and Environmental Science, New Jersey Institute of Technology, Newark, NJ 07102

A mathematical model has been developed to describe biological denitrification in a sequencing batch reactor (SBR). The model assumes noninhibitory kinetics for nitrate reduction, inhibitory kinetics for nitrite reduction, and incorporates the possible toxicity effect of nitrite on the biomass. The model has been successful in qualitatively describing experimental data from a 1200 gallon pilot unit. Extensive sensitivity analyses have been performed to indicate the kinetic parameters that have the greatest impact on reactor performance. Numerical results have also indicated ways of properly selecting the operating parameters (hydraulic residence time, time devoted to fill, fraction of reactor contents to be emptied during draw-down), in order to achieve complete denitrification and avoid high nitrate and nitrite concentrations during operation. The results of the model are currently being applied to the design of a facility to denitrify a high strength 0.42 MGD munitions waste stream.

Sequencing batch reactors (SBR) are known to offer a number of advantages over conventional continuous flow systems (9, 16), such as cycling between anoxic and aerobic periods of operation, effective process and quality control, and the fact that they do not require a separate clarifier. For cases of aerobic biodegradation, mathematical models, verified by experiments, have shown that SBRs can also have a greater volumetric efficiency than continuous flow reactors (3), achieving the same level of treatment, at an equivalent throughput, in a much smaller volume. The experience gained in this earlier work enabled us to modify the equations for a denitrifying application.

[1]Current address: Allied-Signal, Inc., Morristown, NJ 07962-1087

0097-6156/91/0468-0282$06.00/0

Reported Rates for Nitrate and Nitrite Removal

Rates of nitrate and nitrite removal in batch and continuous flow reactors have been reported by various researchers (*2, 5-8, 12*). It has been reported that (when reduction is dissimilatory), nitrite is an intermediate (*11*) which may accumulate in the system (*4,15*). This nitrite build-up is undesirable since it inhibits the overall rate of denitrification (*14*). Nitrogen removal in SBR units has also been reported in the literature (*9, 13, 16*), although these studies involved nitrification/denitrification with activated sludge and municipal wastewaters.

Purpose of the Present Study

The present study is for a SBR designed to denitrify a high strength industrial waste. The intent was to model the bio-denitrification process, verify the model with available experimental data, and use the model to optimally design the unit.

Assumptions Made in Deriving the Model

In deriving the model, the SBR operating cycle was assumed to consist of only three phases, namely, fill, "react", and draw-down, all of which operate anaerobically. The volume variation during the cycle is shown in Figure 1. The inlet volumetric flow rate during the fill phase, and the outlet volumetric flow rate during the draw-down phase are constant, although not necessarily equal to one another. Reaction occurs throughout the cycle; thus, the "react" phase is simply the period during which there is neither an input to, nor an output from the reactor (pure batch mode). Settling is neglected in this model, which is based on kinetics only, and does not incorporate physical phenomena. This means that solids will be lost during draw-down that will have to be made up in growth. Therefore, the model is likely to underpredict field performance. Nitrate and nitrite are assumed to be the only substances exerting rate limitation on the process. Therefore, the carbon source (usually methanol) and phosphorous are in excess. The rate of nitrate reduction is described by Monod's (non-inhibitory) expression, while the rate of nitrite consumption is described by Andrews' (inhibitory) expression. Provision is also made in the model to incorporate the possibility of nitrite exerting a toxic effect (deactivation) on the biomass. It is also assumed that the unit operates under isothermal conditions.

Mathematical Description of the Process

Based on the assumptions described above, the general equations describing bio-denitrification in a SBR are the following:

$$\frac{dV}{dt} = Q_f - Q \tag{1}$$

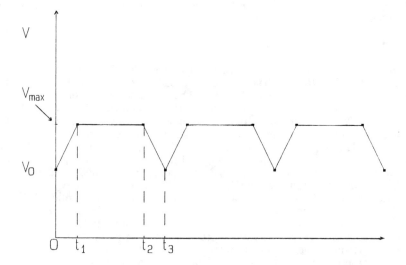

Figure 1.   Reactor contents volume variation with time.

$$\frac{ds}{dt} = \frac{Q_f}{V} (s_f - s) - \frac{b}{Y_1} \mu_1 \tag{2}$$

$$\frac{dp}{dt} = \frac{Q_f}{V} (p_f - p) + \alpha\mu_1 b - \frac{b}{Y_2} \mu_2 \tag{3}$$

$$\frac{db}{dt} = - \frac{Q_f}{V} b + (\mu_1 + \mu_2)b - kpb \tag{4}$$

where,

$$\mu_1 = \frac{\mu_{max} s}{K + s} , \text{ and } \quad \mu_2 = \frac{\mu^* p}{K' + p + p^2/K_i}$$

Equation 1 represents an overall mass balance for the reactor contents, while Equations 2 through 4, represent mass balances for the nitrate, nitrite, and biomass, respectively. The specific form of the equations is different for each phase of the cycle, since some of the terms (those involving volumetric flow rates) may be equal to zero. The last term in Equation 4 represents the loss of (active) biomass due to the possible toxic effect of the nitrite.

Dimensionless Form of the Model

The model equations have been reduced to dimensionless form, in order to decrease the number of parameters. This was done by scaling the parameters and variables as follows :

$$u = \frac{s}{K}, \quad u_f = \frac{s_f}{K}, \quad x = \frac{b}{Y_1 K}, \quad \omega = \frac{K'}{K}, \quad \varphi = \frac{\mu^*}{\mu_{max}},$$

$$\gamma = \frac{K}{K_i}, \quad \eta = \frac{Y_1}{Y_2}, \quad v = \frac{p}{K}, \quad v_f = \frac{p_f}{K}, \quad \rho = \alpha Y_1,$$

$$Q'_f = \frac{Q_f}{Q^*_f \sigma_1}, \quad Q' = \frac{Q}{Q^*_f \sigma_1}, \quad \sigma_1 = \frac{t_1}{t_3}, \quad \sigma_2 = \frac{t_2 - t_1}{t_3},$$

$$\sigma_3 = \frac{t_3 - t_2}{t_3}, \quad V' = \frac{V}{V_{max}}, \quad \beta = \frac{\mu_{max} V_{max}}{Q^*_f \sigma_1}, \quad \delta = \frac{V_o}{V_{max}}$$

$$\varepsilon = kK\mu_{max}, \quad \vartheta = \frac{tQ^*_f \sigma_1}{V_{max}}$$

It can be shown that the time (in dimensionless form) devoted to each phase of the cycle is the following:

Fill: $0 \leq \vartheta \leq (1 - \delta)\sigma_1$ (5)

React : $(1 - \delta)\sigma_1 \leq \vartheta \leq (1 - \delta)(1 - \sigma_3)$ (6)

Draw-down : $(1 - \delta)(1 - \sigma_3) \leq \vartheta \leq 1 - \delta$ (7)

The total time period, τ, of the cycle is equal to $1 - \delta$.
Therefore, Equations 2 through 4 take on the following dimensionless form:

Fill-phase:

$$\frac{du}{d\vartheta} = \frac{u_f - u}{\delta\sigma_1 + \vartheta} - \beta x\lambda_1 \qquad (8)$$

$$\frac{dv}{d\vartheta} = \frac{v_f - v}{\delta\sigma_1 + \vartheta} + (\rho\lambda_1 - \eta\lambda_2)\beta x \qquad (9)$$

$$\frac{dx}{d\vartheta} = \frac{- x}{\delta\sigma_1 + \vartheta} + (\lambda_1 + \lambda_2)\beta x - \varepsilon\beta xv \qquad (10)$$

React and Draw-down phases:

$$\frac{du}{d\vartheta} = - \beta x\lambda_1 \qquad (11)$$

$$\frac{dv}{d\vartheta} = (\rho\lambda_1 - \eta\lambda_2)\beta x \qquad (12)$$

$$\frac{dx}{d\vartheta} = (\lambda_1 + \lambda_2)\beta x - \varepsilon\beta xv \qquad (13)$$

where,

$$\lambda_1 = \frac{u}{1 + u} , \text{ and } \lambda_2 = \frac{\varphi v}{\omega + v + \gamma v^2}$$

It should be mentioned that the volume variation has been incorporated wherever needed in Equations 8 through 13, and thus Equation 1 does not appear in the dimensionless formulation of the model. By making the equations dimensionless, the number of model parameters has been reduced from 17 in the original formulation to 11 in the final one, which substantially reduces the amount of numerical work needed for sensitivity studies.

Special Case of the SBR Operation.   When the fill and draw-down times are insignificant, the SBR becomes a conventional batch reactor, operated in a cyclic fashion. Although the derivation is slightly different in this case, it can be shown that the batch operation is described by Equations 11 through 13, provided that one sets $\beta = 1$ and $\vartheta = t\mu_{max}$.

Reaching the Steady Cycle. After operation over a period of time (the extent of which depends on the start-up conditions), the unit (SBR or batch) reaches the steady cycle, in which all concentrations repeat their temporal variation. This is expressed as:

$$e_{n+1}(\vartheta) = e_n(\vartheta), \quad 0 \leq \vartheta \leq \tau \quad (14)$$

e being any concentration (i.e., u, v, or x). The subscript indicates the number of the cycle. Equation 14 was used in the numerical studies as the criterion (within an accuracy of 10^{-5}) for checking whether the system reached its steady cycle of operation.

Numerical Results and Discussion

The results presented here, are based on Equations 8 through 14. These equations were used in extensive numerical studies in order to reveal all types of behavior of the system, and to compare the theoretical predictions of the model with available experimental data.

Numerical Method and Procedure

The computer simulations were performed by using a 4th-order Runge-Kutta integration routine. No particular problems of numerical stiffness were encountered. In studying the effect of a parameter (e.g., δ) on the system, the following procedure was used: The values of all parameters were fixed, and it was assumed that at time zero the concentrations of nitrate and nitrite in the reactor are zero, while the concentration of the biomass was set at a positive value. The equations were integrated up to the point where the steady cycle (also known as the limit cycle) was reached. The criterion used to decide convergence to the steady cycle was described earlier in this paper. This first integration never took more than 10 to 15 cycles in order to reach the limit cycle. Subsequently, all parameters were kept fixed, except the one under study. The latter (e.g., δ), was slightly varied, and the integration was repeated by using as initial conditions the values at the end of the steady cycle of the previous integration. The system usually reached the new limit cycle after integration for two or three cycles. More integration time may be needed if the value of the parameter under study is significantly altered. Using this approach (known as continuation), the computer time needed for the simulations was kept at low levels. The integrations were performed on a VAX 11/780 computer.

Possible Outcomes and Operating Diagram

Because of the fact that settling of solids is not incorporated in the model, during the draw-down phase a known amount of biomass is lost. Unless that amount of biomass is made-up by growth during the other phases of the cycle, the biomass will eventually wash out of the system and operation will be disrupted. Loss (or washout) of biomass can be prevented if the mean hydraulic residence time is long enough. This is shown in Figure 2. In this diagram, β (a dimensionless measure of the of the hydraulic residence time) and u_f (the dimensionless concentration of nitrate in the untreated waste) have been varied, while for the other parameters the following values were used: $\omega = 1.2$, $\sigma_1 = 0.1$, $\gamma = 0.05$, $\delta = 0.5$, $v_f = 0$, $\varepsilon = 0$, $\varphi = 1.5$, $\rho = 0.25$ and $\eta = 0.6$. This diagram indicates when survival of the biomass has been achieved at steady cycle. For a given value of u_f, β (which can be set by properly picking the volume of the reactor and the feed flow rate during the fill) must be selected so that it falls in Region II in order to guarantee survival of the biomass. If it falls in Region Region I, the biomass will wash out of the system. If β falls in Region III (for a given feed concentration, u_f), the biomass will be either maintained or washed out, depending on how the system is started-up.

Region of Multiple Outcomes

Some further clarification is needed for Region III of the operating diagram. It is known (1) that in continuous culture under inhibitory kinetics, there are conditions of operation that lead to multiple steady states. In fact, the chemostat (based on steady state analysis), can reach three steady states. One in which the culture is washed-out, and two in which the biomass is established in the reactor. Out of the two states of survival, one implies high substrate concentrations (which physically correspond to severe inhibition), and one implies growth under relatively uninhibited conditions (known as the normal steady state). When the stability of these states is studied, it turns out that the wash-out and the normal state of survival are locally stable, while the inhibited state of survival is locally unstable. The notion of local stability implies the following: Suppose that the chemostat operates at the normal state of survival. Minor fluctuations in the values of the variables will decay and the system will again reach the same steady state. On the other hand, severe operational upsets will lead the system to go away from the normal state of survival, and it will in fact reach another locally stable state (i.e., in this case the culture will be washed-out). The inhibited state of survival will never be physically realized (unless one applies proper control action), because the ever existing small fluctuations will cause the system to go away from it. What has been reported for the chemostat case, happens here for the SBR, if the operational conditions are selected in such a way that they fall in Region III of Figure 2. In the SBR case instead of steady states, there are limit cycle (periodical) solutions. In fact, in Region III the system has three solutions: a wash-out solution, and two limit cycles of biomass establishment, one of which is unstable. The unstable limit cycle is in fact undesirable, because it leads to high concentrations of nitrate and nitrite (i.e., low levels of waste treatment). Thus,

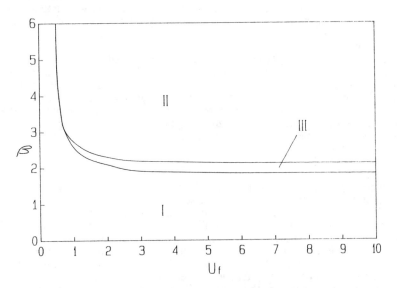

Figure 2. Operating diagram for SBR.

there is no point for applying control action in order to stabilize (and reach) this cycle. Numerically, one can see this unstable cycle if integration of the model equations is performed backwards in time, that is if one sets -t wherever t appears in the model. Whether the system will reach the normal cycle of operation or it will be washed-out, depends on the start-up conditons of the unit, and more specifically on the amount of biomass which is originally present in the reactor. The fact that an SBR can have multiple outcomes depending on the start-up conditons, has been experimentally verified (Baltzis, Lewandowski, and Ko, unpublished results) in a 4-liter unit where a culture of *Pseudomonas Putida* was used to biodegrade phenol under aerobic conditions. If conditions in Region III are selected, operational upsets, or even relatively small fluctuations, may lead to loss of biomass even if the strat-up conditions were properly chosen. Since multiple outcomes have never been reported for non-inhibited biological systems, the existence of Region III is attributed to the inhibitory kinetics assumed for nitrite removal.

A Note on the Settling Phase. It should be added that although in practice there is always a settling phase (which has been neglected here), and thus no substantial loss of biomass during the draw-down phase, biomass (solids) is periodically discharged from the unit. The type of diagram shown in Figure 2 can be used for predicting the amount of biomass that can be discharged without creating problems for the subsequent operation of the unit.

Parameter Sensitivity Studies

The effects of various model parameters have been investigated, and are presented here in a series of diagrams. The parameters can be classified into two categories: operating parameters (i.e., those that can be selected in the different opetating scenarios), and system parameters (i.e., those that depend on the kinetic rate constants, characteristics of the biomass, etc.). Although one can affect the system parameters to some extent (via temperature, pH, etc.), they cannot be freely selected for operation. Since concentrations vary during the cycle, it was decided to compare the effect of different parameters by plotting the maximum concentration of nitrate and nitrite during the steady cycle.

Effects of Operating Parameters

Figures 3 and 4 show the effect of operating parameters. These diagrams have been prepared for the case where $\varphi = 2$, $\gamma = 2$, $\varepsilon = 0$, $\omega = 10$, $\eta = 2$, $\rho = 0.75$, and $u_f = 40$. When $\sigma_1 = 0.5$, $v_f = 2$, $\delta = 0.5$, the effect of β, which is a measure of the hydraulic residence time, is shown in diagrams (a) and (b) of Figure 3. As expected, the nitrate concentration drops as β increases. However, nitrite is produced as a result of nitrate reduction. As diagram (b) of Figure 3 indicates, at low residence times, the maximum nitrite concentration is very high, since there is not enough time for its depletion. Low residence times would lead to possibly acceptable

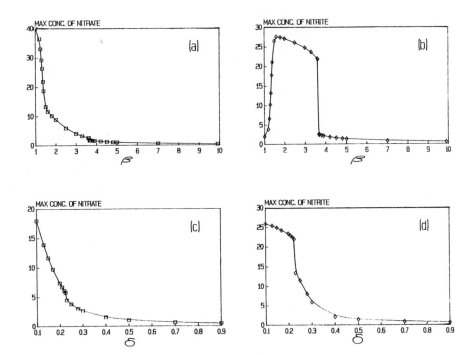

Figure 3. SBR maximum nitrate and nitrite concentration dependence on β and δ.

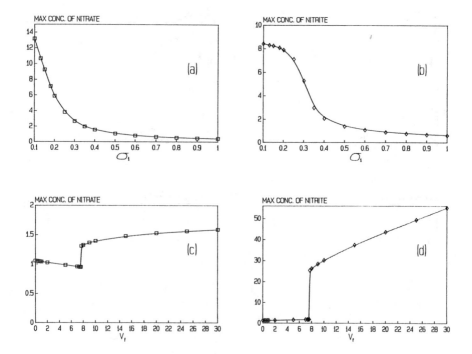

Figure 4.   SBR maximum nitrate and nitrite concentration
dependence on σ_1 and v_f.

nitrate levels, but definitely unaceptable nitrite levels. From Figure 3, a value of $\beta = 5$ seems to be appropriate.

When $\beta = 5$, $\sigma_1 = 0.5$, and $v_f = 2$, the effect of δ, which is the fraction of the reactor contents emptied during draw-down, is indicated in diagrams (c) and (d) of Figure 3. It should be noted here that the maximum concentrations are not the concentrations at the end of the cycle (i.e., those in the effluent). Effluent concentrations may be low, while they are quite high in the reactor during periods of the cycle. This is undesirable, particularly for nitrite, since it is inhibitory and potentially toxic. As the graphs indicate, a value of $\delta = 0.5$ seems to be appropriate.

Diagrams (a) and (b) of Figure 4 show the effect of σ_1, which is the fraction of cycle time devoted to fill, when $\beta = 5$, $\delta = 0.5$ and $v_f = 2$. The indication is that a fast fill should not be used in order to avoid high nitrite concentrations. In fact, it seems that the waste should be fed to the reactor over a period equal to half the cycle time ($\sigma_1 = 0.5$).

Diagrams (c) and (d) of Figure 4 show the effect of v_f, which is the concentration of nitrite in the untreated waste, when $\beta = 5$, $\delta = 0.5$, and $\sigma_1 = 0.5$. As can be seen , values of v_f up to about 8 have almost no effect on either the nitrate or the nitrite maximum concentration. Higher v_f values result in increases of both nitrate and nitrite. This can be attributed to high concentrations at the beginning of the cycle, which inhibit the available biomass. If v_f were greater than 8, either the waste would have to be pre-diluted, or σ_1 would have to be increased. However, in the application for which the model has been developed (munitions waste), v_f will be on the order of 0.1.

Effects of System Parameters

Diagrams (a) and (b) of Figure 5 show the effect of ε, which is a system parameter and a measure of the toxic impact that nitrite may have on the biomass. For low values of ε (up to 0.05), there is almost no effect. As ε increases, the maximum value of nitrate concentration increases up to a value equal to the nitrate concentration in the untreated waste. This implies (as expected) that the toxicity level is so high that no reduction of nitrate occurs. The same things happen for nitrite (Figure 5b). At intermediate values of ε, the nitrite concentration increases because some nitrate is reduced. At high values of ε, the process stops due to complete inactivation of the biomass, and the nitrite concentration is equal to that in the untreated waste. This concentration is assumed low, and thus the drop shown in the diagrams. We anticipate, however, based on experimental data (*10*), that the value of ε will be low.

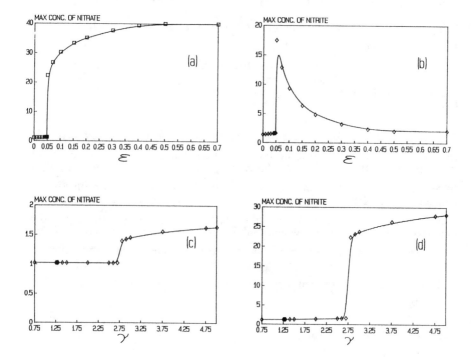

Figure 5.   SBR maximum nitrate and nitrite concentration dependence on ε and γ.

The effect of γ, which is a measure of the inhibition exerted by nitrite on the biomass, is shown in diagrams (c) and (d) of Figure 5. Values of γ up to 2.75 seem to have no effect (inhibition is insignificant), while higher values of γ can have a significant effect on nitrite concentrations. These results are expected from the biology of the system. Except for ε and γ which were varied, the parameter values used in preparing the diagrams of Figure 5 are the same with those used for the diagrams of Figures 3 and 4, with $\beta = 5$, and $\delta = 0.5$.

Experimental Data and the Model

It has been already established (*10*) that the model can, at least qualitatively, predict batch experimental data. An example is given in Figure 6. The data (Figure 6a) are from a 1200-gallon pilot reactor operated in batch mode. Figure 6b shows the model simulation (also in batch mode), for parameter values as follows: $\varphi = 2$, $\gamma = 2$, $\omega = 10$, $\eta = 2$, $\rho = 0.75$, $\varepsilon = 0.$, $\delta = 0.5$, $\tau = 1.1$, $u_f = 40$, and $v_f = 2$. The comparison is very good, given the   complexity of the actual process, and the relative simplicity of the model equations. We have compared many more sets of pilot data, with similar results.

An example of the importance of modeling is given in Figure 7. Figure 7a is the same batch simulation shown in Figure 6b. The actual pilot reactor was run in a discrete batch mode. Figure 7c shows what the effluent concentrations would have been at steady cycle if the reactor had been operated cyclically (with instantaneous fill and draw). This shows that even though the final nitrite concentration in a discrete batch was probably acceptable, it leads to an unacceptable nitrite build-up after several cycles.

By contrast, Figures 7b and 7d show the concentration profiles for a fill period equal to 50% of the total cycle time. For these diagrams $\beta = 5$, $\sigma_1 = 0.5$, while all other parameters are the   same as in the diagram   of Figure 6b. Figure 7d shows that during the steady cycle, nitrate and   nitrite concentrations fall off to zero. In addition, note the differences in scale. The peak concentration in Figure 7b is five times smaller than for the comparable batch case (due to slower feeding), and the steady-cycle peak concentration   in Figure 7d is 20 times smaller than for the batch case (Figure 7c). These predictions are now being tested in the 1200-gallon pilot reactor.

Conclusions and Future Work

It can concluded that a powerful model has been derived for describing the biodenitrification process in a sequencing batch reactor. Comparison with pilot-scale batch reactor performance indicates that the model can effectively descirbe trends in the concentration data. The most important aspect of the model is its ability to indicate   the impact of changes in the operating and design parameters, and the location of optimum conditions. These predictions, if verified, can have important consequences for the design and operation of denitrifying SBRs. We are currently in the

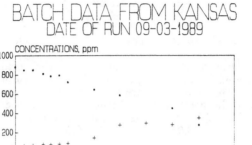

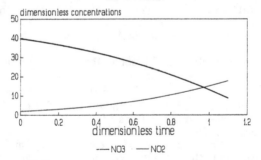

Figure 6. Qualitative comparision of experimental data with model prediction.

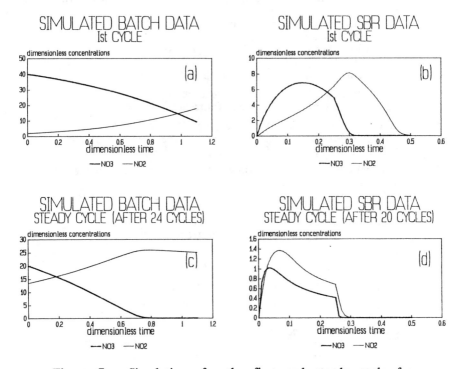

Figure 7. Simulations for the first and steady cycle for batch and SBR operation.

process of experimentally determining the specific kinetic rate constants for a munitions waste application, and will continue to experimentally explore the model predictions using bench-scale reactors.

Notation

b : concentration of biomass in the reactor

e_n : concentration of any quantity in the reactor during the n-cycle

e_{n+1} : concentration of any quantity in the reactor during the n+1-cycle

k : constant associated with the rate of biomass loss due to toxic effect exerted by the nitrite

K : saturation constant in the Monod expression

K' : constant in the Andrews model

K_i : inhibition constant associated with the Andrews model

p : concentration of nitrite in the reactor

p_f : concentration of nitrite in the (untreated) waste

Q : volumetric flow rate of the reactor effluent

Q' : dimensionless volumetric flow rate of the reactor effluent

Q_f : volumetric flow rate of waste input to the reactor

Q'_f : dimensionless volumetric flow rate of waste input to the reactor

Q^*_f : volumetric flow rate of waste input to the reactor during the fill phase

s : concentration of nitrate in the reactor

s_f : concentration of nitrate in the (untreated) waste

t : time

t_1 : time indicating the end of the fill-phase

t_2 : time indicating the end of the "react"-phase

t_3 : time indicating the end of the draw-down-phase, and the duration of the cycle.

u : dimensionless concentration of nitrate in the reactor

u_f : dimensionless concentration of nitrate in the (untrated) waste

V : volume of the reactor contents

V_0 : volume of the reactor contents in the beginning of the cycle

V_{max} : volume of the reactor contents during the "react" phase

V' : dimensionless volume of the reactor contents

v : dimensionless concentration of nitrite in the reactor

v_f : dimensionless concentration of nitrite in the (untreated) waste

x : dimensionless concentration of biomass in the reactor

Y_1 : constant yield coefficient, expressing the amount of biomass produced per unit amount of nitrate reduced

Y_2 : constant yield coefficient, expressing the amount of biomass produced per unit amount of nitrite depleted

Greek Symbols

α : constant, expressing the amount of nitrite produced per unit amount of biomass produced due to nitrate reduction

β : dimensionless measure of the hydraulic residence time

γ : dimensionless inverse inhibtion constant associated with the Andrews model

δ : fraction of the reactor contents emptied at the end of a cycle

ε : dimensionless constant associated with the rate of biomass loss due to toxic effect exerted by the nitrite

η : ratio of the yield coefficients Y_1 and Y_2

ϑ : dimensionless time

λ_1 : dimensionless version of the Monod model

λ_2 : dimensionless version of the Andrews Model

μ_1 : Monod's model

μ_2 : Andrews' model

μ_{max} : maximum specific growth rate in the Monod expression

μ^* : constant associated with Andrews' model

ρ : constant indicating the amount of nitrite produced, per unit amount of nitrate reduced

σ_1 : fraction of cycle time devoted to fill phase

σ_2 : fraction of cycle time devoted to "react" phase

σ_3 : fraction of cycle time devoted to draw-down phase

τ : period of the cyclic operation of the reactor

φ : constant, equal to μ^*/μ_{max}

ω : constant, equal to K'/K

Acknowledgements

The authors gratefully acknowledge the support of the U.S. Army, Production Base Modernization Activity (Robert Goldberg, Project Officer) and the assistance of graduate students Y.S. Ko, J.H. Wang and Z. Shareefdeen in preparing the figures and typing the manuscript.

Literature Cited

1. Andrews, J.F.; "A Mathematical Model for the Continuous Culture of Microorganisms Utilizing Inhibitory Substrates"; *Biotech. Bioeng.* **1968**, *10*, 707.

2. Balakrishnan, S.; Eckenfelder, W.W.; "Nitrogen Relationships in Biological Treatment Processes-III: Denitrification in the Modified Activated Sludge Process"; *Water Res.* **1969**, *3*, 177.

3. Baltzis, B.C.; Lewandowski, G.A.; Chang, S.H.; Ko, Y.F.; Fill-and-Draw Reactor Dynamics in Biological Treatment of Hazardous Waste"; In *Biotechnology Applications in Hazardous*

Waste Treatment; Lewandowski, G.; Armenante, P.; Baltzis, B.; Engineering Foundation: New York, NY, 1989, 111-128.

4. Betlach, M . R.; Tiedje , J.M.; "Kinetic Explanation for Accumulation of Nitrite, Nitric Oxide, and Nitrous Oxide During Bacterial Denitrification"; *App. Env. Microbiol.,* **1981,** *42,* 1074.

5. Christianson, C.W.; Rex, E.H.; Webster, W.M. ; Vigil, F.A.; "Reduction of Nitrate Nitrogen by Modified Activated Sludge"; In *US Atomic Energy Commision,* TID-7517 (pt-1A); **1957,** 264.

6. Dawson, R.L.; Murphy, K.L.; "The Temperature Dependency of Biological Denitrification"; *Water Res.* **1972,** *6,* 71.

7. Francis, C.S.; Callahan, M.W.; "Biological Denitrification and its Application in Treatment of High-Nitrate Wastewater"; *J. Env. Quality* **1975,** *4,* 153.

8. Halt r ich, W.; "Elimination of Nitrate from an Industrial Waste"; In *Proceedings, 22nd Purdue Industrial Waste Conference,* Purdue University, West Lafayette, IN,**1967,** 203

9. Irvine, R.L.; Ketchum, L.H.; Breyfogle, R.E.; Barth, E.F.; "Municipal Application of Sequencing Batch Treatment at Culver, Indiana"; *J. Water Pollut. Control Fed.* **1983,** *55,* 484.

10. Lewandowski, G.A.; Baltzis, B.C.; Sanyal, S.; Radhakrishnan S.; Yamoza, J.; Carrazza, J.; "Engineering Study of a Sequencing Batch Reactor for Denitrification of Munition Wastes"; In *Proceedings, 14th Annual Army Environmental R&D Symposium,* Williamsburg, VA; USATHAMA Report No. CETHA-TE-TR 90055, 1990, 315-327.

11. Mc Carty, P.L; Beck, L.; St. Amant, P.; "Biological Denitrification of Wastewater by Addition of Organic Materials"; In *Proceedings, 24th Purdue Industirial Waste Conference,* Purdue University, West Lafayette, IN,**1969,** 1279.

12. Monteith, H.D.; Bridle, T.R.; Sutton, P.M.; "Industrial Waste Carbon Sources for Biological Denitrification"; *Progress in Water Tech.* **1980,** *12,* 127.

13. Palis, J.C.; Irvine, R.L.; "Nitrogen Removal in Low-Loaded Single Tank Sequencing Batch Reactor"; *J. Water Pollut. ControlFed.* **1985,** *57,* 82.

14. Randall, C.W.; Buth, D.; "Nitrite Build-up in Activated Sludge Resulting from Combined Temperature and Toxicity Effects"; *ibid.* **1984,** *56,* 1045.

15. Requa , D.A.; Schroeder, E.D.; "Kinetics of Packed Bed Denitrification"; *ibid.* **1973,** *45,* 1696.

16. Silverstein, J.A . ; Sc h roeder, E.D.; "Performance of SBR Activated Sludge Processes with Nitrification/Denitrification"; *ibid.* **1983,** *55,* 377.

RECEIVED April 5, 1991

SOLID WASTE MANAGEMENT

Chapter 15

Contaminant Leaching from Solidified–Stabilized Wastes

Overview

Paul L. Bishop

Department of Civil and Environmental Engineering, University of Cincinnati, Cincinnati, OH 45221

The current state-of-the-art of solidification/stabilization (S/S) technologies is reviewed. This includes the legal impetus and basis for use of solidification/stabilization for hazardous wastes or contaminated soils, the principles and chemistry of contaminant immobilization within the waste form matrix, leaching mechanisms, and environmental factors affecting leachability. It is shown that S/S processes can be very effective at immobilizing certain waste materials, but other wastes may not be amenable to these processes.

Stabilization/solidification (S/S) processes have been developed to concurrently eliminate land disposal of liquid wastes and minimize leaching of the resultant solid waste after disposal. These processes are also being used to remediate existing hazardous waste sites by markedly reducing the rate of leaching of pollutants from contaminated soils and debris.

Land disposal of wastes should not be the primary means of waste disposal if other alternatives are available. Waste reduction, recycle and reuse are much superior alternatives. Where this is not possible, destruction or detoxification options should be considered. There will always be some wastes, though, where these options are not viable. Most reuse or destruction operations will result in some residue which cannot be further reduced and which must be disposed of on land. This includes flyash and bottom ash from incineration processes, mixed metal sludges, foundry sands, heavy metal contaminated soils, etc. Direct land disposal of these wastes could lead to potentially serious consequences, however, if contaminants in the waste leach into ground or surface waters. Many of these wastes can be effectively treated by stabilization/solidification processes so as to minimize leaching to environmentally acceptable levels.

0097–6156/91/0468–0302$06.00/0

Much of the impetus for S/S of hazardous wastes has been provided by the Resource Conservation and Recovery Act (RCRA) of 1976, including the 1984 amendments, and the Comprehensive Environmental Response, Liability and Recovery Act (CERCLA) of 1980, later reauthorized in 1986 as the Superfund Amendments and Reauthorization Act (SARA). RCRA deals primarily with the generation, handling, treatment and disposal of hazardous wastes, while CERCLA and SARA established a massive remedial program for the cleanup of existing sites that threaten the environment.

In 1985, under RCRA authority, the U.S. EPA banned the disposal of bulk hazardous liquids into landfills, necessitating solidification of the waste. Stabilization/solidification technologies have been specified by EPA as "best demonstrated available technologies" for a number of waste streams, and some can be used as a basis for "delisting" a waste as hazardous under RCRA.

Under SARA provisions, permanent treatment of contaminated soil and debris is being emphasized rather than the use of nontreatment containment systems such as covers, grout walls and similar methods. A large number of Superfund sites are now using S/S treatment processes for soil treatment.

Stabilization/solidification technology refers to treatment processes that are designed to (1) improve the handling and physical characteristics of the waste (2) decrease the surface area of the waste mass across which transfer or loss of contaminants can occur, and/or (3) limit the solubility of any hazardous constituents of the waste such as by pH adjustment or sorption phenomena.

Stabilization processes attempt to reduce the solubility or chemical reactivity of a waste by changing its chemical state or by physical entrapment. The hazard potential of the waste is reduced by converting the contaminants to their least soluble, mobile or toxic form. Solidification refers to techniques that encapsulate the waste in a monolithic solid of high structural integrity. Solidification does not necessarily involve a chemical interaction between the wastes and the solidifying reagents, but may mechanically bind the waste into the monolith. Contaminant migration is restricted by vastly decreasing the surface area exposed to leaching and/or by isolating the wastes within a relatively impervious capsule [1].

The most important factor in determining whether a particular stabilization/solidification process is effective in treating a given waste is the reduction in the short- and long-term leachability of the waste [2]. Leaching can be defined as the process by which a component of waste is removed mechanically or chemically into solution from the solidified matrix by the passage of a solvent such as water. Resistance to leaching will depend on both the characteristics of the solidified/stabilized waste and on those of the leaching medium it will come into contact with.

This paper discusses the principles of contaminant immobilization in solidified/stabilized wastes, leaching mechanisms from these materials, factors which affect leaching, and models which can be used to predict short- and long-term leaching rates.

Principles of Immobilization

Stabilization/solidification processes employ systems which both solidify the waste mass and eliminate free liquids, and stabilize the contaminant in their least soluble form. The overall objective is to minimize the rate of leaching of pollutants from the resulting waste form. These processes typically involve the addition of binders and other chemical reagents to the contaminated soil or sludge to physically solidify the waste and chemically bind the contaminants into the monolith.

Binder systems can be placed into two broad categories, inorganic or organic. Most inorganic binder systems in use include varying combinations of hydraulic cements, lime, flyash, pozzolans, gypsum and silicates. Organic binders used or experimented with include epoxy, polyesters, asphalt/bitumen, polyolefins (primarily polyethylene and polybutadiene) and urea formaldehyde. Combinations of inorganic and organic binder systems have also been used. These include diatomaceous earth with cement and polystyrene, polyurethane and cement, polymer gels with silicates, and lime cement with organic modified clays [2].

Most immobilization processes currently in use involve hydraulic cements, such as Portland cement, cement kiln dust, flyash, or other pozzolanic materials. Consequently, the focus of this review will be on cement-based processes.

The main components of cement are lime and silicates. Cementation of the mixture begins when water is added, either directly or as part of the waste being immobilized. First, a calcium-silicate-hydrate gel forms, followed by hardening of the material as thin, densely-packed, silicate fibrils grow and interlace. The hydration reactions form a variety of compounds as the cement paste sets, including calcium hydroxides and calcium silicate hydrates. The latter provides the cement's structural stability, while the former supplies large amounts of entrapped alkaline material.

The water-to-cement ratio (W/C) is very important to the properties of the final product. The volume of the cement approximately doubles upon hydration, creating a network of very small gel pores. The volume originally occupied by the added water forms a system of much larger capillary pores. As the water-cement ratio increases, the percentage of larger pores increases, substantially increasing the permeability of the waste form and increasing the potential for contaminant leaching. A W/C of 0.48 by weight will fully hydrate the cement, leaving some free pore water, gel water and air voids. Above this W/C, permeability increases rapidly, which could lead to increased leaching rates. Because of economic reasons, though, these low water-binder ratios are usually not feasible for waste immobilization. Very low permeability is sacrificed for a decrease in the amount of binder required.

A number of factors affect the degree of immobilization, or fixation, of constituents in the waste. Major factors include solubility minimization through pH or redox potential control; chemical reaction to form carbonate, sulfide or silicate precipitates; adsorption, chemisorption; diadochy (substitution in the calcite crystal lattice); and encapsulation.

Not all wastes can be effectively treated by solidification/stabilization technologies. The major category of wastes for which immobilization is applicable are those which are essentially all inorganic. Cement and pozzolan-based waste forms rely heavily on pH control for pollutant containment. Cement-based waste forms typically have a pore water pH of 10-12 because of the excess lime present in the pores. These high pH values are usually desirable for heavy metal immobilization because most metal hydroxides have minimum solubility in the range of 7.5-11. Some metals, though, are amphoteric and have higher solubility at both low and high pH. These metals may be soluble at the high pH of the pore water (see Figure 1). Other contaminants, such as anions (arsenate, selenite, etc.), may be more soluble at high pH than low. Metals may also precipitate as carbonates, silicates or metal sulfides [3,4].

Metal immobilization is primarily dependent on the extent of solubilization of precipitated metals. This is governed by the solubility product, K_{s0}. Therefore, the solid metal concentration in the waste does not affect the concentration in the pore water which can leach out; only the pore water (leachant) composition will govern the amount of metal which will leach. It is only the solubilized fraction which can diffuse out of the waste form into the surrounding environment.

Cement and pozzolan-based systems rely heavily on hydroxide formation for metal containment, but other factors can come into play. Shively et al. [5] demonstrated that even after the alkali was leached from cement-based waste forms, lead and chromium leaching was much lower than would be expected from metal hydroxide solubilities. In this case, the metals were probably bound into the silica matrix itself. Cote [3] also found differences between calculated hydroxide solubility-pH curves and those determined empirically.

A number of agents which may be present in the waste may interfere with the binding systems and lead to decreased immobilization of hazardous constituents. Some of these are inorganic (certain metals, sulfates, etc.), while others are organic (oil, grease, HCB, TCE, phenol, etc.) [6,7]. For example,

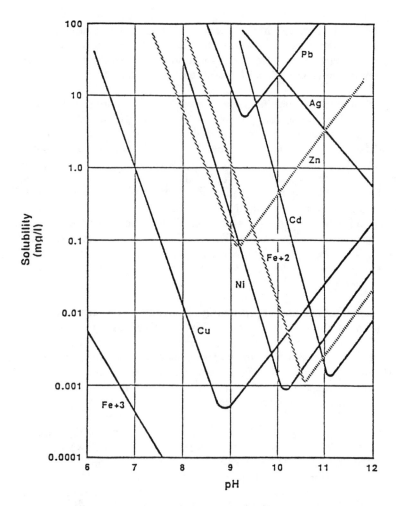

Figure 1. Solubilities of metal hydroxides as a function of pH.

some metals may temporarily inhibit setting of cement-based processes; chlorides may decrease durability; oils, greases and other nonpolar organics may inhibit setting and decrease long-term durability. Sulfates are a particular problem because they cause the expansive compound etringite to form.

Long-term immobilization of a contaminant incorporated into a cement matrix depends primarily on the ability of the matrix to maintain its integrity. Durability refers to the resistance of the matrix to chemical and physical interactions in the environment. All compounds of cement hydration are relatively insoluble in neutral water with the exception of $Ca(OH)_2$. Lime will leach easily, leaving a much more porous structure. Acids can dissolve the matrix of hydrated cement, releasing many of the bound metals. Freeze-thaw and wet-dry cycles can cause fracturing of the matrix, resulting in increased liquid-solid interfaces where leaching can occur. Little research has been conducted on the influence of these factors on long-term containment of heavy metals.

Very little research has been reported on immobilization of organics in waste forms, and what has been reported is often contradictory. Several researchers have reported chemical reactions between organic waste and binder, resulting in immobilization, but most researchers report that these positive results may actually be due to sorption effects, volatilization of organics or dilution by reagent chemicals. Much more research is needed on the stabilization/solidification of organics, because most principally inorganic wastes which are suited for stabilization/solidification also contain appreciable quantities of organics which may leach.

Organic constituents tend to retard cementitious reactions, inhibiting the formation of a solid monolithic mass. Also, the organic components may be easily leached from the waste form. Recently, research has been conducted into the use of organically modified clays in order to overcome these difficulties [8]. When these clays are mixed with cement-based stabilization agents, they reportedly adsorb and retain organic pollutants while solidifying organic wastes into a stable mass with low leaching potential.

The modified organophilic clays are made by mixing quaternary ammonium ions with the clay. The $[R_4N]^+$ ions substitute for metal ions present between the layers of alumina and silica in the clay minerals. This yields clays that have both organic and inorganic properties. Introduction of these ions increases the interplanar distance between clay plates allowing organics to penetrate, and makes the polarity of the stationary phase more compatible to that of the organic waste to be stabilized [9,10]. Preliminary studies indicate that chemical bonding between the organophilic clay binder and certain organic wastes may occur, but it is too early to tell the long-term fate of these complexes. It is possible that a unique binding mixture may be required for each organic compound encountered.

Leaching Mechanisms

A solidified waste is a porous solid at least partially saturated with water. The pore water in the solid is in chemical equilibrium with the solid phase. When the solid is exposed to leaching conditions, equilibrium is disturbed. The resulting difference in chemical potential between the solid and the leaching solution causes a mass flux between the solid surface and the leachant. This in turn causes concentration gradients that result in bulk diffusion through the solid [11,12]. Figure 2 depicts the leaching mechanisms in effect. Transport can either be by diffusion of metal ions from the solid matrix surface into the bulk aqueous phase, or by dissolution into the water in matrix pores and microfractures and then diffusion out. Consequently, the porosity and integrity of the waste form is of major importance.

For any constituent to leach, it must first dissolve in the pore water of the solid matrix. The amount of dissolution which occurs is dependent on the solubility of the constituent and the chemical makeup of the pore water, particularly its pH. Under neutral pH leaching conditions, the leaching rate is controlled by molecular diffusion of the solubilized species. Under acidic conditions, however, the rate will also be governed by the rate of penetration of hydrogen ions into the solid matrix, since this establishes the speciation and solubility of the contaminants present. Acid attacks pozzolanic-based paste through permeation of pore structure and dissolution of ions that must diffuse back through a chemically altered layer to enter solution. Acid consumes most of the calcium hydroxide in the leached layer and leaves a highly porous structure. Diffusion across this layer can be considered as a steady-state process since the leached layer provides little resistance to diffusion. At the leaching front, diffusion of hydrogen ions proceeds as if the unleached medium is infinite and dissolution reactions occur simultaneously in the pores. Proton transfer reactions are usually very fast with half-lives less than milliseconds. Hence, the dissolution reactions can be treated as diffusion-controlled fast reactions. The whole process then can be described as steady-state diffusion across the leached layer and unsteady-state diffusion controlled fast reactions in the porous leaching front [13].

Figure 3 depicts theoretical concentration gradients produced in the waste form during leaching [12]. The H^+ ions in the penetrating leachant react with metal hydroxides in the waste form, solubilizing the metal and reducing the H^+ concentration. The result is a leached layer where H^+ is essentially totally consumed. The soluble metal concentration in the pore water peaks at the leaching front. There is a gradient for the metal to diffuse out to the surface, but it can also move further back into the matrix where it reacts with the excess alkalinity and reprecipitates. Thus a narrow zone behind the leaching front is denser than the bulk solid.

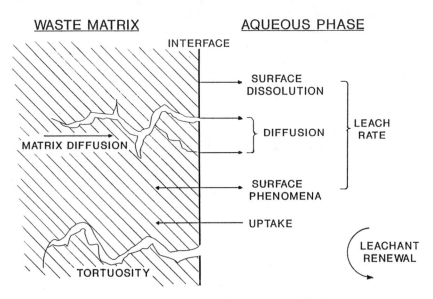

Figure 2. Schematic showing leaching mechanisms from a waste product. (Reprinted with permission from ref. 12. Copyright 1989 American Society for Testing and Materials.)

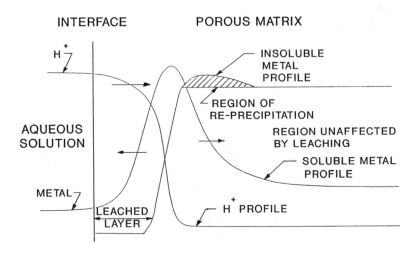

Figure 3.   Concentration gradients during leaching.   (Reprinted with permission from ref. 12.   Copyright 1989 American Society for Testing and Materials.)

Recent research in our laboratories confirms these findings [14]. Measurements of pH and heavy metals across the profile at various stages of leaching demonstrate the leached layer, leaching zone and reprecipitation zone. It has been shown that the leaching zone is very narrow, usually less than 10μm deep. There is little or no change in pH from that of the leachant all the way through the leached zone to the leaching front, indicating that all available alkalinity is leached from the leaching zone as the leaching front progresses.

Once solubilized, the constituent is transported from the solid matrix through the leached zone to the leaching solution by molecular diffusion. The flux of the constituent within the solid can be described by Fick's first law:

$$J = -D \frac{dC}{dz} \tag{1}$$

where:

C   = concentration of the constituent
D   = diffusion coefficient
J   = flux
z   = distance

A semi-infinite medium diffusion model with uniform initial concentration and zero surface concentration can be used to interpret the kinetic data generated from serial batch leaching tests [15]. The equation takes the form

$$\frac{\Sigma a_n}{A_o} \frac{V}{S} = 2 \left(\frac{D_e}{\pi} \right)^{0.5} t_n^{0.5} \tag{2}$$

where a_n = contaminant loss during leaching period n (mg)
A_o = initial amount of contaminant present in the specimen (mg)
V   = volume of specimen (cm^3)
S   = surface area of specimen (cm^2)
t_n = time to end of leaching period n (sec)
D_e = effective diffusion coefficient (cm^2/sec)

The American Nuclear Society recommends use of a series of seven batch leaching tests in order to determine the effective diffusion coefficient in the Godbee and Joy model [16]. They suggest that the results be presented as a "leachability index", LX, equal to the average negative logarithm of D_e.

$$LX = \frac{1}{7} \log \sum_{1}^{7} \left(\frac{1}{D_e} \right) \tag{3}$$

This index can be used to compare the relative mobility of different contaminants on a uniform scale that varies from about $5(D_e = 10^{-5}$ cm^2/s, very mobile) to 15 ($D_e = 10^{-15}$ cm^2/s, immobile) [17].

Leachate generation is an extremely complex process. The free alkalinity present in the pozzolanic-based paste maintains a high pH environment and limits the metal leachability of fixed wastes. Calcium hydroxide, which is produced by the hydration reactions of the binder, provides most of the buffering capacity. The leaching model shown above, however, does not include the factor of acid strength of the leachant and cannot describe the rate of movement of leaching front into the waste solid.

Cheng and Bishop [13] have shown that the rate of advance of the leaching front can be expressed as steady-state diffusion across the leached layer. Figure 4 depicts the cumulative amount of calcium leached and Figure 5 shows the penetration distance into the waste matrix versus the square root of time for two leachant acetic acid strengths (5 and 15 meq/g solids). The plots are linear, as would be expected for diffusion-based processes. The leachant strengths differ by a factor of 3.0, but the ratios of the amount of leaching and the penetration distances differ by factors of 1.65 and 1.95, respectively. This can possibly be explained by comparing the free hydrogen ions arising from dissociation of the diffusing acetic acid. A 15 meq/g acetic acid leachant contains approximately 1.73 times more free hydrogen ions than the 5 meq/g leachant. This is a very close agreement, considering the complex, heterogeneous nature of the waste matrix. Thus, leachant acid strength may be very important in determining the rate and extent of leaching of solidified/stabilized wastes. This concept is currently not included in any leaching models, however.

Factors Affecting Leachability

There are a number of factors which can affect the leachability of a particular solidified/stabilized waste form. Table 1 presents a summary of some of the more important ones.

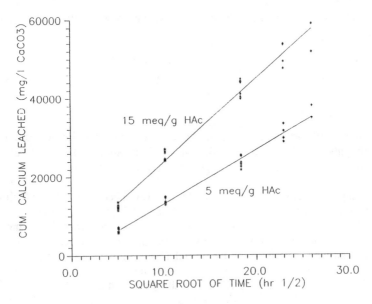

Figure 4. Cumulative calcium hardness leached versus square root of time during leaching of portland cement solidified/stabilized wastes using two strengths of acetic acid leachant.

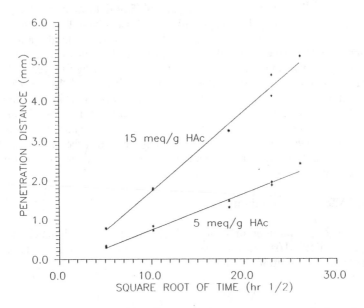

Figure 5. Acid penetration distance versus square root of time during leaching of portland cement solidified/stabilized wastes using two strengths of acetic acid leachant.

Table 1. Factors That Affect Leaching from Solidified/Stabilized Wastes

Waste Form Factors
    Contaminant binding mechanisms
    Alkalinity
    Surface-to-volume ratio
    Porosity and pore tortuosity
    Durability

Leachant Factors
    Composition (pH, acidity, E_h, chelating potential, etc.)
    Leachant volume to waste form surface area ratio
    Flow rate
    Temperature

Obviously, the composition of the waste form determines the physicochemical properties and the leaching mechanisms. Every effort should be made to minimize the potential for leaching by improving the quality of the waste form. Alkalinity is needed in the final product to maintain metals in their most insoluble form and to buffer against acid dissolution. One of the principle factors governing diffusion of soluble metals from the waste form is the solid surface-to-volume ratio (see equation 3). The larger the monolith, the smaller the surface-to-volume ratio and the smaller the potential for leaching. The role of internal fractures in determination of the applicable surface-to-volume ratio to use has not as yet been determined. Porosity and tortuosity, which is a measure of the path length for a diffusing substance to reach the surface of the waste product through winding and convoluted pores, governs to a large extent the rate of diffusion to be expected. A highly porous matrix will have a higher effective diffusion coefficient for a particular contaminant than a less porous one, while wastes with a large tortuosity factor will have reduced D_e. The "effective" diffusion coefficient in equation 3 modifies the true diffusion coefficient for the contaminant to account for these variations in porosity and tortuosity. Waste durability is very important because if an initially intact and acceptable monolith weathers poorly over time, the porosity and surface-to-volume ratios will increase markedly, resulting in increased leachability.

The leachant composition determines the reactions which will occur within the waste form. Acid dissolution, oxidation-reduction reactions and metal complexation can all occur, depending on the chemical composition of the leachant. Increased leachant volume to surface area ratios and flow rates will increase leaching because diffusing substances will be removed from the monolith surface more rapidly and concentration gradients in the solid will be greater. Temperature is not normally a factor for buried waste forms once the exothermic heat of reaction of the cement has dissipated. Increases in temperature result in increases in all reaction rates, including those involved in leaching, which can be described by the Arrhenius equation.

Summary

Stabilization/solidification processes are being used to minimize the potential for groundwater pollution from land disposal of hazardous wastes. Many variations are used, but most rely on pozzolanic reactions to chemically stabilize and physically solidify the waste. Portland cement alone or in combination with fly ash, cement kiln dust, lime or other ingredients is the principal solidifying agent used.

Stabilization/solidification processes are very effective at immobilizing most heavy metals present in sludges, contaminated soils and other wastes. They are not as effective at immobilizing toxic organic materials. Organically modified clays are now being evaluated as an additive to S/S processes in order to adsorb and retain these organic pollutants in the solidified waste form.

The environmental acceptability of stabilization/solidification processes will depend on the long-term ability of the waste form to retain contaminants. This will be governed by the chemical binding mechanisms involved and by the durability of the waste form. Many S/S processes have been developed which can pass regulatory leaching tests, but these tests do not indicate the potential for leaching after long-term environmental exposure. Wide spread acceptance of stabilization/solidification processes will be hampered until the long-term durability of the waste form can be demonstrated.

Literature Cited

1. Cullinane, M., Jones, L. and Malone, P., _Handbook for Stabilization Solidification of Hazardous Wastes_. U.S. EPA, EPA/540/2-86/001, 1986.

2. Poon, C., "A Critical Review of Evaluation Procedures for Stabilization/Solidification Processes," _Environmental Aspects of Stabilization and Solidification of Hazardous and Radioactive Wastes_, ASTM STP 1033, American Society for Testing and Materials, Philadelphia, 1989, pp. 114-124.

3. Cote, P., _Contaminant Leaching from Cement-Based Waste Forms Under Acidic Conditions_, Ph.D. Dissertation, McMaster University, Hamilton, Ontario, 1986.

4. Bishop, P., "Leaching of inorganic hazardous constituents from stabilized/solidified hazardous wastes," _Hazardous Wastes and Hazardous Materials_, 1988, vol. 5, pp. 129..

5. Shively, W., Bishop, P., Brown, T. and Gress, D., "Leaching Tests of Heavy Metals Solidified and Stabilized with Portland Cement," _Journal Water Pollution Control Federation_, 1986, vol. 58, pp. 234-241..

6. Wiles, C., "A review of solidification/stabilization technology," *Journal of Hazardous Materials*, 1987, *vol. 14*, pp. 210.

7. Jones, L., *Interference Mechanisms in Waste Solidification/ Stabilization Processes*. Final report for U.S. EPA, IAG No. SW-219306080-01-0, 1988.

8. Alther, G., Evans, J. and Pancoski, S., "No Feet of Clay," *Civil Engineering*, 1990, *vol. 60*, pp. 60-61.

9. Soundararajan, R., Barth, E. and Gibbons, J., "Using an Organophilic Clay to Chemically Stabilize Waste Containing Organic Compounds," *Hazardous Materials Control*, 1990, *vol. 3*, pp. 42-45.

10. PEI Associates, Inc., "Use of Organophilic Clays for Organic Stabilization," unpublished report to U.S. EPA, Cincinnati, OH, 1990.

11. Conner, J., *Chemical Fixation and Solidification of Hazardous Wastes*. New York: Van Nostrand Reinhold, 1990.

12. Cote, P., Bridle, J. and Benedek, A., "An approach for evaluating long-term leachability from measurement of intrinsic waste properties," *Hazardous and Industrial Solid Waste Testing and Disposal*, ASTM STP 933, American Society for Testing and Materials, Philadelphia, 1989, pp. 63-78.

13. Cheng, K. and Bishop, P., "Developing a kinetic leaching model for solidified/stabilized hazardous wastes," *Journal of Hazardous Materials*, in press.

14. Cheng, K. and Bishop, P., unpublished data, University of Cincinnati, Cincinnati, OH.

15. Godbee, H. et al., "Application of mass transport theory to the leaching of radionuclides from solid waste," *Nuclear and Chemical Waste Management*, 1980, *vol. 1*, pp. 29.

16. American Nuclear Society, "Measurement of the Leachability of Solidified Low-Level Radioactive Wastes by a Short-Term Procedure," 1986.

17. Cote, P. and Hamilton, D., "Leachability comparison of four hazardous waste solidification processes," *Proceedings of the 38th Annual Purdue Industrial Waste Conference*, 1983, *vol. 38*, pp. 221.

RECEIVED April 5, 1991

Chapter 16

Importance of Soil–Contaminant–Surfactant Interactions for In Situ Soil Washing

R. C. Chawla[1], C. Porzucek[2,3], J. N. Cannon[1], and J. H. Johnson, Jr.[1]

[1]School of Engineering, Howard University, Washington, DC 20059
[2]Los Alamos National Laboratory, Los Alamos, NM 87545

Organic pollutants can bind to soil matter by various types of physical and chemical interactions—hydrophobic bonding, hydrogen
bonding, ion exchange, charge transfer, etc. They enter the environment from spills, leaks, leaching and other sources. They are
emplaced in soils through capillary action or by the physical and
chemical forces of sorption. The displacement and recovery of organic pollutants by in situ soil washing may use aqueous surfactant solutions. In a surfactant-assisted soil-washing operation, soil-
contaminant-surfactant interactions, and the surrounding soil environment, all have major effects on the pollutant mobilities. The
various soil-contaminant-surfactant properties, and their effects on
the removal of organics, are reviewed. The interactions between
soil/aquifer systems, contaminants, and surfactants are discussed
in the context of the fate of contaminants in the subsurface system.

Remedial action plans for cleanup of contaminated soils generally fall into
two categories. They are either above ground or in situ activities. The above
ground techniques require excavation of the contaminated zone followed by
some treatment such as bioremediation, soil washing, or incineration. The in
situ techniques are used on contaminated sites that are either very large, very
inaccessible, or both. One example of an inaccessible site is a gasoline spill that
occupies the soil beneath an airplane hangar. Another example is a leak from
a pipe that is buried 100 feet below the ground.

For many inaccessible sites, in situ cleanups are the only practical treatment methods. However, in situ cleanups are more difficult than above ground
cleanups because it is not possible to have complete control over critical soil
remediation process parameters (e.g., temperature, pH, and the spatial flow of
fluids). Consequently, in situ cleanups are riskier than above ground actions.

---

[3]Current address: Dow Chemical Texas Operations, Freeport, TX 77541

0097–6156/91/0468–0316$07.50/0

The Chemical Countermeasures Program, sponsored by U.S. Government agencies like the Environmental Protection Agency (EPA), the Department of Defense, and the Department of Energy, is designed to evaluate in situ remediation processes for the cleanup of contaminated soils (*1,2*). Sites that are candidates for these cleanup methods should have the following characteristics:

1. contaminants are spread over large volumes (e.g., 100 to 100,000 m³ at a depth of 1 to 10 m),
2. contaminant concentrations are low (i.e., less than 10,000 ppm),
3. contaminants can be removed or immobilized by aqueous chemical solutions injected into the contaminated site, and
4. the hydraulic conductivity of the contaminated soil is greater than 10^{-4} cm/s.

Two types of in situ cleanup technologies have been tested. First, acids and chelating agents have been evaluated for the in situ remediation of soils containing hazardous heavy metals. Secondly, surfactant washing techniques have been considered for the in situ remediation of soils contaminated by hazardous organics.

Chawla et al. (*3*) have critically reviewed past work on surfactant-assisted, in situ soil washing for the removal of hazardous organics. Laboratory, pilot, and field-scale studies have been conducted to investigate and evaluate this remedial action technique. Initial laboratory studies were conducted at the Texas Research Institute (TRI) (*4,5*). They showed the effectiveness of this technique in recovering gasoline from contaminated sands. For example, one set of experiments showed that surfactants could be used to mobilize residual gasoline in the vadose zone. Once mobilized, the gasoline migrated downward to the water table where it was recovered. A second set of experiments tested different methods for application of the surfactant solution. The best case results from both types of experiments suggested that about 80% recovery of residual gasoline from the sands was possible.

Surfactant-assisted, in situ soil-washing studies have also been conducted by the Science Applications International Corporation, Inc. (SAIC) (*1,2*). These studies showed the effectiveness of this technique in removing polychlorinated biphenyls, Murban crude oil, and phenols from doped soil samples using mixtures of nonionic, biodegradable surfactants. Results from core flood experiments suggested that about 68–88% of the pollutants could be removed. After these successful lab experiments, SAIC conducted a second set of experiments using contaminated soil from a fire fighting area of an Air National Guard Base (*6*). The soil was contaminated with petroleum derived hydrocarbons (JP-4 jet fuel), chlorinated solvents (1,1,1-trichloroethane, dichloromethane, trichloroethylene, and chloroform), and a fire-fighting foam. In these tests they were able to remove 90–95% of all contaminants (i.e., aliphatic, aromatic, and unresolved hydrocarbons).

Following these lab studies, a field test was conducted by Mason & Hanger-Silas Mason Company at the same Air National Guard Base (7,8). Aqueous surfactant solutions were applied to the contaminated soil surface and allowed to percolate downward. This type of surfactant solution application should have drained contaminants from the vadose zone down toward the water table, but results showed essentially no contaminant movement. The failure was attributed to the development of preferential flow paths that bypassed contaminated areas of the soil. These preferential flow paths developed after many years of rainfall accompanied by a redistribution of fine-sized, soil particles.

More recent studies were performed to evaluate this technology for remediating soils at a railroad-tie treating plant in Laramie, Wyoming. Soil contaminants included creosote oil, polynuclear aromatics, and pentachlorophenol. Since this site was severely contaminated, a multistep remediation process was adopted. Initially, contaminants were pumped directly out of the ground, similar to a primary oil recovery operation. This action was followed by a waterflood for secondary contaminant recovery. Then a surfactant flood, designed by MTA Remedial Resources, Inc., was implemented (9). The formulation consisted of a surfactant, alkali, and polymer mixture. While their laboratory studies had shown greater than 98% recovery of creosote from contaminated soils with this surfactant wash, pilot tests at the railroad-tie treating facility only yielded a 67% reduction in contaminant concentrations. However, the overall recovery from primary pumping, secondary waterflooding, and the tertiary surfactant flood was 94%.

Another recent implementation of this technology occurred at an industrial site in Florida which was contaminated with a viscous oil (100 cp at 20 °C). The oil existed as a free layer (0.01 to 1.5 feet thick) floating on the water table over a 40,000 ft^2 area. Their contaminant layer was separated from the drinking water supply aquifer by only a semipermeable silt layer. Again, MTA Remedial Resources designed a surfactant system for an in situ soil wash (10). Due to the location of the contaminant and its proximity to the drinking water aquifer, the surfactant system consisted of "chemicals used in water treatment and as food additives" and a "biodegradable polymer". Field results showed that these washing solutions were able to remove up to 75% of the trapped oil.

In 1988, a joint study was initiated between Howard University and Los Alamos National Laboratory to determine the feasibility of using surfactants for in situ remediation of contaminated soils and aquifers. The first step of the study was to identify the mechanisms behind the transport of contaminants in soils and aquifers. More specifically, the mechanisms of interaction between contaminants, soils, and ground water were identified. This study was followed by an investigation into the effects of surfactants on each mechanism. The results were used to explain successes and failures of previous laboratory and pilot-scale field studies of surfactant-enhanced, in situ soil washing (3,11).

The design of effective in situ remedial action plans requires a thorough understanding of physical chemistry and hydrogeology. The most common in situ remedial action plans for ground water cleanup use the pump-and-treat philosophy. This philosophy treats the contamination problem like a "black box." Liquid is pumped from the aquifer and treated on the surface. Pumping is continued until an acceptable level of aquifer contamination is reached. Recent studies show, however, that contaminants can take decades to dissolve in the ground water (*11,12*). This makes pump-and-treat plans slow and expensive.

In order to design effective in situ remedial action plans, it is necessary to understand the interactions between contaminants, soil, and ground water. Here we describe the relevant physical properties of these components and their interactions with each other and with aqueous surfactant solutions. Explanations of sorption/desorption, capillary forces, and mass-transfer limited solubilization are given along with the influence of surfactants on each. Other mechanisms that influence the transport and retardation of contaminants in soil and aquifer systems are also discussed. Multiple processes are responsible for the transport of contaminants in a soil and aquifer system. The most effective remedial action plans will be those whose designs are based on an understanding of these fundamental transport phenomena.

The Contamination Problem

In order to define the mechanisms of contaminant transport, it is necessary to develop a physical picture of a contaminant discharge into the environment. For example, spills or releases of nonaqueous phase liquids (NAPLs) come from sources such as leaky storage tanks and pipes, and spills on the ground. The typical movement of contaminants following a release is illustrated in Figure 1. As the contaminants move, they are acted upon by various forces in the soil.

Initially a contaminant, referred to as a NAPL, moves downward through the vadose zone because of the influence of gravity. Some of it is held trapped in the vadose zone by capillary forces. This NAPL is called a "residual" and it exists as small blobs and pendular rings in the soil pore structure (*13*). Experimentally, Hoag and Marley (*14*) found residual gasoline concentrations in the vadose zone to be about 14% in coarse sands with average particle diameters around 2.2-mm, and as high as 60% in fine sands with average particle diameters about 0.26-mm.

The NAPL continues to migrate downward through the vadose zone until it encounters the water table, also shown in Figure 1. At this point, a lighter-than-water NAPL will spread out and form a lens on the water table. On the other hand, a heavier-than-water NAPL will continue to migrate downward through the aquifer until it hits an underlying rock layer and forms a NAPL pool. In both cases, NAPL will dissolve into the aquifer and create a contaminated ground water supply.

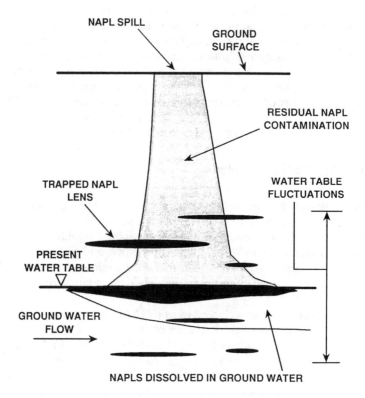

Figure 1. Migration of a Contaminant Spill

Due to natural fluctuations in aquifer recharge and discharge or pumping, the height of the water table will fluctuate throughout the year. This results in a distribution of trapped, lighter-than-water, NAPL lenses at different depths in the aquifer and vadose zone. The NAPL lenses trapped below the water table were emplaced there when the water table fell. As the water table rose again, the buoyancy forces exerted on the NAPL phase were insufficient to overcome the capillary forces that held the NAPL trapped in the pores. The end result is a NAPL trapped below the water table.

A similar situation occurred to emplace NAPLs in the vadose zone. Here, the water table fell and never recovered back to its original height. In this case, the capillary forces were larger than the gravity forces and the NAPL became trapped.

In addition to NAPL dissolution into the ground water, it will migrate away from the original contamination zone due to the natural hydraulic gradient of the aquifer. These NAPLs will be pushed along as a separate phase by the viscous forces of the ground water flow. However, they may encounter soil pores that exert sufficient capillary forces to entrap them. Such trapped NAPLs will no longer exist as lenses, but instead as ganglia and blobs. Wilson and Conrad (*13*) estimated residual NAPL saturations of 10% to 50% in saturated (aqueous) soil systems. Residual NAPLs also pose a long-term contamination problem because they slowly dissolve into the ground water.

From this description of NAPL movement, it is clear that NAPLs can exist in any of several forms in a soil/aquifer system. They may occur as a residual phase, or as trapped lenses, or as dissolved species. The exact disposition of a particular contaminant depends on the specific physical properties of a surfactant, contaminant, and soil. It also depends on the interactions between these species as described below.

Soil/Aquifer-Contaminant-Surfactant Interactions

In a surfactant-enhanced, in situ soil-washing operation, surfactant, contaminant, and the soil/aquifer system are the three interacting components. Here the term soil/aquifer system is used to mean soil in general. This includes the saturated as well as unsaturated (vadose) zone. Each component interacts with the other two as illustrated by the triangular diagram of Figure 2. In this diagram, each of the three components occupies an apex. The interactions between any two components are listed along the line connecting the two respective apexes. These component-component interactions can be physical, chemical or biological in nature.

Soil/Aquifer System Properties.   Properties of the soil/aquifer system that are important in the transport of contaminant and surfactants are shown in Fig-

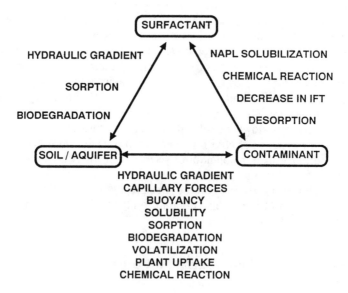

Figure 2. Interaction Triangle for Soil/Aquifer-Contaminant-Surfactant

ure 3. Temperature is important because surfactant and contaminant solubilities are functions of temperature. Species sorption properties are also temperature dependent. Therefore, it is important to know the temperature of the soil/aquifer system before a surfactant-enhanced soil washing system is designed.

Soil/aquifer system pH is also a critical variable. The pH will determine the surface charge on any minerals present in the soil/aquifer system. It will also affect the configuration of any humic materials (e.g., coiled or uncoiled) (15,16). Both of these changes will in turn quantitatively affect the amount of surfactant or contaminant sorbed.

In addition to pH, organic carbon content of the soil/aquifer system has been shown to correlate well with the amount of contaminant sorbed onto the soils (17–20). Therefore, it is necessary to have an idea of the location and amount of organic carbon.

The particle size distribution of the soil/aquifer system is also important. Soils with larger average particle sizes, say greater than 62-μm, are usually sandy and have little or no sorptive capacity. On the other hand, soils with smaller particles (e.g., average diameters less than 2-μm), usually contain more clay minerals. These clay minerals have a sorptive capacity for contaminants. Also, soils with small particles tend to be less permeable to fluid flow.

In soil/aquifer systems that are saturated with water, the porosity, permeability, and pressure all affect the flow field of the ground water. Ground water tends to follow paths that lead through more permeable sections of the soil. It also tends to follow paths through soil sections that have higher porosity. Finally, ground water will travel the fastest from areas of high pressure to areas of low pressure. Therefore, it is helpful to have an idea of the permeability, porosity, and pressure distribution in a soil/aquifer system before designing an injection and recovery well system for remediation. Using this map it may be possible to design a system that maximizes the flow through the contaminated zone and also recovers fluids that have been swept through this zone.

Finally, the moisture content of the soil/aquifer system is a critical variable. For remedial action design purposes it is necessary to know if the contamination problem exists in the vadose zone (low moisture content) or the aquifer (saturated zone). Flow in the vadose zone is more unpredictable when compared to the saturated zone (aquifer) because of the presence of air. On the other hand, flow in the aquifer can be predicted well if one has an idea of the permeability and pressure distribution in the aquifer.

Contaminant Properties. In order to remove or clean up a contaminant one must define the contaminant and its physical properties. These properties will give some idea as to the mobility of the contaminant and how difficult it will be to remove from the soil/aquifer system. Important contaminant properties are shown in Figure 4.

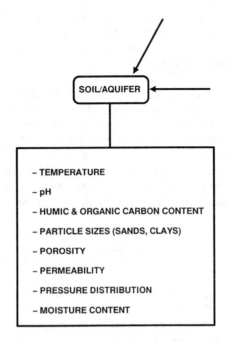

Figure 3. Physical Properties of Soil/Aquifer

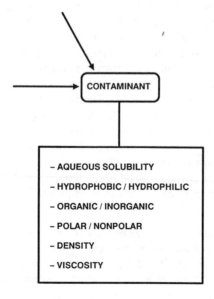

Figure 4. Physical Properties of Contaminant

The aqueous solubility of the contaminant will determine if the contaminant is a major threat to the ground water. A highly water-soluble contaminant will spread into the ground water and increase the zone of contamination much faster than a contaminant that is only slightly water soluble. On the other hand, a slightly water-soluble contaminant poses a long-term threat to the ground water because it will take longer to dissolve.

Nonpolar contaminants are those with little or no net dipole moment. Such contaminants are hydrophobic and hence only slightly water soluble. On the other hand, polar contaminants have a net dipole moment. These contaminants are hydrophilic and hence highly water soluble. The transport properties of polar and nonpolar contaminants in the soil/aquifer system will be different.

It is also important to know if the contaminant is organic or inorganic since the remediation schemes differ for these two categories. In addition, the density of the contaminant must be known. As mentioned previously, a contaminant that is denser than water will lie predominantly beneath the aquifer while a less dense contaminant will be found floating on top. Finally, it is important to know the contaminant viscosity. If the viscosity is too high, it may not be possible to use a hydraulic gradient for site remediation.

Surfactant Properties.   Surfactant properties that are important in the remediation process are shown in Figure 5. A surfactant must increase NAPL solubilization in the aqueous phase and reduce interfacial tension between the aqueous and NAPL phases. In general, surfactants can be cationic, anionic, or nonionic. Since ionic surfactants have a tendency to flocculate clays, the nonionic surfactants at first seem preferable. However, laboratory experiments should be performed using soil/aquifer samples and different types of surfactants to determine potential flocculation problems.

One successful pilot-scale implementation of surfactant-enhanced soil washing used nonionic and ionic surfactant washes in series. However, an alkali preflush was used to reduce the amount of exchangeable calcium and magnesium from the soil surfaces (9). In any event, the surfactant should be biodegradable and nonbiocidal so that it cannot harm the environment if it is injected and not recoverable.

Soil/Aquifer System and Contaminant Interactions.   Soil/aquifer systems and contaminants interact by many mechanisms as shown in Figure 2. Among these is the natural hydraulic gradient of the aquifer. This hydraulic gradient acts to push the NAPL through the soil matrix. The net result is a transport of NAPL away from the point of entry, thus enlarging the contamination zone. On the other hand, contaminant transport by hydraulic gradients can be used to remediate contaminated soils. Pumping wells can be installed down gradient from the contaminant source and used to induce an artificial hydraulic gradient in the aquifer. Contaminants and water are then drawn toward the pumping wells and recovered.

Capillary force is a second interaction mechanism which is measured quantitatively by the capillary pressure. Mathematically, it is the pressure difference between the NAPL and water phases, and is described using the Young & Laplace relation, equation 1 below (21). In writing this form of the Young & Laplace equation, we have assumed a hemispherical NAPL/water interface. This is the simplest case, but will allow easy estimation of the capillary pressure.

$$P^{cap} = P^{NAPL} - P^{aq} = \frac{2\gamma_{ow}}{r} \qquad (1)$$

Here, P^{cap} is the capillary pressure, P^{NAPL} is the pressure in the NAPL, P^{aq} is the pressure in the aqueous phase, γ_{ow} is the NAPL/water interfacial tension, and r is the radius of curvature of the hemispherical interface.

Physically, the capillary pressure is a measure of the force necessary to deform a curved interface so as to allow it to pass through a constriction. The situation is illustrated in Figure 6 which shows a NAPL drop trapped in a pore. The natural hydraulic gradient in the aquifer contains insufficient energy to deform the interface at r_2 and allow the droplet to pass through the constriction.

Capillary forces are most significant for silt or clay-sized particles with a grain diameter of less than 0.075-mm. Equivalently, these are particles that will pass through a Tyler 200 mesh screen.

The Young & Laplace equation can be used to estimate the order-of-magnitude of the capillary forces that hold NAPL drops trapped in silt or clay-sized particles, and the hydraulic gradient required to push the NAPL drop through the constriction shown in Figure 6. Let P_1^{aq} and P_2^{aq} be the aqueous phase pressures on the up and down gradient sides of the droplet, respectively. Let r_1 and r_2 be the radii of curvature of the interfaces as shown in Figure 6. The hydraulic gradient necessary to push the droplet through the constriction, ΔP, is then given by equation 2.

$$\Delta P = P_1^{aq} - P_2^{aq} = P_2^{cap} - P_1^{cap} = 2\gamma_{ow}\left(\frac{1}{r_2} - \frac{1}{r_1}\right) \qquad (2)$$

A typical assumed value for a NAPL/water interfacial tension is 30 dyne/cm. Further assuming a cylindrical pore geometry, and a hemispherical NAPL/water interface, r_1 and r_2 become the pore radii up and down gradients of the NAPL, respectively. Representative values for r_1 and r_2 may then be selected for clay-sized particles, and a few values representing the fine sands, the next larger particle size classification. Berg (22) provides a relationship for estimating pore throat diameters from particle diameters for rhombohedral packing of spheres which is also helpful in selecting useful r_1 and r_2 values. For example, r_1 should vary from 0–25-μm and r_2 from 0.5–10-μm. Pore radii

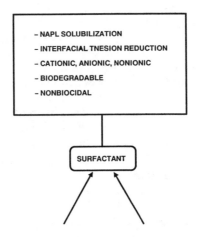

Figure 5. Physical Properties of Surfactant

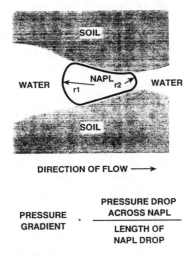

DIRECTION OF FLOW ⟶

$$\text{PRESSURE GRADIENT} = \frac{\text{PRESSURE DROP ACROSS NAPL}}{\text{LENGTH OF NAPL DROP}}$$

Figure 6. NAPL Drop Trapped in a Soil Pore Constriction

values less than 5.8-μm correspond to silt or clay-sized particles while those between 5.8- and 32.3-μm correspond to fine-sand-sized particles.

Using such values, the pressure gradients that are needed to push a NAPL drop one-cm in diameter through a given pore constriction can be estimated. Figure 7 shows the results in parametric form. Looking at the case where r_1 is 5-μm and r_2 is 0.5-μm, a pressure drop of 10,800 kPa/m (480 psia/ft) is required to dislodge the NAPL droplet. This pressure drop is many orders of magnitude larger than the typical natural gradient of 0.13 kPa/m (0.006 psia/ft) that is observed in aquifers which have no large elevation changes. This is certainly the worst case seen in Figure 6. However, even when r_2 and r_1 differ by as little as a factor of two, the required pressure drop is still very large compared to the typical natural gradients in aquifers. Thus, these calculations suggest that the natural gradients in aquifers are insufficient to mobilize NAPLs that are trapped as droplets in pore constrictions of silt or clay-sized particles.

Similar calculations for fine-sand-size particles are also helpful as indicated by the curves for r_2 at 5- and 10-μm in Figure 7. In these cases, the pressure gradients to dislodge a one-cm diameter drop were on the order of 0.1 to 0.2 kPa/m and are much closer to the natural gradients present in horizontal aquifers. Therefore, for particles in this size classification, the natural aquifer gradients may be sufficient to mobilize NAPLs that are trapped by capillary forces alone.

The calculations just presented are only order-of-magnitude estimates. The real subsurface environment will have a pore size distribution. Variations of an order-of-magnitude in grain size are common between adjacent layers in an alluvial environment (12). Although simple, our calculations show how large capillary forces can be.

In one sense capillary forces are good because they tend to hold the contaminant immobilized. On the other hand, many NAPLs are slightly water soluble, a few ppm to a few hundred ppm. Therefore, these immobilized NAPLs dissolve into the ground water, in accord with their water solubilities, and are then transported along with the ground water. This phenomenon increases the size of the contaminated zone. To make the situation worse, some evidence suggests that the dissolution process is mass transfer limited (12). This means that dissolved NAPL concentrations in ground water are less than their solubilities. The result is an increased dissolution time (i.e., time period required for a droplet to dissolve) and a greater long-term contamination problem for the ground water.

Capillary and buoyancy forces influence contaminant movement in the vertical direction in the soil/aquifer system. The NAPL can be thought of as an object partly immersed in a fluid. The fluid is air when the NAPL is in the vadose zone and the fluid is water when the NAPL is in the saturated zone. Archimedes' principle states that a body wholly or partly immersed in a fluid is buoyed up with a force equal to the weight of the fluid displaced by the body (23). In opposition to buoyancy is the gravity force that pulls the sub-

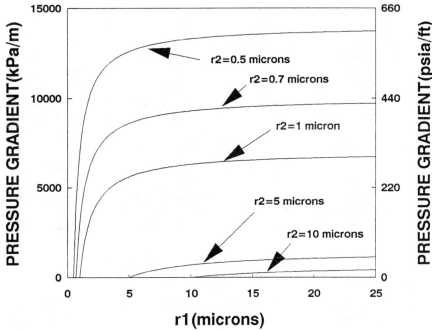

Figure 7. Pressure Gradient Required to Overcome Capillary Forces

merged object in a downward direction. The net force is given by the difference in buoyancy and gravity forces. This net force is directed either upward or downward in a vertical plane. If this net force causes the NAPL droplet to rise through the soil matrix, many pore constrictions will be encountered. At some point the NAPL drop may encounter a pore constriction that is very small. In this case the net upward force is much smaller than the capillary force necessary to push the drop through the pore constriction and the drop is trapped in the pore. The end result is an immobilized NAPL phase. A similar phenomenon occurs for NAPLs that have a net force directed in a downward direction; Berg provides a more general discussion of this interaction (22). In either case, the interaction of capillary and buoyancy forces results in NAPL emplacement in pores of the soil/aquifer system.

Sorption represents another important interaction mechanism between the soil/aquifer system and contaminants. Depending on the forces of attraction, contaminant sorption to soil particles may be classified as physical, chemical, or electrostatic. Physical sorption is characterized by dipole-dipole interactions with relatively low heats of sorption in the range of 1–2 kcal/mole. Chemical sorption (chemisorption) represents the other extreme. It is characterized by covalent and hydrogen bonding with differential heats of sorption that can be as high as 50–100 kcal/mole. The last sorption category is electrostatic. Electrostatic interactions are due to ion-ion and ion-dipole interactions.

Two additional properties affect the soil/aquifer system-contaminant sorption interaction. These are the soil-cation exchange capacity and the surface area of the soil particles. Cation exchange capacity values of various soils constituents are 2–6 meq/100 g for oxides and hydroxides, 80–150 meq/100 g for montmorillonite, and 200–400 meq/100 g for organic matter. Surface areas range from 7–30 m^2/g for kaolinite to 500–800 m^2/g for organic matter (24). From these values of cation exchange capacity and soil surface area, it appears that organic carbon content will have a larger effect on sorption of contaminants when compared to the clay mineral content.

The sorption mechanism between the soil/aquifer system and the contaminant must be quantified in order to design effective in situ remedial action schemes. Simply put, it is necessary to estimate the amount of contaminant sorbed onto the soil particles. This sorbed contaminant will be more difficult to mobilize and recover when compared to contaminant that exists as a separate nonaqueous phase liquid.

A tremendous amount of work dealing with the sorption, transport and fate of all classes of organic compounds in subsurface soils has appeared in the literature (18,25–27). In particular, mathematical models have been used to correlate contaminant sorption with the aqueous phase contaminant concentration. Some of the more common mathematical models to predict sorption of contaminants onto soil particles are shown in Table I. These mathematical equations relate the equilibrium solid phase concentration, x/m (μg of contaminant

Table I. Sorption Models

Freundlich	x/m	$= K_p C^{1/n}$
Linear	x/m	$= K_p C$
Langmuir	x/m	$= (x/m)_{max} K_p C / (1 + K_p C)$

x = mass of contaminant sorbed, μg
m = mass of solid sorbent, g
C = aqueous phase equilibrium concentration, mg/mL
n = empirical constant, $n \geq 1$

sorbed per g of soil or sorbent), to the equilibrium liquid phase concentration, C (mg of contaminant/mL of liquid).

The parameter K_p is the distribution coefficient. This coefficient is a measure of the strength of sorption and can be thermodynamically correlated to the Gibbs' free energy and the temperature. The Freundlich model is very general and fits most situations. In some situations it reduces to the simpler forms. It is linear if $n = 1$ and the Langmuir model results from the assumptions of constant differential heat of sorption and monolayer coverage. For hydrophobic organic compounds at low concentration, sorption isotherms (equilibrium plots of x/m versus C at constant temperature) are linear. The value of K_p depends on the properties of the contaminant and the soil. K_p is a strong function of the organic carbon fraction of the soil, f_{oc}, and the octanol-water partition coefficient, K_{ow} (26). K_{ow} is a measure of the hydrophobicity of a compound.

A normalized sorption coefficient is used in many sorption correlations. This normalized coefficient, K_{oc}, is defined based on a soil that is 100% organic carbon. The mathematical definition is given in equation 3.

$$K_{oc} = \frac{K_p}{f_{oc}} \tag{3}$$

K_{oc} varies widely for different contaminants (e.g., 1 cm^3/g for 2-propanol, 30,000 cm^3/g for oil, 200,000 cm^3/g for dioxin, and 5×10^6 cm^3/g for grease)

The sorption coefficient, K_p, is correlated with K_{ow} for polynuclear aromatics and chlorinated hydrocarbons (26), polynuclear aromatics (17), and halogenated alkenes and benzenes (18). These correlations are presented in Table II. Their utility is that they can be used to estimate the amounts of contaminants that sorb onto soil particles.

To estimate contaminant sorption, two physical parameters are needed: K_{ow} and f_{oc}. The K_{ow} value for a contaminant can typically be found in the literature (12,28). The f_{oc} value for a particular soil can be estimated or measured. With these values of K_{ow} and f_{oc} an appropriate correlation for K_{oc} is selected from Table II and a K_p value is calculated from equation 3. Then an appropriate sorption model is selected from Table I and the amount of contaminant sorption is calculated.

Table II. K_{oc} Correlations

Correlation	Ref	Application
$\log K_{oc} = \log K_{ow} - 0.21$	(26)	polynuclear aromatics & chlorinated hydrocarbons
$\log K_{oc} = \log K_{ow} - 0.317$	(17)	polynuclear aromatics
$\log K_{oc} = 0.72 \log K_{ow} + 0.49$	(18)	halogenated alkenes & benzenes

The amount of contaminant sorbed onto soil particles is an important parameter in the design of any in situ remedial action plan. It is important to know what percentage of the contaminant will be sorbed on the soil particles versus the amount left either as a separate phase or dissolved in the ground water.

Biodegradation is another interaction mechanism that may occur between contaminants and soil/aquifer systems. Microorganisms that are native to a soil/aquifer system, or that are acclimated to a contaminant and grow by natural selection and mutation, might be able to use the contaminant as a food source. If this happens, the contaminant will be transformed into a different species that may be more or less harmful than the original contaminant. The species that is formed depends on several variables such as: (1) the contaminant, (2) the microorganisms, (3) nutrient additions or availabilities, and (4) an aerobic or anaerobic environment.

Biodegradation transforms the contaminant into a different species with different physical properties. In some cases, biodegradation can transform a molecule into a species that is nonbiodegradable (29). In other cases it can create a species that is also toxic. For example, one biodegradation pathway for trichloroethylene (TCE) produces vinyl chloride (30), another toxic substance. However, in the best cases of complete biodegradation under aerobic conditions, contaminants are transformed into carbon dioxide and water.

NAPLs may also interact with soil/aquifer systems through the mechanisms of vaporization and diffusion in the vadose zone. Vaporization is due to the vapor-liquid phase equilibrium for NAPLs that are in contact with air. Vaporized NAPLs diffuse through the vadose zone in the directions of decreasing concentrations. Also, the transport of vaporized NAPLs in a vadose zone will occur much faster than the corresponding NAPL transport in the water saturated zone because the NAPL vapor phase diffusion coefficients are several orders-of-magnitude greater than the corresponding liquid phase diffusion coefficients. Typical vapor phase diffusion coefficients are on the order of 10^{-1} cm^2/s while those of the liquid phase are on the order of 10^{-5} cm^2/s (31). Thus, vaporization is a transport mechanism that increases the extent of the contaminated zone.

Contaminants and soil/aquifer systems may also interact through plant root uptake. Plants absorb minerals through their roots and can also absorb contaminants in soils via this same mechanism. This causes contaminants to become part of the plant and results in their spread to the food chain.

Finally, contaminants and soil/aquifer systems can react and form a different species. Biodegradation falls into this category, but contaminants can also react with other species that may already be present in the soil/aquifer system (e.g., humic materials and other contaminants). The products of these reactions are new species with different physical and chemical properties than the reactants. Their interactions with soil/aquifer system and with surfactants must then be investigated.

Interaction Mechanisms between Surfactants and Contaminants.   Surfactants and contaminants interact through the mechanism of enhanced solubilization. When a surfactant is added to water, it increases the aqueous phase solubilities of NAPLs. On the one hand, this increased solubility results in contaminated ground water. On the other hand, the contaminated ground water can be recovered using a recovery or pumping well. The recovered ground water can then be treated for contaminant removal; the cleaned water can be recharged to the aquifer. While this cleanup strategy has the potential to increase pollutant dispersion, it is important for many otherwise inaccessible sites. The surfactant mobilizes the contaminant so that it can be recovered.

A surfactant does increase the aqueous phase solubilities of NAPLs, but dissolution rates are also important. Experimental evidence indicates that NAPL dissolution is mass transfer limited (*32*). However, Hunt et al. (*12*) present a model for NAPL dissolution kinetics which considers a spherical NAPL droplet suspended in a flowing aqueous stream and estimates the time required for the NAPL droplet to dissolve. Their model is based on a mass balance around a spherical NAPL droplet. The rate of change in the droplet mass is set equal to the rate of dissolution of the NAPL phase. The rate of dissolution is then expressed as a function of a mass transfer coefficient and the diameter of the NAPL drop. The resulting mass balance is an ordinary differential equation which can be solved to give the diameter of the drop as a function of time. The time at which the diameter goes to zero is defined as the droplet dissolution time. Relevant model parameters include the droplet diameter, NAPL solubility, and the flow velocity of the aqueous stream.

This dissolution model was modified to estimate surfactant effects (e.g., increased NAPL solubilities). For a given water flow velocity, the model dissolution times were computed for progressively increasing NAPL solubilities. The results are plotted in Figure 8 which illustrates the dissolution times for a spherical, 10-cm droplet of TCE. Dissolution times for other size TCE droplets are given by Porzucek (*11*).

The parametric curves in Figure 8 are for different values of the water flow

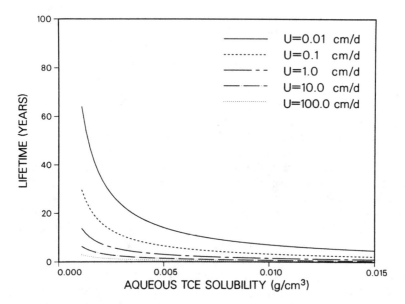

Figure 8. Trichloroethylene Droplet Lifetimes in an Aquifer as a Function of Aqueous Phase TCE Solubilities and Ground Water Flow Velocities (10 cm droplet). (Reprinted from Ref. 8.)

velocity, U. The top curve, $U = 0.01$ cm/day, shows dissolution times for a droplet in a very slowly moving aquifer. The curve for $U = 0.1$ cm/day (1.2 ft/year) is a realistic upper bound on naturally occurring ground water flow in the absence of large changes in aquifer elevation. The three remaining parametric curves are for larger ground water flow rates. These are more typical of aquifers that experience large changes in elevation or have an artificially induced hydraulic gradient.

The droplet lifetimes that correspond to a TCE solubility of 0.0012 g/cm^3 are for systems without surfactant. As the surfactant concentration increases, the TCE solubility increases and the droplet lifetimes decrease. For the case where $U = 0.1$ cm/day, surfactant decreases the dissolution time from 30 years to approximately five years and gives a quantitative measure of the magnitude of the surfactant effect. It also suggests that pump-and-treat remediation plans that don't use surfactants may require long-term strategies.

Surfactants and contaminants can also interact by chemical reaction mechanisms. These mechanisms are not really desirable because they will keep a surfactant from performing its job of increased solubilization and lowering the interfacial tension. In general, laboratory studies should be used to eliminate candidate surfactants that could possibly react with contaminants.

Surfactants and contaminants also interact through the mechanism of lowering interfacial tension. Surfactants that are injected into the ground water will eventually contact NAPL drops or lenses that are held in place by capillary forces. They will then lower the interfacial tension between the aqueous and nonaqueous (i.e., contaminant) phases. In common tertiary oil recovery experiments, for example, surfactant formulations lowered the interfacial tension from approximately 35 dyne/cm to about 0.001 dyne/cm. Similar reductions are expected from surfactant-contaminant interactions although data are not now available. This decreased interfacial tension reduces the capillary forces to the point where it might be possible to dislodge the trapped NAPL drops by the natural pressure gradient in the soil/aquifer system. From the previous discussion, the pressure gradient needed for displacement of a one-cm NAPL droplet trapped in silt-sized particles was 10,800 kPa/m. Using surfactants, it may be possible to lower the required pressure drop to 0.1 kPa/m, a more realizable value that could naturally exist in an aquifer. Once the NAPL drop is mobilized, it can be recovered using a strategically located pumping well.

Surfactants and contaminants also interact by the mechanism of desorption since surfactants are capable of removing contaminants from soil particle surfaces. For strongly sorbed (i.e., chemisorbed) contaminants, however, a given surfactant may not be able to desorb all the contaminant from the soil. Thus, laboratory screening experiments for surfactants are essential to determine those formulations which are most effective.

Another factor to consider, in addition to the desorption equilibrium, is the desorption rate. This rate is an important consideration because subsurface remediation problems are expected to be mass transfer limited. These desorption rate experiments can be carried out in batch mode using doped soil samples, a surfactant solution, and a containment device such as a test tube.

Doped soil samples are placed in a test tube and surfactant solution is added. The tubes are then agitated and allowed to settle. Samples of the aqueous solution are then drawn from the test tube and analyzed for contaminant concentration. Figure 9 shows results of typical batch experiments from our studies using TCE-doped soils and 0.5 vol% Triton X-100, a nonionic surfactant. The soil used was a mixture of 80% Ottawa sand and 20% commercial top soil. Results are plotted as the cumulative percentage TCE desorbed as a function of time. Data are shown for three different TCE-contaminated soils. The first contaminated sample contained 12 μg TCE/g of soil while the second and third samples contained 25 and 34 μg TCE/g of soil, respectively. Equilibrium is reached in less than six hours and 80–95% of the original TCE on the soil is desorbed. Trichloroethylene is a slightly hydrophilic compound (up to 1200 mg/L dissolve in water) and it is not surprising that such high desorption rates and extent of desorption are achieved. A more hydrophobic compound on a soil with greater organic content is expected to show slower desorption rates and decreased percentages of contaminant removed. In such cases, the choice of surfactant is critical.

Soil/Aquifer System and Surfactant Interactions.   In a surfactant-enhanced, in situ soil-washing scheme, the hydraulic gradient of a pump is used to force flow aqueous surfactant solutions through the contaminated zone. Thus, the soil/aquifer system and surfactant interact because of the hydraulic gradient imposed by the pump. This hydraulic gradient not only causes flow of the aqueous surfactant solutions, it also creates mixing between the surfactant and contaminant phases. This mixing is desirable and hence larger hydraulic gradients are attractive. Unfortunately, large hydraulic gradients are not easily generated in soils because the soil matrix cannot withstand these pressures without fracturing and losing integrity. Therefore, relatively low hydraulic gradients must be used for in situ soil-washing schemes, practical values are typically around 0.13 kPa/m.

Surfactants dissolved in water will interact with the soil/aquifer system by sorption. By definition, surfactants are surface active species and like to sorb onto surfaces. More specifically, surfactants will sorb onto clay minerals and associate with dissolved humic materials that exist in soils. This sorption is influenced by the solution pH, cationic strength, and temperature.

Finally, surfactants interact with the soil/aquifer systems through the mechanism of biodegradation. Native microorganisms present in a soil/aquifer system may be capable of biodegrading a surfactant. Under aerobic conditions and complete mineralization, this process eventually converts the surfactant into CO_2 and water. Moreover, it is desirable to use a biodegradable surfac-

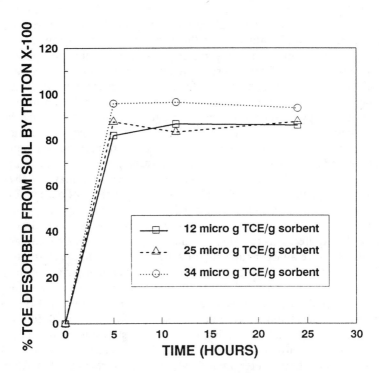

Figure 9. Cumulative Desorption from TCE-Contaminated Soil Contacted with 0.5 vol% Triton X-100 Surfactant

tant so that any surfactant left unaccounted for will simply degrade into harmless components. On the other hand, if the surfactant degrades too quickly, it will be ineffective at cleaning the soil. Thus, biodegradation tests should be performed using the selected surfactant and native microorganisms from the contaminated site. These tests will show how long a surfactant will survive before biodegradation starts.

Several factors impede the application and flow of surfactant in a soil/aquifer system. Heterogeneities in soil type, porosity, and permeability will alter the fluid flow in soils and add uncertainty to the flow path of surfactant solutions. These factors make it difficult to obtain a uniform, predictable, areal sweep of injected surfactant solution through the desired soil/aquifer area.

In addition to heterogeneities in soil physical properties, soil pore blockage will also affect the flow of surfactant solutions. Such blockages can occur if emulsions or flocs form between surfactants and contaminants or surfactants and the soil/aquifer system. It should be possible to avoid such undesirable phase behavior by performing proper laboratory phase studies using the surfactants and contaminants. However, surfactant phase behavior is known to be a strong function of pH, temperature, cationic strength, and organic concentration (*33,34*). Unfortunately, these are variables that are not completely controllable at most in situ remediation sites. Therefore, it may not always be possible to avoid undesirable phase behavior and subsequent pore blockage.

Pore blockage can also occur by mobilization of fines (e.g., clay-sized particles). Fines can be mobilized and redistributed by the natural hydraulic gradient of the aquifer. Structurally, the fines consist of organic carbon and clay minerals and have a larger capacity for sorption when compared to other soil particles. Therefore, a large portion of a contaminant can sorb onto the fines.

Once the fines have redistributed in the soil/aquifer system, they create preferential flow paths which bypass pores that are blocked by the fines. Surfactant solutions that have been injected into the soil/aquifer system will follow these preferential flow paths because they offer less resistance than the more tortuous paths that go through the fines. When the fines segments of the soil are bypassed due to these preferential flow paths, the surfactant is ineffective in cleaning up the soil/aquifer system.

Summary

Soil/aquifer system-contaminant-surfactant interactions are important in the design of in situ remedial action plans for contaminated sites that are either very large, very inaccessible, or both. Important physical properties of the soil/aquifer system, contaminant, and surfactant are identified and discussed as they pertain to further spread of contamination or cleanup. Interactions between these components and the influence of surfactants are also important.

The mechanisms of sorption/desorption, capillary forces, and mass transfer-limited solubilization are particularly important.

Acknowledgements

This work was partially supported by Los Alamos National Laboratory under Contract No. 9-X58-8080U-1. We are thankful to Dr. Michael Ebinger of Los Alamos National Laboratory for his help in this study and Alexander Helou of Howard University for performing the laboratory studies.

Literature Cited

1. Ellis, W., Payne, J., Tafuri, A., and Freestone, F. In *Proc Hazardous Material Spills Conf*, pages 116–124, Nashville, TN, 1984.
2. Ellis, W., Payne, J., and McNabb, G. *Treatment of Contaminated Soils with Aqueous Surfactants*. Final Report EPA/600/2-85/129, Hazardous Waste Engineering Research Laboratory, Office of Research and Development, U.S. EPA, Cincinnati, OH, NTIS, Springfeld, VA, PB86-122561, 1986.
3. Chawla, R., Diallo, M., Cannon, J., Johnson, J., and Porzucek, C. In-situ treatment of soils contaminated with hazardous organic wastes using surfactants: a critical analysis. In Muralidhara, H., Ed., *Solid/Liquid Separation: Waste Management and Productivity Enhancement, 1989 Int Sym*, pages 355–367, Battelle Press, 1989.
4. *Underground Movement of Gasoline on Groundwater and Enhanced Recovery by Surfactants*. API Publication 4317, Texas Research Insititute, 1979.
5. *Test Results of Surfactant Enhanced Gasoline Recovery in a Large-Scale Model Aquifer*. API Publication 4390, Texas Research Institute, 1982.
6. McNabb, Jr., G., Payne, J., Ellis, W., Kirstein, B., Evans, J., Harkins, P., and Rotunda, N. Chemical countermeasures application at the Volk Field site of opportunity. Internal Report to EPA, 1985.
7. Nash, J. Field studies of in-situ soil washing. Final Report to EPA on Contract No. 68-03-3203, 1986.
8. Nash, J. and Traver, R. In *Proc Hazardous Material Spills Conference*, Nashville, TN, 1984.
9. Sale, T. and Pitts, M. Chemically enhanced in situ soil washing. In *Proc NWWA/API Conf on Petroleum Hydrocarbons and Organic Chemicals in Groundwater: Prevention, Detection and Restoration*, pages 487–503, Houston, TX, Nov. 15-17 1989.
10. Pouska, G., Trost, P., and Day, M. In *Proc 6th National RCRA/Superfund conference and Exhibition*, pages 423–430, New Orleans, LA, 1989.

11. Porzucek, C. *Surfactant Flooding Technology for In Situ Cleanup of Contaminated Soils and Aquifers — A Feasibility Study*. Report LA-11541-MS, Los Alamos National Laboratory, 1989.
12. Hunt, J., Sitar, N., and Udell, K. Nonaqueous phase liquid transport and cleanup 1. Analysis of mechanisms. *Water Resources Res*, 24(8):1247–58, **1988**.
13. Wilson, J. and Conrad, S. Is physical displacement of residual hydrocarbons a realistic possibility in aquifer restoration?   In *Proc NWWA/API Conf Petroleum Hydrocarbons and Oranic Chemicals in Ground Water — Prevention, Detection and Restoration*, pages 275–298, Houston, TX, Nov. 5-7 1984.
14. Hoag, G. and Marley, M. Gasoline residual saturation in unsaturated uniform aquifer materials. *J Env Eng*, 112(3):586–604, **1986**.
15. Carter, C. and Suffet, I. Binding of DDT to dissolved humic materials. *Env Sci Tech*, 16(11):735–740, **1982**.
16. West, C. *Dissolved Organic Carbon Facilitated Transport of Neutral Organic Compounds in Subsurface Systems*. PhD thesis, Rice University, Houston, TX, 1984.
17. Means, J., Wood, S., Hassett, J., and Banwart, W. Sorption of polynuclear aromatic hydrocarbons by sediments and soils. *Env Sci Tech*, 14(12):1525–29, **1980**.
18. Schwarzenbach, R. and Westall, J. Transport of nonpolar organic compounds from surface water to groundwater, laboratory sorption studies. *Env Sci Tech*, 15(11):1360–67, **1981**.
19. Karickhoff, S. Organic pollutant sorption in aquatic systems. *J Hydral Eng*, 110(6):707–735, **1984**.
20. Garbarini, D. and Lion, L. The influence of the nature of soil organics on the sorption of toluene and TCE. *Env Sci Tech*, 20(12):1263–69, **1986**.
21. Adamson, A. *Physical Chemistry of Surfaces*. John Wiley & Sons, Inc., 1982.
22. Berg, R. Capillary pressures in stratigraphic traps. *Am Ass Pet Geol Bull*, 59(6):939–956, **1975**.
23. Halliday, D. and Resnick, R. *Fundamentals of Physics*. John Wiley & Sons, Inc., 1974.
24. Bailey, G. and White, J. Factors influencing the adsorption, desorption and movement of pesticides in soils. *Residue Rev*, 32:29–92, **1970**.
25. Weber, W. Sorption processes and their effects on contaminant fate and transport in subsurface systems. 1990 Lecture, Association of Environmental Engineering Professors, 1990 Distinguished Lecture Series also submitted (with P.M. McGinley and L.E. Katz) to *Water Res*, Feb. 1990.
26. Karickhoff, S., Brown, D., and Scott, T. Sorption of hydrophobic pollutants on natural sediments. *Water Res*, 13(3w):241–248, **1979**.
27. Means, J. and Wood, S.   Sorption of amino- and carboxy-substituted

polynuclear aromatic hydrocarbons by sediments and soils. *Env Sci Tech*, 16(2):93–98, **1982**.

28. Leo, A., Hansch, C., and Elkins, D. Partition coefficients and their uses. *Chem Rev*, 71(6), **1971**.

29. Karveta, L., Chung, H., Guin, K., Shebs, W., and Smith, L. Ultimate biodegradation of an alcohol ethoxylate and a nonylphenol ethoxylate under realistic conditions. Annual Meeting Soap and Detergent Association, Boca Raton, Florida, Jan. 27-31 1982.

30. Fisher, M. *Transport and Fate of Organic Chemicals in Bandelier Tuff at Los Alamos National Laboratory Chemical Waste Site, New Mexico*. Master's thesis, University of Texas, El Paso, 1988.

31. Bird, R., Stewart, W., and Lightfoot, E. *Transport Phenomena*. John Wiley & Sons, Inc., 1960.

32. Sitar, N., Hunt, J., and Udell, K. Movement of nonaqueous liquids in groundwater. In *Proc Geotechnical Practice for Waste Disposal '87*, pages 205–223, Geotechnical Division, American Society of Civil Engineers, Ann Arbor, MI, 1987.

33. Healy, R. and Reed, R. Physicochemical aspects of microemulsion flooding. *Soc Pet Eng J*, 14(3):491–501, **1974**.

34. Healy, R., Reed, R., and Stenmark, D. Multiphase microemulsion systems. *Soc Pet Eng J*, 16(3):147–160, **1976**.

RECEIVED April 5, 1991

Chapter 17

Mechanisms of Mobilization and Attenuation of Inorganic Contaminants in Coal Ash Basins

Shingara S. Sandhu[1] and Gary L. Mills[2]

[1]Clafin College, Orangeburg, SC 29115
[2]Savannah River Ecology Laboratory, University of Georgia,
Drawer E, Aiken, SC 29801

A study was undertaken to evaluate the field effectiveness of technology applied to the disposal of ash generated by coal-fired power plants located at the Savannah River site. Sediment cores, collected from ash basins of varying ages, were sectioned and analyzed to collect information for elucidating the modes of mobilization and attenuation of several major and minor nutrients. There was significant translocation of most elements from the upper to the lower horizons of the impounded ashes where attenuation occurred. This effect was most pronounced in the old basin, with a well-established plant community. Low pH, generated by decomposing organic matter was considered partially responsible for solublizing these elements; however, the complexation by organic ligands may also have been important. The leached components remained in solution until the percolating ground water reached a zone of higher pH and redox conditions, resulting in the attenuation of several elements by precipitation, coagulation, and surface reactions. Further downward migration of ash-derived elements can be expected as the lower horizons of ash basins become ion saturated and acidified.

The utility industry produces more than 7.5×10^7 t of solid waste annually of which only 2.0×10^7 t (26.67%) is put to practical use. Available estimates indicate that solid wastes generated by the utility industry will double by the turn of the century. The electrical utilities reported (1) that 70% of the current ash production is sluiced to disposal ponds where it comes in contact with the aquatic environment posing contamination potential for ground and surface waters.

Fly ash is generated at the Savannah River Site (SRS) by routine operations of a four unit, 4×10^8 btu hr^{-1}, coal burning power plant that has been operating since 1952. In 1987, the plant burned 2×10^5 t of coal, averaging 1.9% sulfur. The ash content of the coal is currently 10.4%, a decrease of 14% from the mid-seventies. Prior to 1976 about 67% of the ash was collected by mechanical cyclones. The remaining ash (33%) went up the stacks. Since 1976, electrostatic precipitators have been in operation to remove fly ash from stack emissions with greater than 99% efficiency. The ash particulates collected by the electrostatic precipitators are smaller in size than those trapped by the cyclone collectors and contain high concentrations of many trace metals (2).

0097–6156/91/0468–0342$06.75/0
© 1991 American Chemical Society

Several studies *(3-9)* have examined the chemical composition and water chemistry of the ash basin water at the SRS as well as the stream system which receives overflow discharged from the basins. One of these studies *(9)* also examined the changes in water chemistry within the ash basin water system during normal seasonal cycles and compared these changes to those occurring in nonimpacted impoundments. However, the mobility and fate of metals in the ash basin sediments formed by the accumulation of ash in the impoundments has not been reported. The older ash basins were abandoned when filled and presently have been colonized by vegetation which has accumulated high levels of several toxic and non-toxic metals *(10)*. The abandoned ash basins have aged and weathered for several years; thus, the partitioning of the trace metals among solid phase components and their leachability from the ash sediments may have changed. Quantitative rates for long-term leaching of fly ash contaminants from utility waste disposal sites are necessary to accurately predict the potential for ground water contamination. This study reports the release, mobilization and attenuation of inorganic contaminants from ash basins of various ages which are located at the Department of Energy's Savannah River Site (SRS) in South Carolina. The study also describes the mechanisms involved in the leaching and attenuation of the elements in the ash basins.

Materials and Methods

Study Area. The basins received ash sluiced from power production facilities in D-area at SRS, which have been in service since plant start up in 1952. D-area (Figure 1) has three basins which are adjacent to each other. The primary ash basin and secondary ash basin are active disposal sites and receive 5.07×10^4 m^3 of ash generated by the power plant. About 4.9×10^9 L yr^{-1} of water is used to transport the ash. The sluiced water gravity flows alternately into the small receiving basins at sites A or F and subsequently moves into the primary basin where sedimentation of ash takes place. The supernatant water, still quite turbid, flows into the secondary ash basin. Overflow water from the secondary disposal basin, along with smaller sized or less dense suspended ash particulates drains into a small swamp which in turn discharges into Beaver Dam Creek, which is a tributary of the Savannah River. The dimensions of the primary ash disposal basin are 100m x 100m x 3.0 m and that of the secondary ash disposal basin, 100m x 50m x 3.0 m. The primary basin has a calculated retention time of 39 days and the secondary ash disposal basin a retention time of 22 days *(9)*. Though the latter had several feet of water, it contained no vegetation except peripheral grasses, wax myrtle etc., and was inaccessible. However, several varieties of vegetation, including grass and bushes, covered half of the primary ash disposal basin.

The sampling sites, New South One (NS-1) and New South Two (NS-2), had a thick and healthy growh of wax myrtle, and an undergrowth of dog fennel, cattails etc. The sampling sites, New North One (NN-1) and New North Two (NN-2), were dominated by several varieties of aster, cattails, dog fennel, rabbit tobacco and dandelion, and also showed a growth of sea myrtle and wax myrtle. Swamp cottonwood, black willow and red maple were also scattered in the primary ash disposal basin. At the time of the study, vegetation was observed in the half of the primary ash disposal basin which was under several feet of water.

The old ash basin (488-D), located to the north of the secondary ash disposal basin (Figure 1), adjoined a coal burial basin. It had a thick

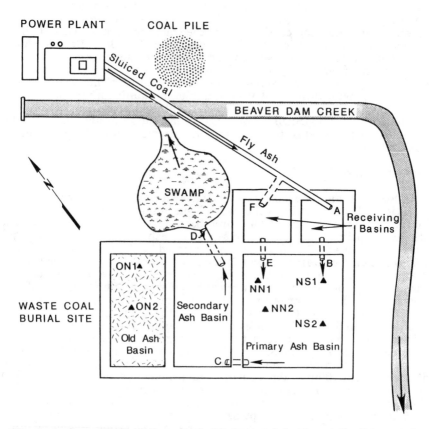

Figure 1. Sampling locations in the D-Area ash basins at the Savannah River Site.

growth of common swamp vegetation, including swamp cottonwood, sycamore, black willow, wax myrtle, pokeweed, and elderberry. Its surface was covered with a thick layer of organic matter, in various stages of decomposition. All the ash basin sites were surrounded by above-ground containment which was over two meters high.

Sampling location. After a visual survey of Area B, each basin (488-D and 488-ID) was divided into several small lots. Final site selection of each ash disposal basin was based on the representativeness of the area and its accessibility for sampling throughout the project period. Two sampling sites (ON-1, ON-2) were selected in the old ash disposal basin. The four sites, selected in the primary ash disposal basin were designated NN-1, NN-2, NS-1, and NS-2. Four surface water sampling sites, A, B, C, and D, were also selected. Site A consisted of ash sluiced water, and sites B, C and D of water just entering the primary ash disposal basin, the secondary ash disposal basin and the swamp (Figure 1), respectively.

Sampling. Several profiles in the old and new ash basins were exposed to a depth of 0.75 m to examine the ash stratification. Several cores of undisturbed ash from each site were obtained using a Wildco soil coring apparatus with a PVC core liner. Sample cores were 90 cm long and 5.12 cm in diameter. Since no stratification or horizon development in the ash basins was observed, the ash from each column was separated arbitrarily into categories with dimensions of 0-7 cm, 7-15 cm 15-30 cm, 30-60 cm, and 60-77 cm. It was impractical to remove the ash cores below the above mentioned depths due to excessive moisture.

The ash samples were transferred to 500 mL centrifuge bottles, weighed, and sequentially extracted with deionized (Milli-Q, Millipore Corp) water, 1.0 M $MgCl_2$ (pH 7), 0.04 M NH_2OH; HCl in 25% HOAc, and 30% H_2O_2 adjusted to pH 2 with HNO_3 *(11)*. The samples were centrifuged to separate the liquid from the solid. An aliquot of supernate from each sample was passed through 0.45 mm membrane filters (Millipore Corp., type HA), acidified with 0.5 mL of ULTREX nitric acid, and stored in acid washed polyethylene bottles for elemental analyses. Procedural blanks were generated by passing deionized water through the same processes. The nutritional level *(12)*, mechanical composition, and porosity of ash sample were also determined.

The mechanical analysis *(13)* and porosity determinations *(14)* were carried out with the composite samples prepared by mixing equal amounts of ash from NN-1 with NN-2, NS-1 with NS-2 and ON-1 with ON-2. Surface water samples from stations A, B, C, and D were collected in the summer of 1987 and 1988 to evaluate the characteristics of sluiced water and the changes it underwent during its passage through the ash disposal basins. The water quality parameters, temperature, pH, and conductivity and the amount of dissolved oxygen were determined using an in situ Water Quality Analyzer (Hydrolab Corp., Model 8000). All the elemental analyses were conducted with a Hitachi Model 180-80 Zeeman Effect Atomic Absorption Spectrophotometer using Zeeman background correction equipped with a GA-3 graphite atomizer *(15)*. Zirconium acetate treated, pyrolytically coated cuvettes *(16)* and nickel nitrate additions were used to improve arsenic analysis. Calcium was determined in the air-acetylene flame after the addition of $LaCl_3$. Finally, the matrix effects for all elements were corrected by using the method of standard additions.

Results and Discussion. The American Society of Testing and Materials *(17)* broadly classifies fly ash into C and F groups. Though a detailed elemental or mineralogical analysis for the ash under study was not conducted, the available information suggested that the ash at SRS basins in the D-area belonged to the C group, as it contained considerable amount of calcium and the fresh water suspension was distinctly alkaline (Table I). It was also necessary to establish the similarities and dissimilarities between the parent materials forming the ash sediments at the study sites. Consequently, a set of ash samples from the ON and NS basins were isotopically analyzed by Chempet Research Corporation *(18)*. The analysis indicated that Sr isotopic compositions of the ON and NS samples were not significantly different indicating a common origin of ash at the two sites. Thus, based on ASTM tests (Methods E-178 and D-277) and Sr isotope analysis, it was concluded that all samples embraced a common origin of parent ash material.

Table I. Characteristics of sluiced fly ash water (N = 4)

| Characteristics | Station | | | |
	A	B	C	D
pH	8.98	7.58	7.41	7.05
Conductivity, mmho/cm	0.28	0.59	0.43	0.40
Dissolved oxygen, mgL^{-1}	4.80	4.50	7.45	5.95
Temperature, C	35	35	31	30
Ca, µg g^{-1}	16.5	21.4	17.1	15.7
Mn, µg kg^{-1}	1.76	2.41	1.89	1.84
Fe, µg kg^{-1}	80.9	3.55	19.3	16.8
Ni, µg kg^{-1}	5.00	6.67	5.00	8.33
Cd, µg kg^{-1}	0.30	0.13	0.64	0.64
As, µg kg^{-1}	183	6.89	46.0	64.9
Cr, µg kg^{-1}	7.28	1.15	3.38	3.38

Sluiced Fly Ash Water. The dissolution of ash components begins on contact with water and the solubility of trace metals and other components associated with power plant solid waste residues depends on many factors, especially the pH. The ash slurry entering the receiving pond had pH values between 7.0 and 8.98 (Table I). The average pH of the sluiced water decreased as it moved from the receiving pond (Site A) through the primary and secondary ash disposal basins to the swamp (Site D). The pH changes were probably associated with the chemical equilibrium that existed between calcium oxide, calcium carbonate and atmospheric carbon dioxide *(19)*. The data presented in Table 1 shows that the sluiced fly ash water is rich in calcium. Thermodynamic considerations dictate that during the combustion process of pulverized coal in the furnace of the power plant, calcium is oxidized to CaO which on contact with moisture and atmospheric CO_2 changes to $Ca(OH)_2$, $CaCO_3$, and $Ca(HCO_3)_2$ respectively, according to the following equations:

$$CaO + H_2O \leftrightarrows Ca(OH)_2 \qquad (1)$$

$$Ca(OH)_2 + CO_2 \leftrightarrows CaCO_3 + H_2O \qquad (2)$$

$$CaCO_3 + H_2O + CO_2 \leftrightarrows Ca(HCO_3)_2 \qquad (3)$$

At Site A, $Ca(OH)_2$ (pH = 12.4 for saturated solution) dominated. However, as the water moved through the ash basin system, dynamic equilibria was established between atmospheric CO_2, $Ca(OH)_2$, $CaCO_3$ (pH 9.4 for saturated $CaCO_3$) and $Ca(HCO_3)_2$ (20), which affected the pH of the system at particular locations. The characteristics of the sluiced ash water, especially its pH, can play an important role in the weathering of ash, as well as in the release, mobilization, and attenuation of inorganic contaminants in the ash basin system. These characteristics were not very conducive for the dissolution and mobilization of most of the trace metals in ash. The concentration of every trace element (Table I), except arsenic in water, was below the standard prescribed for drinking water *(21)*.

Physical and Chemical Characteristics of Ash Columns. Composite samples for particle-size analysis and porosity determinations were prepared by mixing ash from the New North site (NN) with the corresponding depth ash samples from the New South site in the ratio of 1:1. The data on the mechnical compositions in Table II show that no significant difference in porosity appeared to exist among samples, indicating that no further compaction occured with aging. However, both sites were very porous and water movement within the ash columns was unimpeded. It was interesting to note a significantly (R^2 = -0.98) decreasing organic matter gradient with depth in the ON ash basin. The surface layer of the Old North site contained a thick layer of organic matter in various stages of decomposition making it more acidic than the surface horizons of the NN and NS sites. In addition, water-soluble organic compounds derived from the decomposing litter moved down through the ash columns with the percolating water until they reached an ash environment with changed redox and pH conditions which led to its attenuation by precipitation or adsorption. The organic compounds are potential vehicles for the mobilization and transport of trace metals from surface to lower horizons *(22)*.

The ash-like soil samples were analyzed by the method developed by the United States Department of Agriculture *(23)* to obtain the textural classes (Table II). The analysis indicated that there was no significant difference in the fractions of particle size distribution at both study sites which were dominated by silt sized particles. The percentage of coarse (sand), medium (silt), and fine (clay) particles did not change significantly with the changing depth. The ON sampling site, with its thick vegetation, contained higher amounts of organic matter and lower levels of silt in comparison to the NN and NS sites which were being filled at the time of this study.

In addition to fossilized organic material, coal also contains small amounts of clay, silt, and sand sized minerals. During the combustion of coal in the power plant furnace, some of this material is likely to condense

Table II. Physical Characteristics of ash (composite samples)

Site	Depth cm	Density, g/cc			Organic Matter	Gravel	Percentage of the total				Structural Classes
		Bulk	Particle	Porosity			Sand	Silt	Clay		
NS-NN	0-7				2.95	0.11	13.58	72.05	9.20		Silt Loam
NS-NN	7-15	0.78	1.86	58.1	2.53	2.84	20.28	63.20	8.03		Silt Loam
NS-NN	15-30				1.78	2.57	47.50	34.56	7.80		Sandy Loam
NS-NN	30-60	0.81	2.19	65.3	1.99	3.29	23.12	60.53	6.35		Silt Loam
ON	0-7				5.15	2.61	40.39	38.72	14.16		Loam
ON	7-15	0.7	1.81	61.4	4.25	0.13	33.89	52.57	9.97		Silt Loam
ON	15-30				3.78	1.20	29.20	54.82	8.76		Silt Loam
ON	30-60	0.68	1.79	62.1	2.60	7.91	31.13	51.15	7.64		Silt Loam
ON	60-77				1.25	17.80	28.23	47.44	9.56		Loam

as part of the ash. A 1981 study *(24)* had reported that the top stratum samples of fly ash exposed to the sluice pond were tan, brown, and orange in color and that the fine ash particles in the upper stratum were enriched with iron due to the oxidation. Contrary to the above findings, the present study did not show any well developed stratification or coloration. All ash columns at each site, to a depth of about 70 cm, were found to contain only gray colored material, even though a considerable concentration of iron was found (Table III) on the surface layers of these columns.

Ash samples obtained from various horizons were analyzed *(23)* for total elemental concentrations to ascertain the fertility and toxicity levels of elements needed to support plant growth (Table III). The data also provided insight into the activities of several major elements which could have played important roles in the transport and attenuation of several ash-derived elements. The data indicated that there was a significant translocation of most elements from the upper to lower soil horizons. This pattern was evident in ash columns from the Old North site which had been subjected to natural weathering for about 30 years. At this site, most of the leached material had been deposited at a depth of 30-60 cm at the expense of the sub-surface horizon (7-15 cm). It is conceivable that elemental leaching also took place from the surface ash layer (0-7 cm), which was later replenished with additional elements released by the decomposing organic matter. The element migration process was also noticeable to some extent in the ash profiles of the NS site. However, the NN site, which contained fresh ash deposits, showed virtually no translocation of elements.

Estimating transport and fate of solutes in the ash deposits is a complicated process because no single model can describe the mobilization and attenuation of chemical components for such systems. The changes in redox and pH conditions of the ash deposits, probably more than any other factors, were responsible for the release of metals and non-metals. These elements were mobilized to a depth determined by the nature of the element and the changing environmental conditions.

At the time of the study, the primary ash disposal basin was receiving fresh sluiced ash which probably affected the translocation of elements in the ash columns at this site. The surface layers were also constantly being replenished with fresh deposits of soluble and insoluble materials. However, the old ash basin had been an inactive site for the last 30 years, its ash deposit had undergone weathering and thus projected a more realistic picture about the potential element mobilization and subsequent contamination of ground water.

The data in Table III suggests interesting mobilization and attenuation patterns for different elements in the old ash basin. It appears that P, Ca and Al were mobilized mainly from the 7-15 cm horizon and deposited at the 30-60 cm layers. A simple mathematical model dv/dt (v = depth in centimeters, t = time in years) applied to the mobilization of elements in the old basin provided a mobility rate of less than 2 cm/year. According to a report by Warren and Dudas *(25)*, the mobility of these elements in ash columns followed a similar pattern, where the constituents were initially mobilized from artificially weathered ash, and later retained by the ash residues below. The translocated material formed secondary mineral precipitates of iron oxyhydroxides, calcium carbonate, and aluminum silicate through a variety of mechanisms and were responsible for precipitation and absorption of most trace elements.

In the primary ash basin there appeared to be some migration of iron to the lower horizons. However, in the old ash basin, iron distribution

Table III. Chemical Characteristics of Ash

Location							Concentration, $mg\ kg^{-1}$								
Site	Cm	pH	P	K	Na	Ca	Mg	Al	Pb	Mn	Fe	B	Cu	Zn	Mo
NS	0-7	6.7	33.2	82.0	63.2	485	48.4	545	0.8	8.5	143	4.7	7.9	1.8	0.4
NS	7-15	6.4	41.1	44.0	46.0	287	26.9	456	0.5	3.8	153	3.4	8.6	1.7	0.4
NS	15-30	6.5	52.6	52.7	39.8	247	23.1	377	0.5	2.9	126	2.7	7.9	1.6	0.2
NS	30-60	7.2	139	64.8	43.7	416	--	454	0.5	4.8	196	2.4	7.9	2.2	0.3
NN	0-7	6.0	67.8	86.0	249.9	497	67.9	577	0.7	5.0	156	5.9	4.5	1.1	0.5
NN	7-15	6.4	35.9	74.6	155.4	155	56.8	630	0.7	3.1	165	5.5	5.0	1.6	0.4
NN	15-30	6.5	34.3	66.0	126.3	350	46.4	466	0.5	2.9	165	4.4	4.1	1.5	0.3
NN	30-60	7.3	71.4	67.8	195.9	601	68.9	779	0.7	4.6	264	9.6	5.3	2.7	0.4
ON	0-7	5.8	62.2	106.8	34.9	861	46.4	468	0.4	13.7	117	1.3	0.7	3.2	0.2
ON	7-15	5.6	15.9	95.8	22.2	487	38.5	422	0.3	3.1	101	1.3	1.4	0.5	0.2
ON	15-30	5.6	46.7	119.7	68.4	721	54.4	504	0.7	4.2	89	2.8	2.1	0.9	0.3
ON	30-60	6.3	54.0	145.1	95.8	1847	171.3	63	0.8	8.9	84	4.5	1.6	1.5	0.3
ON	60-77	7.3	29.6	162.4	108.3	640	240.0	313	0.7	8.1	85	4.3	1.5	1.2	0.2

decreased with depth ($R^2 = -0.85$). Manganese followed the iron distribution pattern in the primary ash basin, but in the old ash basin it accumulated at a depth of 30-77 cm after being leached from the subsurface layer (7-15 cm). The manganese appears to be replenished in the surface layer (0-7 cm) by the decomposing organic litter. The zinc mobilization pattern was similar to manganese.

Sodium, potassium, magnesium and, to some extent, boron showed similar mobilization and attenuation patterns in the old as well as in the primary ash basins. The highest accumulation of these elements was found at a depth of 60-77 cm in the old ash basin. The primary ash basin did not show significant moblization of these elements. The translocation of sodium, potassium, and boron ions through the ash columns was limited primarily by the depth of downward movement of percolating water This study differed from Warren and Dudas' conclusion *(25)* that boron is not attenuated in ash, as the data in Table III reflect a significant mobilization and attenuation pattern for boron. This situation would probably change if there was uninhibited water movement through the ash columns.

The solubility of calcium, aluminum, and phosphate in the ash environment was pH controlled (pH 5.8) making it conducive for the slow dissolution and mobilization of these elements *(26)*. The most active zone of chemical reaction and eluviation appeared to be the subsurface ash layer (7-15 cm). The solubility product constants (log K_{sp}) for $Fe_2 O_3 \cdot 3H_2O$, $Al_2 O_3 \cdot 3H_2 O$, phosphate, and calcium carbonate in neutral aqueous systems are very low *(26)*. It is conceivable that the interaction of several elements took place in the zone of eluviation (30-77 cm) leading to their attenuation as coprecipitates of calcium, aluminum, and phosphate.

Chemistry of Minor Elements. The four trace elements - cadmium, nickel, chromium and arsenic - selected for in-depth study are on the Priority List of Pollutants of the United States Environmental Protection Agency (USEPA). The salts of these elements are highly toxic. In fact, nickel, chromium and arsenic are suspected carcinogens *(27)*. Each one of these elements depicts a different chemistry and behavior *(28)*.

Cadmium: The speciation of cadmium shows that it was mainly associated with the dissolved phase, which accounted for 62.8 to 94.4 percent of the total (Table IV; Figure 2). There was significant eluviation from the subsurface layer (7-15 cm) the zone of active chemical and microbial activity - and attenuation in the 15-30 cm horizon at the ON site. The surface layer had been enriched by the deposition of cadmium through the recycling, growth, and decay of vegetation. Cadmium exists only in a divalent form *(28)* which is presumably stable ($Cd^{+2} + 2e \rightarrow Cd$, $E° = -0.4026$ v) in the environment. It does not undergo rapid transformation due to minor redox and pH fluctuations, and because of its relatively low boiling point (765°C) is easily volatilized during combustion and subsequently condenses onto the layers of particles after leaving the zone of combustion.

Cadmium, along with other +1- and +2-valent elements, is concentrated *(1, 29)* in the glass phase which undergoes a slow hydrolysis reaction releasing cadmium into the surrounding aqueous environment of ash. The pH of 5.6 observed in ON subsurface layer appeared conducive for the release of cadmium to the bulk solution. Similar trends for the dissolution of cadium were reported (30) for a pH range of 4.0-8.5. Though various cadmium compounds such as $Cd(OH)^+$, $Cd(OH)_2$ (aq), $Cd(OH)_3^-$, $CdCo_3$ (aq), and CdS (aq) prevail if pH conditions are distinctly alkaline

Table IV. Concentration (μg/kg) of Cadmium in ash columns

Site	Depth(cm)	Dissolved	Exchange	Fe/Mn	Organic	Sum
NS	0-7	115	1.08	47.5	16.8	180
NS	7-15	190	1.32	45.0	16.9	253
NS	15-30	282	2.26	40.0	12.8	338
NS	30-60	496	2.98	35.6	14.6	550
NN	0-7	452	2.05	42.5	10.3	507
NN	7-15	1250	2.14	42.2	15.2	1310
NN	15-30	488	0.97	37.5	6.8	541
NN	30-60	310	0.73	37.5	8.3	356
ON	0-7	2535	2.77	32.7	6.1	2577
ON	7-15	222	0.45	25.0	4.1	252
ON	15-30	1162	0.39	22.5	3.5	1189
ON	30-60	180	0.58	32.5	6.2	211
ON	60-77	156	0.11	67.7	9.4	232

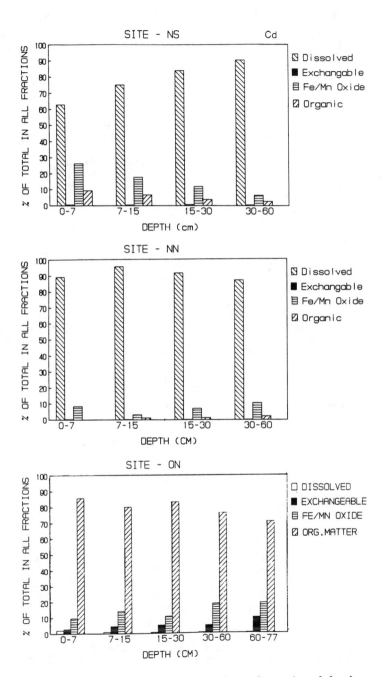

Figure 2. Distribution of Cd among various phases in ash basin sediments.

(pH = 9.5 or over), they were not expected at a pH 5.6 (acidic) observed at the ON study site. The concentration of cadmium in the dissolved phase was high and poses a potential groundwater pollution problem in the future.

Nickel: Table V shows the distribution of nickel in the various ash phases. The organic phase contained the major concentrations of nickel ranging from 537 µg/kg (lowest) to 10144 µg/kg. Strong correlations (ON, $R^2 = 0.95$; NN, $R^2 = 0.87$) between the concentrations of nickel and the amount of organic matter were found in various ash horizons. Small amounts of nickel were also associated with Fe/Mn and exchangeable phases. Together they accounted for less than 30% of the nickel whereas organic matter contained 75% to 87% of the element (Figure 3). The fraction of nickel observed in dissolved phase was negligible.

The data collected showed a great contrast between the behavior of cadmium and nickel in the ash environment. Nickel is stable in a divalent state, though under strongly reducing conditions it may also be found as Ni^{+3} and like cadmium is also slow in responding to redox changes ($Ni^{+2} + 2e \rightarrow Ni$, $E° = 0.25V$). Although a past study linked nickel and many other first transition series metals in the magnetic spinal phase of ash (24), the present study did not support these findings.

The reason for incorporation of nickel in the organic phase at the NN and NS sites was not evident, as the fresh ash deposits of NN did not contain significant amounts of vegetation litter. The nickel in ash at these sites, must have been incorporated into the organic carbon which formed black coatings on minerals during the combustion of coal in the furnace. The fresh ash samples collected from the power plant in the D-area were found to contain an appreciable amount of carbon (31) [Rear Past Ash = 2.01% C, Precipitated Ash = 2.2% C, Cyclone Ash = 24.33% C.].

Chromium: The chromium concentrations associated with various phases of ash deposits (Table VI) varied from 13437 µg/kg to 839 µg/kg in the organic phase to 1078 µg/kg to 4502 µg/kg in the Fe/Mn phase. The distribution of chromium (Figure 4) showed very interesting patterns. The surface ash deposits of NS and NN contained over 84% of chromium in the Fe/Mn phase. However, in the lower horizons at these sites, the fractions of chromium associated with this phase appeared to be decreasing with depth or age of ash.

The NN and NS subsurface horizons contained higher concentrations of Fe (Table III) than the surface layers. Several studies (24, 25) have indicated the association of chromium with the iron-rich mineral phase of fresh ash, a hypothesis supported by the present study. However, the chromium associated with the mineral phase appeared to be easily released. The pH conditions in the ash basins were weakly (pH = 6.0) to mildly (5.6) acidic and thus conducive to the slow dissolution of various minerals (glass, silicates, carbonates, etc.) which had chromium incorporated in the matrix. As chromium was released through the various physico-chemical activites, it was complexed by the organic compounds (humic acid, low mol. wt. organic acids, etc.) released by plant litter decomposition. The redox and pH conditions prevailing in the ash basin were such that chromium existed in the cationic form in the initial contact with water during ash disposal.

Although chromium exists in several states (+2, +3, and +6) (19, 28), Cr (III) and Cr (VI) are more common in the natural environment. The former exists as a cation and the latter as an anion; their valence is

Table V. Concentration (µg/kg) of Nickel in ash columns

Site	Depth(cm)	Dissolved	Exchange	Fe/Mn	Organic	Sum
NS	0-7	8.0	391	838	2720	3958
NS	7-15	40.5	375	657	2962	4036
NS	15-30	15.5	401	905	4789	6110
NS	30-60	11.6	221	540	537	6151
NN	0-7	33.9	141	730	4958	5864
NN	7-15	87.8	372	1315	10145	11920
NN	15-30	36.3	366	962	1726	3092
NN	30-60	16.3	128	555	1953	2652
ON	0-7	150.7	221	697	6164	7224
ON	7-15	44.8	233	712	4036	5028
ON	15-30	28.9	230	468	3594	4322
ON	30-60	20.3	259	992	4074	5347
ON	60-77	15.4	503	974	3520	5113

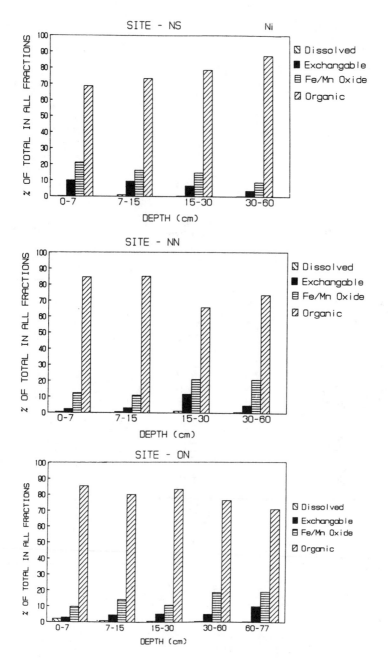

Figure 3. Distribution of Ni among various phases in ash basin sediments.

Table VI. Concentration (μg/kg) of Chromium in ash columns

Site	Depth(cm)	Dissolved	Exchange	Fe/Mn	Organic	Sum
NS	0-7	12.0	11.1	4042	839	4905
NS	7-15	65.7	13.4	4340	4789	2908
NS	15-30	10.5	5.9	3597	4999	86131
NS	30-60	17.3	2.2	2557	7009	9586
NN	0-7	32.4	10.6	2855	4487	7385
NN	7-15	94.3	23.8	3444	5995	9558
NN	15-30	23.1	49.2	2855	8486	11414
NN	30-60	22.4	19.0	255	8174	10773
ON	0-7	178.8	19.1	1078	6317	7593
ON	7-15	63.8	10.2	1668	8800	10542
ON	15-30	36.5	13.7	1523	10703	12277
ON	30-60	19.9	15.3	1818	12985	14838
ON	60-77	22.3	17.8	4502	13438	17980

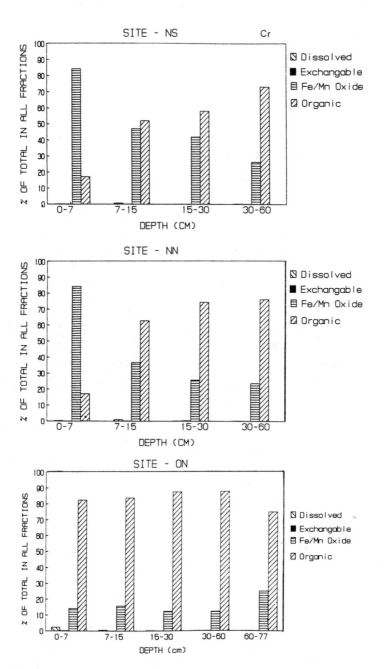

Figure 4. Distribution of Cr among various phases in ash basin sediments.

controlled by pH and redox conditions of the system, especially by the presence of small amounts of Fe (II) and Mn (IV) O_2 *(32)*. Acidic conditions (pH = 2 or less) are conducive to the rapid transformation of Cr (III) to Cr (IV) ($Cr^{+6} + 3e \rightarrow Cr^{+3}$, $E° = 1.10$ V), whereas the reverse is true for basic conditions (19, 28), though reaction kinetics for the latter is slow. Recent studies *(19, 32)* have indicated that chromium (III) exists as Cr $(OH)_3$ (aq), over a pH range of 6.5 to 10.5. This explains the fact that a very small fraction of chromium was found associated with the dissolved phase (log K_{sp} = -30.18 for Cr $(OH)_3$). Under the present set of field conditions in the ash basins, Cr (VI) probably did not exist, thus precluding the possibility of immobilizing chromium as barium chromate, calcium chromate, lead chromate etc.

Arsenic: The concentration and fraction of total arsenic associated with various ash phases are given in Table VII and Figure 5, respectively. Its concentration with the Fe/Mn phase is very high (2195 µg/kg to 8107 µg/kg). The second major fraction of arsenic was associated with organic matter, whereas the dissolved and exchangeable phases contained a very small fraction of arsenic. The amount of arsenic found in the NS and NN ash basins was almost 2.5 times the amount found in the ON ash basin.

One study among the several leaching studies *(24, 25, 30)* of ashes at various pH levels found toxic levels of arsenic in the leachate obtained at pH 4. In the human environment, arsenic is normally found as As^{+3} and As^{+5} *(28)* in the cationic as well as in the anionic form. The equilibrium between these two states is determined by the pH and redox conditions of the system *(19)*. Under acidic conditions, As^{+3}, which is more toxic than As^{+5}, is likely to dominate (As $O_4^{-3} + 2H^+ \rightarrow As O_3^{-3} + H_2O$, $E°$ 0.559 V) in the system. Arsenic sublimates at a low temperature (610°C) and thus it is likely to evaporate easily during the combustion process of coal in the furnace and to recondense on the iron or silicate minerals which have been generated in the furnace. Significantly, the pH and redox conditions in the ash basins were conducive for the presence of As^{+3} which probably existed as $H_3 As O_3$.

Conclusions

Higher concentrations of decomposing organic matter in the surface horizon in a 30 year old, inactive ash basin were responsible for the release and transport of metals by lowering the pH, reducing the Eh, and by the production of organic chelates. Metals leached from the ash deposited in the surface horizon were immobilized in the lower horizons in the basin sediments where attenuation took place by precipitation, coagulation, and surface reactions.

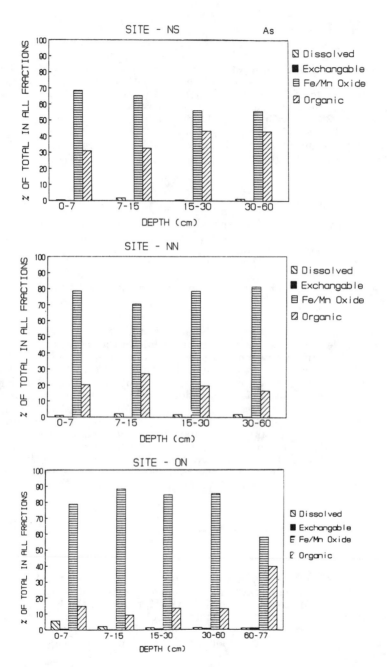

Figure 5. Distribution of As among various phases in ash basin sediments.

Table VII. Concentration (µg/kg) of Arsenic in ash columns

Site	Depth(cm)	Dissolved	Exchange	Fe/Mn	Organic	Sum
NS	0-7	57.6	5.46	8108	3660	11831
NS	7-15	195.1	4.17	7560	3795	11554
NS	15-3(	84.2	3.73	7362	5680	13130
NS	30-60	145.6	17.72	6862	5300	12326
NN	0-7	75.8	6.35	6068	1565	7715
NN	7-15	167.9	6.94	5444	2100	7719
NN	15-30	112.8	6.78	5172	1295	6587
NN	30-60	167.4	24.60	7015	1435	8648
ON	0-7	193.1	19.32	2690	510	3412
ON	7-15	129.6	12.06	5275	555	5954
ON	15-30	51.2	22.50	3185	510	3768
ON	30-60	37.1	24.75	2936	450	3447
ON	60-77	37.5	44.59	2196	1505	3783

Acknowledgments

This research was supported by a grant from the U.S. Department of Energy, HBCU Program (DE-FG09-865R.15170). Manuscript preparation was supported by contract DE-AC09-76SROO-819 between the University of Georgia and the U.S. Department of Energy. We thank Ms. Patricia Davis for her help in preparing the manuscript.

Literature Cited

1. Muraka, I. P.; and McIntosh, D. A. Solid-Waste Enviromental Studies: Description, Status and Available Results. Electric Power Research Institute, Palo Alto, CA 94304, 1987, EA-5322-SR.
2. Horton, J. H., Dorsett, R. S.; and Cooper, R. E. Trace elements in the terrestrial environment of coal-fired power plant, DOE report, DP-1475, E. I. DuPont de Nemours and Company, Savannah River Laboratory, Aiken, SC, 1977, pp. 49,
3. Dressen, D. R., Gladney, E. S.; Owens, J. W.; Perkins, B. L.; Wienke, C. L.; and Wangen, L. E. Comparison of trace elements extracted from fly ash levels found in effluent waters from a coal-fired power plant. Environ. Sci. Technol., 1977, 11, 1017.
4. Cherry, D. S.; and Guthrie, R. K. Mode of elemental dissipation from ash basin effluent. Water, Air, Soil Pollut., 1978, 9, 408.
5. Chu, T. Y.; Ruane, R. J.; and Krenkel, P. A. Characterization and reuse of ash pond effluents in a coal-fired power plant. J. Wat. Pollut. Control Fed. Nov., 1978, 2494-2498.
6. Evans, D. W.; and Giesy, J. P., Jr. Trace metal concentrations in a stream-swamp system receiving coal ash effluent. In Ecology and Coal Resources Development, Walie, M. K.; Ed., Proc. Intern. Cong., Energy and Ecosystem, 2, Grand Fork, ND, 1978, pp. 782-789.
7. Evans, D. W.; Alberts, J. J. and R. A. Clark III. Reversible ion exchange fixation of cesium-137 leading to mobilization from reservoir sediments. Geochem. Cosmochem. Acta., 1983, 47, 1041-1048.
8. Theis, T. L.; and Wirth, J. L. Sorptive behavior of trace metals on fly ash in aqueous systems. Environ. Sci. and Technol., 1977, 11, 1096-1101.
9. Alberts, J. J., Newman, M. C.; and Evans, D. W. Seasonal variations of trace elements in dissolved and suspended loads for coal ash ponds and pond effluents. Water, Air, soil Pollut., 1985, 26, 11-129.
10. Babcock, M. F.; Evans D. W.; and Alberts, J. J. Comparative uptake and translocation of trace elements from coal ash by Typha latifolia. The Science of Total Environment, 1983, 28, 203-206.
11. Tessier, A., Campbell, P. G. C.; and Bisson, M. Sequential extraction procedure for the speciation of particulate metals. Anal. Chem., 1979, 51, 844-851.
12. Rich, C. J. Elemental Analysis by Flame Photometry, In Methods of Soil Analysis Part 2, Black, C. A.; Ed.-in-Chief, American Society of Agronomy, Inc., Madison, WI, 53711, 1965, pp. 849-864.

13. Day, Paul R. Particle Fraction and Particle Size Analysis, In *Method of Soil Analysis, Part I,* Black, C. A.; Ed.-in-Chief; American Society of Agronomy, Inc., Madison, WI, 53711, 1965, pp. 545-566.

14. Blake, G. R. Particle Density and Bulk Density, In *Methods of soil Analysis, Part I,* Black, C. A., Ed.-in-Chief; American Society of Agronomy, Inc., Madison, WI, 53711, 1965, pp. 371-377.

15. *Hitachi Applications and Software Manuals,* NSI Hitachi Scientific Instruments, Mountain View, CA, 1982, pp 20-25.

16. Vickrey, T. M.; Harrison, G. V.; and Ramelow, G. J. Treated graphite surfaces for determination of tin by graphite furnace atomic absorption spectrometry. *Anal. Chem.,* 1981, *53,* 1573-1579.

17. *American Society for Testing and Materials, Fly ash and raw or calcined natural pozzolan for use as a mineral admixture in Portland Cement concrete.* Part 14, Annual Book of ASTM Standards, 1982, pp. 381-384.

18. Hurst, R. W. Sr isotope analyses of D-Area ash basins. Chempet Research Corporation, Moorpark, CA, 93021, 1988.

19. Nebergall, W. H.; Schmidt, F. C.; and Holtzclaw, H. F., Jr. *College Chemistry with Qualitative Analysis,* D. C. Health and Company, Lexington, MA., 4H edition, 1982, pp. 814-818, 868-870.

20. *Handbook of Chemistry and Physics;* Weast, Robert, C., Ed.; 57th Edition; D-135, CRC Press, 18901 Cranwood Parkway. Cleveland, Ohio 44128, 1977, Table F 90.

21. United States Environmental Protection Agency, Water Quality Criteria: Availability of documents; Notice. Federal Register, 1985, *50,* 145, pp. 30784-30796.

22. Douglas, G. S.; Mills, G. L.; and Quinn, J. G. Organic copper and chromium complexes in the interstial waters of Narragansett Bay sediments. *Marine Chem.,* 1986, *19,* 161-174.

23. Brady, N. C., *The Nature and Properties of Soils,* McMillan Publishing Co. 9th Ed.; 1984, p. 44.

24. Hulett, L. D., Weinberger, A. J.; Ferguson, N. M.; Northcut, K. J.; and Lyon, W. S. Trace Element and Phase Relation in Fly Ash, Oak Ridge National Laboratory, Analytical Chemistry Division, Oak Ridge, TN 37830, 1981, EA-1822, Research Report 1061.

25. Warren, C. J.; and Dudas, M. J. Mobilization and Attenuation of Trace Elements in an Artificially Weathered Fly Ash. Electric Power Research Institute, Palo Alto, CA 94304, 1986, EA-4747.

26. Lindsay, W. L. and Printing Vleak, P. L. G. In *Phosphate Minerals, Minerals in Soil Environment,* Third Printing; Dixon, J. B. and Weed, S. B. Co-editors; Soil Science Society of America, Madison, WI, 1982, pp. 662-670.

27. *United States Environmental Protection Agency,* Second Annual Report on Carcinogens, 1981, NTP 81-43, pp. 31-36, 81-87, and 163-168, December.

28. Windholz, M., *The Merck Index,* Merck and Co., Inc., Rahway, NJ. Tenth Edition, 1983.

29. Davison, R. L.; Natusch, D. F. S.; Wallace, J. R.; Evans, C. A., Jr. Trace metals in fly ash: dependence of concentration on particle size. *Environ. Sci. Technol.,* 1974, *8,* 1107-1118.

30. Grisafe, D. A.; Angino, E. E.; and Smith, S. M. Leaching characteristics of a high-calcium fly ash as a function of pH: a potential source of selenium toxicity. *Appl. Geochem.,* 1988, *3,* 601-608.

31. Sandhu, S. S.; and Mills, G. L. Kinetics and Mechanisms of the Release of Trace Inorganic Contaminants to Ground Water from Coal Ash Basins of the Savannah River Plant, Project Report; Prepared for the U.S. Department of Energy (USDOE) Savannah River Operation Office, Aiken, SC, **1989**, Contract Number DE-FG 09-86-SR-15170.
32. Thermodynamic and Kinetic Data for Aqueous Chromium, Electric Power Research Institute, Palo Alto, CA 94303, **1986**, RP 2485-03.

RECEIVED April 5, 1991

Chapter 18

Enhanced Metal Mobilization Through Leachants Containing Acetate Ion

Richard D. Doepker

Spokane Research Center, U.S. Bureau of Mines, Spokane, WA 99207

Soils close to metal smelters often have elevated concentrations of lead, cadmium, copper, zinc, and other heavy metals. Lead in wind-blown dust from the countryside adjacent to a smelter has been cited as a major source for the high levels of lead found in blood of children in these areas. Lead in soil is generally immobile and persistent. Initial studies have been carried out at U.S. Bureau of Mines (USBM) laboratories to examine the possibility of mobilizing metal contaminants, such as cadmium, copper, and lead, within the top foot of soil and then redepositing and stabilizing them at depths greater than 16 in. The use of acetate ion as a leachant for cadmium, copper, lead, and zinc has been investigated. Column and slurry test procedures have been used to evaluate the effects of soil depth, acetate ion concentration, leachate pH, and total soil metal concentrations on specific metal extraction efficiencies. A 0.5 F ammonium acetate solution at a pH of 4.5 can effectively mobilize cadmium and partially mobilize zinc and lead. Only small amounts of copper from the top 2.5 to 5 cm of contaminated soil could be mobilized with acetate leachants.

Residents around many urban smelters have drawn attention to the hazards of high lead concentrations in the soil, especially to young children (1). Testing of surface soil near secondary lead smelters has revealed lead concentrations as high as 11,500 mg/kg on residential land and 51,000 mg/kg on industrial land (2-4). Even though there may be several sources for lead in dust, the immediate source is usually soil (5-6). As a general rule, because inhalation and ingestion are the specific pathways leading to elevated blood lead levels, elevated lead in soil (and thereby dust) produces elevated blood lead levels in children exposed to that soil (7).

Lead in soil that has originated from an atmospheric rather than a geological source is generally rather immobile and persistent. The retention of heavy metals by soil has recently been discussed in detail by Evans (8). As pointed out, metals may be retained through adsorption reactions involving the formation of either outer-sphere or inner-sphere complexes with the surfaces of mineral and organic constituents. Metals also may be retained through precipitation reactions, leading to the formation of new secondary mineral phases such as oxides, carbonates, phosphates, and sulfates. Thus, metal-contaminated soil is a complex system involving many different thermodynamic and kinetic factors affecting the transport and/or stability of metals in soil.

In a previous study at USBM laboratories (9), a 0.1 formal (F) sodium acetate leachant was shown to enhance lead leaching from columns of silver mine tailings by a factor of 10 over a 0.1 F potassium nitrate leachant. The pH of both leachants was approximately 6. A reduction of the formal concentration of acetate to 0.01 reduced the mobilization of lead to nearly that of deionized water. In addition, leachate concentrations for other species, such as zinc, nickel, and cadmium, also were enhanced through acetate complexation.

The extraction of metals from basalt by humic acids resulted in increased metal dissolution rates (10). Different binding sites were observed by Hering and More (11) for strong complexes of copper and calcium with humic acid. In a baseline study of 10 soils, Korte et al. (12) found that the important factor describing the amounts of elements leached were the total metals originally present. The influence of fatty acids in the leachant showed that lead, nickel, and zinc were mobilized while copper accumulated in the soil (13). In the absence of fatty acids, precipitation and adsorption of the metals took place. Experiments with EDTA suggested that the rate of complexation by EDTA to metals may be greatly decreased by EDTA's binding with calcium and magnesium in hard water (14).

The purpose of this investigation is to examine the possibility that certain heavy metal contaminants in natural soils could be mobilized through the formation of metal acetate complexes. In so doing, it might be possible to transport these metals from the surface down to the subsurface of the soil, thereby reducing the metal-laden dust hazard.

Materials and Methods

Contaminated Soil. The soil samples used in this study were collected at five locations around a smelter complex where lead, silver, zinc, and gold had been processed. The smelter ceased operations in the early 1980's. Because all five samples were collected within a 1-km radius of the complex, soil properties were assumed to be similar, differing only in metal content. There was no soil characterization study conducted as a part of this investigation other than a chemical analysis.

Chemicals. All chemicals in this study were commercially available, analytical-grade reagents used without further purification. The deionized water was produced in the laboratory through distillation (Barnstead glass still) and then deionized with a Barnstead NANOpure II Demineralizer (18.3 megohm/cm). (Reference to specific equipment or trade names does not imply endorsement by the U.S. Bureau of Mines.) The acetate leachate used was prepared by weighing out the appropriate amount of ammonium and/or sodium acetate and adjusting the pH with glacial acetic acid. The volume of the resulting solution was then adjusted to compensate for the additional acetate from the acetic acid. These solutions were stored in carboys.

Slurry Test Procedure. One hundred grams of soil was placed into 27 500-cm^3, high-density polyethylene rectangular bottles (Nalgene). One hundred milliliters of leachant (1.0 M sodium acetate, pH = 5.0 or 7.0; 0.1 M sodium acetate, pH = 5.0 or 7.0; or 0.01 M potassium nitrate) was added to each bottle. The bottles were placed on a variable-speed, reciprocating shaker for 6 hours. The samples were then filtered through No. 42 Whatman filter paper. Part of the filtrate was further acidified for metal analyses while the second portion was reserved for anion analyses.

Column Test Equipment and Methods. Columns 35 cm long were constructed from 7.6-cm outside diameter (OD) acrylic columns with 0.32-cm-thick walls. A solid 6.98-cm-OD acrylic rod was cut into 5-cm lengths, drilled, and tapped to accept a 3/8-in polyethylene tube/pipe fitting and cemented into the end of the acrylic column. A qualitative, medium-pore filter paper was placed on a perforated Nalgene plate fitted at the base of the column over the opening to the polyethylene fitting. The air-dried soil sample was then introduced into the column.

Initial leaching of the air-dried soil column was carried out with 300 or 500 cm^3 of acetate leachant, depending on column depth. The volume of the leachate collected was used to determine pore volume of the soil column (i.e., leachant volume − leachate volume = pore volume). The resulting pore volumes (in cubic centimeters) averaged between 30 and 40 pct of the weight of the soil (in grams) for all soils tested in this study. Each subsequent leaching was carried out with leachants of either 1 or 1/2 pore volume. Leachates from each leaching were collected, and pH, conductivity, and volume were routinely determined in addition to metals.

Analytical Methods and Equipment. Leachable metal concentrations in the soil were determined through a modification of the method of Kuryk et al. (*15*). One gram of soil was placed into a 500-mL, rectangular polyethylene bottle, and 5 mL of concentrated nitric acid, 2 mL of concentrated hydrochloric acid, and 25 mL of deionized water were added. The bottles were capped and placed into a mechanical laboratory oscillation shaker for 4 hours. The samples were then gravity filtered through No. 42 Whatman filter paper and washed with acidified (nitric acid) water. The filtrate and washings were combined and diluted to 100 mL in a volumetric flask.

Metal analyses for the soil filtrates and column leachates were carried out with a Perkin-Elmer Plasma II inductively coupled plasma (ICP) spectrometer. Selected anion analysis was conducted with a Dionex 4000i ion chromatograph (IC) equipped with an AS4A (anion) separation column.

Results and Discussion

Concentrations of metals in a slurry leachate from a contaminated soil sample are given in Table I.

Table I. Slurry Metal Extraction by Acetate Ion from Contaminated Soil, Milligrams per Kilogram of Soil

	SO_4	Al	Ca	Cd	Cr	Cu	Fe	Mg	Mn	Ni	Pb	Zn
							Element					
Soil assay.........	1860	4880	1780	61	1	333	10300	503	1060	4	8810	1190
SODIUM ACETATE LEACHANT												
Formal. 1, pH 5:												
Amount...........	399	24.8	314	35.2	0.1	37.2	38	30.5	52	0.95	2140	316
Pct extracted...	21.5	0.5	17.7	57.2	7.6	11.2	0.4	6.1	4.9	23.7	24.3	26.6
Formal. 0.1, pH 5:												
Amount...........	79	0.42	175	11.6	ND	3.38	1.43	17.4	5.52	0.29	190	113
Pct extracted...	4.2	0	9.8	19	ND	1	0	3.5	0.5	7.2	2.2	9.5
Formal. 1, pH 7:												
Amount...........	791	4.62	226	30.3	0.07	22.5	7.61	23	5.99	0.39	1610	148
Pct extracted...	42.5	0.1	12.7	49.3	5	6.7	0.1	4.6	0.6	9.6	18.3	12.5
Formal. 0.1, pH 7:												
Amount...........	136	0.3	136	7.63	ND	1.39	1.16	14.9	3.02	0.17	102	64
Pct extracted...	7.3	0	7.7	12.5	ND	0.4	0	3	0.3	4.2	1.2	5.4
POTASSIUM NITRATE LEACHANT												
Formal. 0.01, pH 6:												
Amount...........	27.2	0.38	92.1	4.78	0.01	0.42	0.93	10.6	2.94	0.19	8.9	61.5
Pct extracted...	1.5	0	5.2	7.8	1	0.1	0	2.1	0.3	4.9	0.1	5.2

ND No data.
Formal. Formality.

At pH 5 and an acetate concentration of 1.0 F, 57 pct of the cadmium, 11 pct of the copper, 24 pct of the nickel, 24 pct of the lead, and 27 pct of the zinc were removed with a single 1-to-1, solid-to-acetate slurry extraction. At pH 7 (1.0 F acetate), the values were reduced to 49, 7, 10, 18, and 13 pct, respectively. These same removal values with 0.1 F acetate and a pH of 5 yielded 19, 1, 7, 2, and 9 pct for cadmium, copper, nickel, lead, and zinc, respectively, and demonstrated the importance of the effects of the acetate concentration on this extraction. In the case of the control, potassium nitrate, metal removals were reduced to below 8 pct. Of these five metals, cadmium and lead were least effected by pH changes between 5 and 7.

The leachability of cadmium, copper, nickel, lead, and zinc from columns of contaminated soil with 0.1 F ammonium acetate was examined using columns containing 1.3 kg, 1.1 kg, and 0.9 kg of soil. The results of this test protocol are given in Table II. The initial leachant pH of 6.8 and 0.1 F acetate used for the first four leachings (a total of 3.5 to 3.8 pore volumes) was followed by six additional leachings (6 pore volumes) of the same leachate adjusted to a pH of 5.11. Table II reports the total milligrams extracted as a function of total leachant pore volumes used. Also reported in Table II are the total milligrams of each element available based on the leachability assay of the soil used.

As is evident in this preliminary column test series, cadmium, nickel, and zinc were mobilized with 0.1 F acetate even at a pH of 6.8. The percentage of metals extracted (Table II) increased with increased column depth, although the metal content remaining in the soil after leaching (Table III) points out that most of the metal extracted was removed from the upper 3 to 6 cm of the soil column.

A second set of six columns, in which three contained 500 g of soil (11.4 cm column depth) and three contained 800 g (18 cm column depth), was leached with 0.5 F acetate solution at a pH of 5.5. The results of these tests are given in Table IV. Distilled water leaching (2 pore volumes) of the columns after 7 to 7-1/2 pore volumes of acetate reduced the metal dissolution to near the detection limits of the ICP.

After the leachings of the soil columns, nearly all the cadmium, 58 to 60 pct of the nickel, 56 to 57 pct of the lead, and approximately 52 pct of the zinc were removed. Of that total, removal of 90 pct of the cadmium, 66 pct of the nickel, 54 pct of the lead, and 73 pct of the zinc occurred during the first 3 pore volumes of extraction. The extraction efficiency with the 0.5 F acetate leachant was independent of column depth.

The metal distribution within one of the columns containing 800 g of soil (16.5 cm column depth) was determined by cutting the column into seven sections and assaying the sections. The results of this column division (Table V) clearly show that primary leaching of copper and lead occurred in the top 6 cm of the column while zinc dissolution occurred throughout the column, although to a greater degree from the top 6 cm. To further examine the effects of leachant pH, soil metal concentration, and total leachant volume on extraction efficiency, three soil samples containing different levels of metal contamination were collected. Six columns were filled (650 g) with each soil. Two columns of each soil were leached with 0.5 F ammonium acetate of pH 6.96, 5.52, or 4.52. A total of 18 columns containing three levels of metal contamination were leached with

Table II. Leachability of Lead-Contaminated Soils Using 0.1 M Ammonium Acetate

Total pore Volume	pH	Conduc-tance, mmho	Element, total mg extracted						
			Ca	Cd	Cu	Mg	Ni	Pb	Zn
SOIL COLUMN, 1,300 g									
Leachant pH 6.8:									
0.5..............	5.01	0.329	3.76	0.32	0	0.43	0.01	1.19	3.59
1.5..............	4.58	2.91	>56	12.8	1.75	15.4	0.33	87.6	112
2.5..............	4.95	6.17	>108	30.1	4.83	34	0.8	252	253
3.5..............	5.34	8.86	>166	35.6	6.07	38.7	0.94	326	297
Leachant pH 5.11:									
4.5..............	5.56	8.75	>198	38.1	6.84	40.4	1.01	366	316
5.5..............	5.74	10.39	>225	40.3	7.65	41.3	1.06	405	334
6.5..............	5.9	11.31	>251	42.3	8.25	42.3	1.12	444	347
7.5..............	5.86	12.24	>288	45.7	9.2	44.3	1.19	495	366
8.5..............	5.53	11.89	>316	49.1	10	46.1	1.27	557	389
9.5..............	5.72	11.7	>344	53.3	10.9	48.4	1.38	625	416
Total avail., mg..	NA	NA	2310	79.9	433	653	5.2	11400	1550
Pct extracted....	NA	NA	>14.9	66.7	2.52	7.4	26.5	5.48	26.8
SOIL COLUMN, 1,100 g									
Leachant pH 6.8:									
0.65..............	4.62	0.491	7.74	0.74	0.06	0.93	0.02	3.8	7.61
1.65..............	4.64	3.58	>56	15.5	2.38	18.3	0.4	124	135
2.65..............	5.28	5.98	>104	24.3	3.65	27.6	0.64	201	207
3.65..............	5.79	9.37	>134	26.6	4.19	29.7	0.69	230	222
Leachant pH 5.11:									
4.65..............	6.01	8.96	>154	27.8	4.59	30.8	0.72	248	230
5.65..............	6.04	11.06	>176	29.3	4.97	31.5	0.75	276	240
6.65..............	6.22	11.76	>201	30.9	5.33	32.2	0.77	306	248
7.65..............	7.76	9.52	>216	31.9	5.63	32.9	0.8	318	252
8.65..............	7.86	9.36	>223	32.3	5.77	33.5	0.81	321	254
9.65..............	8.22	8.69	>227	32.4	5.95	34	0.82	322	255
Total avail., mg..	NA	NA	1958	68	367	553	4	9691	1309
Pct extracted....	NA	NA	>11.6	47.9	1.6	6.2	18.5	3.3	19.5
SOIL COLUMN, 900 g									
Leachant pH 6.8:									
0.8..............	4.54	1.48	32.4	3.76	0.54	4.33	0.1	25.8	34.7
1.8..............	4.89	4.88	>76	14.2	2.32	15.5	0.36	129	123
2.8..............	7.06	8.41	>95	15.4	2.51	17.2	0.4	138	131
3.8..............	7.42	8.75	>98	15.5	2.57	17.6	0.4	140	131
Leachant pH 5.11:									
4.8..............	7.96	6.83	>109	15.8	2.65	19.1	0.41	141	133
5.8..............	7.7	8.57	>127	16.3	2.78	21.3	0.42	145	135
6.8..............	7.76	11.73	>160	17.4	2.99	23.4	0.43	149	137
7.8..............	7.82	9.95	>176	18	3.21	24.9	0.43	152	139
8.8..............	7.73	11.73	>186	18	3.43	26.5	0.44	152	139
9.8..............	5.27	10.89	>189	18.5	3.49	26.8	0.44	156	141
Total avail., mg..	NA	NA	1602	55	300	452	3.6	7929	1071
Pct. extracted....	NA	NA	>11.8	33.4	1.2	5.9	12.2	2	13.2

NA Not applicable.

Table III. Element Analysis of 1,300-g Soil Column after Acetate Leaching

Element, mg/g soil	Column interval, cm						
	Top-1.3	3.8-5.1	7.6-8.9	11.4-12.7	15.2-16.5	19.1-20.3	23-bottom
Ca........	1.25	1.51	1.54	1.39	1.89	1.72	1.93
Cd........	0.026	0.035	0.046	0.044	0.066	0.076	0.070
Cu........	0.277	0.320	0.344	0.352	0.356	0.375	0.373
Mn........	0.981	1.05	1.14	1.02	1.13	1.6	1.52
Pb........	8.04	8.84	9.62	9.48	10.3	10.5	9.87
Zn........	0.967	1.26	1.37	1.32	1.56	1.69	1.68

Table IV. Effects of Column Depth on Metal Leachability of Contaminated Soil Using Acetate Ion

Leachant	No. of pore vol.	pH	Ca	Cd	Cu	Mg	Mn	Ni	Pb	S	Zn
			Element, total mg extracted								
COLUMN DEPTH 4-1/4 in, PORE VOLUME 220 cm³											
NH4OAc, 0.4 M,	0.4	5.19	75.5	10.4	2.81	8.25	2.07	0.23	219	7.37	62
plus NaOAc,	0.9	5.4	144	19.6	6.45	14.1	5.33	0.44	522	19.1	120
0.1 M,	1.4	5.46	183	25.4	9.44	16.8	8.69	0.56	784	25.3	162
pH 5.45.......	1.9	5.55	210	27.4	11.8	17.8	12.3	0.62	956	35.8	185
	2.4	5.67	236	29.3	14	19.4	25.1	0.71	1140	47.5	208
	2.9	5.64	258	30.7	16.2	21.1	39.2	0.79	1320	58.3	227
NH4OAc, 0.5 M,	3.4	5.22	283	31.7	18.7	23.5	59.3	0.88	1501	70.6	247
pH 5.52.......	3.9	5.26	305	32.4	20.8	25.3	76.3	0.96	1670	81.5	263
	4.4	5.5	320	32.8	22.3	26.5	89.8	1.01	1780	93.6	273
	4.9	5.62	332	33.1	23.6	27.2	98.8	1.05	1890	107	280
	5.4	5.69	341	33.3	24.6	27.6	105	1.08	1970	118	286
	5.9	5.73	348	33.5	25.6	28.1	110	1.11	2070	131	291
	6.4	5.66	354	33.6	26.6	28.5	114	1.13	2160	141	296
	6.9	5.59	359	33.8	27.7	29.1	119	1.15	2250	152	300
	7.4	5.71	364	33.9	28.9	29.7	123	1.18	2340	160	305
Deionized	7.9	5.75	368	34.1	29.9	30	126	1.18	2420	167	308
water.........	8.4	5.7	375	34.1	30.5	30.4	128	1.19	2480	173	311
	8.9	5.74	376	34.2	30.6	30.5	128	1.2	2480	176	312
	9.4	6.5	376	34.2	30.7	30.5	129	1.2	2490	183	312
Tot. avail., mg	NA	NA	890	34.3	167	251	530	2	4400	310	595
Pct extracted..	NA	NA	42.3	99.6	18.4	12.1	24.3	59.8	56.5	59	52.4
COLUMN DEPTH, 7 in, PORE VOLUME 360 cm³											
NH4OAc, 0.4 M,	0.3	5.16	102	14.7	3.66	11	2.82	0.31	260	2.67	86.4
plus NaOAc,	0.8	5.35	215	30.1	9.26	21.8	8.39	0.64	715	20.7	181
0.1 M,	1.3	5.42	284	40.6	14.1	26.5	14.1	0.85	1130	26.8	254
pH 5.45.......	1.8	5.51	330	44.2	18.1	28.4	19.7	0.95	1420	43.9	295
	2.3	5.64	364	47.5	21.7	30.9	41.3	1.1	1720	62.6	331
	2.8	5.64	395	49.7	25.3	33.8	63.8	1.24	2010	79.6	362
NH4OAc, 0.4 M,	3.3	5.18	430	51.2	29.3	37.7	99	1.37	2320	98.3	393
pH 5.52.......	3.8	5.24	461	52.2	32.7	40.5	127	1.49	2590	116	417
	4.3	5.5	484	52.8	35.2	42.3	147	1.57	2780	136	433
	4.8	5.6	501	53.3	37.2	43.3	161	1.63	2950	158	444
	5.3	5.66	513	53.6	38.9	44	170	1.68	3080	178	453
	5.8	5.67	523	54	40.6	44.7	177	1.73	3240	198	461
	6.3	5.63	530	54.3	42.2	45.4	184	1.76	3390	214	468
	6.8	5.56	537	54.5	43.8	46.1	190	1.79	3540	228	475
	7.3	5.68	544	54.8	45.5	46.8	195	1.82	3690	241	482
Deionized	7.8	5.72	549	55	47.3	47.3	200	1.84	3860	251	488
water.........	8.3	5.67	553	55.1	48.1	47.7	202	1.85	3940	260	492
	8.8	5.73	554	55.2	48.2	47.8	203	1.86	3945	264	493
	9.3	6.32	555	55.2	48.3	47.8	203	1.86	3950	279	493
Tot. avail., mg	NA	NA	1424	54.9	267	402	848	3.2	7040	496	952
Pct extracted..	NA	NA	38.9	100	18.1	11.9	23.9	57.9	56.1	56.3	51.8

NA Not applicable.

Table V. Element Analysis of 800-g Soil Column after Acetate Leaching

Element, mg/g soil	Column interval, cm						
	Top-1.3	1.3-3.8	3.8-6.4	6.4-8.9	8.9-11.4	11.4-14.0	14.0-16.5
Ca........	0.941	0.959	1.15	1.39	1.98	1.64	1.78
Cd........	<0.005	<0.005	<0.005	<0.005	0.005	0.006	0.006
Cu........	0.18	0.21	0.27	0.30	0.30	0.30	0.33
Mn........	0.714	0.758	0.856	0.900	1.33	1.22	1.40
Pb........	2.82	3.67	3.74	4.39	6.87	7.48	8.37
Zn........	0.78	0.73	0.88	0.93	1.16	1.22	1.16

three different pH, 0.5 F ammonium acetate leachants. The results of these leachings are discussed below.

Effects of Acetate (Leachant) pH. Test columns of the soils were leached with deionized water, producing leachates ranging from pH 4.2 to 4.8. The pH of the leachates obtained over 4 pore volumes of extractions is given in Table VI.

In Figure 1, the percentage of available cadmium, lead, and zinc removed from soil sample M is reported as a function of acetate leachant pH. Although all three metals were more completely extracted as acetate pH decreased, the effect was greatest with lead, zinc, and cadmium, in that order. It can be assumed that greater efficiencies may be obtained when leachant pH is below 4, but one must assume that there is an acidity limit above which the soil being leached is permanently affected. Soil sample M contained 7.63 mg/g of lead, 1.27 mg/g of zinc, and 0.051 mg/g of cadmium.

Effect of Soil Metal Content. The removal of cadmium, lead, and zinc as a function of soil metal content is shown in Figure 2. The leachant used was 0.5 F ammonium acetate at a pH of 4.52. For all three metals (cadmium, lead, and zinc), removal efficiency after leaching with 4 pore volumes were greatest for the soil with the lowest metal content. It is interesting to note that for the initial 1/2-pore volume extraction, both cadmium and lead produced a greater removal efficiency where the soil contained the highest amount of these metals. It would appear that saturation did not develop in columns of approximately 12.5 cm soil depth. Furthermore, it was possible that these metals, in particular lead and cadmium, were present in the soil in two or more different forms, i.e., as free metals, as oxides, or as complexes with soil components.

Effect of Total Leachant Volume on Removal Efficiency. One set of columns was leached repeatedly with 1/2-pore volume portions of ammonium acetate at a pH of 5.52 for a total of 4-1/2 pore volumes. This leaching was then followed by four 1/2 pore volumes of deionized water. The results obtained for cadmium, copper, lead, and zinc from this test protocol are reported in Figure 3. As seen, the replacement of acetate leachate in the pore water by deionized water effectively stopped metal dissolution. This decrease in leachate metal concentration was accompanied by a similar decrease in ICP carbon concentrations, presumably caused by the removal of the residual acetate ion. With the exception of lead, the removal efficiencies were nearly at a maximum before the water leachant became effective.

A single column of soil sample M that had been leached with 4.5 pore volumes of 0.5 F ammonium acetate of pH 4.52 and then leached by 2 pore volumes of deionized water was cut into 1.3-cm sections. Each section was then assayed to determine the remaining cadmium, copper, lead, and zinc. Figure 4 shows the results. Cadmium was detected at values near the ICP detection limit, so it was assumed to be nearly completely removed. This agreed with results based on the analysis of the leachates obtained during leaching. A major portion of the copper and lead removal occurred between depths of 0 to 3 cm, while zinc dissolution occurred throughout the whole column.

Table VI.   Leachate pH as Function of Leachant (O.5 F Ammonium Acetate) pH and Soil Metal Concentration

pH	Pore volume							
	0.5	1.0	1.5	2.0	2.5	3.0	3.5	4.0
	SOIL SAMPLE H							
6.96....	5.22	5.77	6.05	6.16	6.26	6.34	6.43	6.41
5.52....	5.16	5.56	5.76	5.78	5.79	5.77	5.70	5.61
4.52....	4.43	4.75	5.06	4.98	4.93	4.71	4.66	4.61
	SOIL SAMPLE M							
6.96....	4.78	5.29	5.66	5.79	5.90	5.99	6.03	6.03
5.52....	4.73	5.17	5.49	5.55	5.61	5.63	5.57	5.52
4.52....	4.30	4.63	4.83	4.77	4.75	4.70	4.63	4.60
	SOIL SAMPLE L							
6.96....	4.52	5.05	5.51	5.71	5.87	5.98	6.03	6.06
5.52....	4.10	4.40	4.65	4.68	4.71	4.68	4.63	4.69
4.52....	4.10	4.40	4.65	4.68	4.71	4.68	4.63	4.69

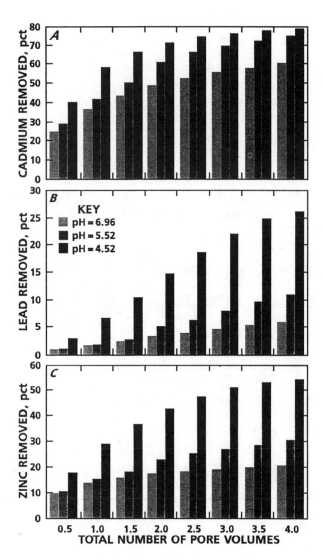

Figure 1.   Percentage of metals removed as function of acetate
leachant pH, soil sample M, 0.5 F ammonium acetate leachant.

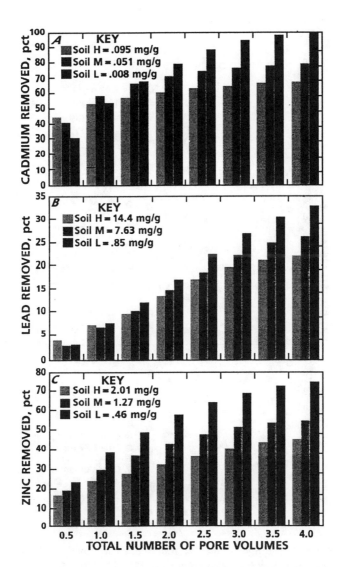

Figure 2.   Percentage of metals removed as function of soil metal content, 0.5 F ammonium acetate leachant at pH of 4.52.

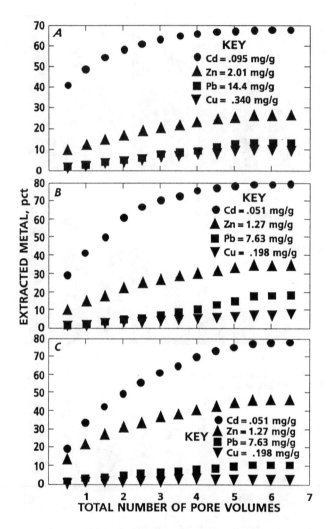

Figure 3. Percentage of metals removed as a function of total leachant volume, 0.5 F ammonium acetate leachant at pH of 5.52. A, Soil H; B, soil M; C, soil L.

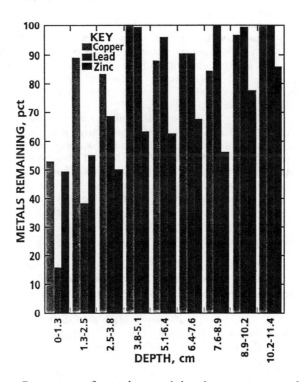

Figure 4.   Percentage of metals remaining in acetate-treated column.

Summary and Conclusions

The use of 0.5 F ammonium acetate leachant with a pH below 6 effectively mobilized cadmium in contaminated soil. At the same time, lead and zinc were partially mobilized. A reduced leachant pH greatly enhanced dissolution, even though the natural soil pH greatly influenced the final leachate pH. Analyses of columns after leaching with 4 to 8 pore volumes of acetate leachant demonstrated that lead dissolution was restricted to the upper few centimeters of soil. Cadmium, lead, and zinc dissolution and elution from soil columns can be effectively stopped through the addition of deionized water. Removal efficiencies increased with decreasing soil metal loads.

This study suggests that cadmium, lead, and zinc may be substantially reduced in the upper few centimeters of soil, which in turn would reduce the likelihood that the metal would be carried in wind-blown dust. Furthermore, this study suggests that acetate leachants, in conjunction with soil flushing and groundwater extraction procedures (16), may be used to enhance the removal of cadmium, lead, and zinc from sites contaminated with these metals.

Although this study suggests that mobilization of cadmium from contaminated soil is possible and that lead and zinc may be reduced in the upper few inches of contaminated soil through acetate ion leaching, it does not address the demobilization or redeposition of these metals into deeper subsoil. Furthermore, the effect on soil viability after treatment with the leachants proposed here has not been considered. These topics and others that address ways to maximize the removal efficiency through various application protocols have been left for continuing research.

Literature Cited

1. Brunekreef, B. *Sci. Total Environ.* **1984**, *vol. 38*, pp. 79-123.
2. Rinn, R.J; Linzon, S.N.; Stokes, P.M. In *Proc., Conference on Trace Substances in Environmental Health*, Univ. of Missouri, Rolla, MO, **1986**, pp. 308-321.
3. Stokes, P.M. "A Survey of Heavy Metals in Soils, Water, and Vegetation from Tannery, Kingston, Ontario"; Report to ARCOM Development, Ltd., Toronto, Ontario, **1977**, 19 pp.
4. *The Missouri Lead Study: An Interdisciplinary Investigation of Environmental Pollution by Lead and Other Heavy Metals from Industrial Development in the New Lead Belt of Southeastern Missouri*; Wixson, B.G. (Ed); Nat. Sci. Found., Washington, DC, *vol. 1*, **1977**, 541 pp.; NTIS PC A24/MF A01.
5. Barltrop, D.; Strehlow, C.D.; Thornton, I.; Webb, J.S. *Post-Graduate Medicine*; **1975**, *vol. 51*, pp. 801-804.
6. Thornton, I.; Culbard, E.; Moorcroft, S.; Watt, J.; Wheatlye, M.; Thompson, M. *Environ. Technol. Letters*; **1985**, *vol. 6*, pp. 137-144.
7. Roberts, T.M. In *Proc., International Conference on Heavy Metals in the Environment*, Toronto, Ontario, **1975**; Elect. Power Res. Inst., **1977**, *vol. 2*, pp. 503-552.
8. Evans, L.J. *Environ. Sci. Technol.*; **1989**, *vol. 23*, pp. 1046-1056.
9. Doepker, R.D. *Information Circular 9183*, U.S. Bureau of Mines, **1988**, *vol. 1*, pp. 210-219.

10. Singer, A.; Navrot, J. *Nature*, **1976**, *vol. 262*, pp. 429-481.
11. Hering, J.G.; Morel, F.M.M. *Environ. Sci. Technol.*, **1988**, *vol. 22*, pp. 1234-1237.
12. Korte, N.E.; Skopp, J.; Niebla, E.E.; Fuller, W.H. *Water, Air and Soil Pollution*, **1975**, *vol. 5*, pp. 149-156.
13. Loch, J.P.G.; Lagas, P.; Haring, B.J. *Sci. Total Environ.*, **1981**, *vol. 21*, pp. 203-213.
14. Hering, J.G.; Morel, F.M.M. *Environ. Sci. Technol.*, **1988**, *vol. 22*, pp. 1469-1478.
15. Kuryk, B.A.; Bodek, I.   "Santhanam"; EA 4215 (research project 2485-4), **1985**, Arthur P. Little, Cambridge, MA, 316 pp.
16. Sturges, S.G., Jr.; McBeth, Paul, Jr.; Pratt, Randy C.   Pres. at *1990 ACS National Meeting*, **1990**; CH2M Hill, **1990**, unpublished.

RECEIVED April 5, 1991

Chapter 19

Three-Dimensional Soil Venting Model and Its Applications

J. F. Kuo[1], E. M. Aieta[1], and P. H. Yang[2]

[1]James M. Montgomery, Consulting Engineers, Inc., Pasadena, CA 91109
[2]Simulation Sciences, Inc., Fullerton, CA 92633

In-situ soil venting has emerged as a promising remedial technology for soils contaminated with volatile or semi-volatile organics. A three-dimensional numerical model is developed to predict subsurface air pressure and velocity profiles during soil venting. The applications of the model include estimating the zone of influence, air flow rate, and pressure drawdown induced by a venting system. Estimated values of these parameters are needed for system design. The paper provides practical examples of how to use the model. These examples show the effects of air supply wells and injection wells on the zone of influence.

Soil contamination by volatile or semi-volatile organics from leaky underground storage tanks has recently been recognized as a major environmental problem. In-situ soil venting, also known as vapor stripping or vacuum extraction, has emerged as a broadly accepted remediation technique for this problem. The technique uses vacuum pump(s) to pull air from contaminated soil zones through a network of vapor extraction wells. The induced air flow sweeps the pore space and disrupts the equilibrium among the contaminants in the moisture phase, in the soil vapor phase, and on the soil surface. The process enhances volatilization of the contaminants into the air stream from both the soil surface and the moisture phase. It also replenishes oxygen within the contaminated zone and may accelerate biodegradation of contaminants by indigenous microorganisms.

The available literature indicates that studies of the venting technique for remediation of contaminated soils started in the early 1980's. Thornton and Wootan presented the results of an experimental study, sponsored by American Petroleum Institute, in 1982 (1). Since then, the literature has reported quite a few controlled experimental studies and field venting operations (2-6, 15, 18-20). However, the reported information is somewhat limited (7) and the design of vacuum extraction in these field operations was done by empirical and site-specific approaches (8). Therefore, there are few guidelines available in the literature for the optimal design, installation and operation of soil vapor extraction systems (20).

Application of the in-situ soil venting technique is conceptually simple. However, the performance of a venting system is controlled by complex physical, chemical, and biological processes. Understanding these processes is necessary to answer the two most commonly asked questions for design: what is the expected radius of influence and what are the effluent concentrations as a function time? Radius of influence determines the number and locations of the wells and sizes of the vacuum pumps. Estimates of effluent concentrations are necessary for selecting and sizing the off-gas treatment system and for predicting the duration of the project. This paper presents a model to predict the radius of influence. The model can potentially be expanded to evaluate effluent concentrations, but that application is beyond the scope of this study.

Before start-up of a full-scale remediation program, people often conduct venting tests to refine information derived from empirical sources. Pressure drawdown data at the extraction well and a few other locations are collected during the tests. These data are then used to develop a rough estimate of the radius of influence. However, without an appropriate analytical tool (analogous to Theis curves for aquifer pump tests) these data alone may not yield useful information for design. To alleviate that problem, a few one- or two-dimensional numerical models have been developed *(3,7,10-14)*. The 2-D models can be used to determine the pressure and velocity profiles of a single-well system. They are relatively easy to use and require short computation time; however, their applications to radially asymmetric geological formations and multiple-well systems are limited.

Recently, two system design modifications have been applied to enhance the performance of venting systems. One modification is the installation of passive air supply wells to minimize the "short-circuit" problem, which refers to the vertical air flux from the ground surface *(2,5,15)*. Another modification is the injection of hot air into the subsurface in an attempt to enhance volatilization of less volatile organics, such as diesel and Stoddard solvents. In these applications, extraction/supply/injection wells produce a complex 3-D reduced pressure zone in the affected subsurface.

There are several problems which can occur without careful design of these multiple-well systems. A short-circuit of flow may occur between an extraction well and an air supply well, and, in turn, some dead spots with little or no air flow will form. For the cases of hot air injection, controlling the flow path of the injected hot air is both economically and environmentally important. Without complete understanding of the air flow pattern, part of the injected hot air may flow away from the capture zone of the extraction wells and carry contaminants with it. The effect of vapor supply wells has been analyzed to a limited extent by a 2-D model, in which venting wells were treated as a system boundary *(14)*. A 3-D model is more appropriate to study these multiple-well systems.

This paper describes a 3-D computer model which provides better understanding of air flow patterns during soil venting. The discussion begins with an equation which describes the air flow. The equation is solved numerically to determine pressure and velocity profiles using a finite-difference scheme in the model. The model can predict the radius of influence, flow rate, and pressure drawdown that are pertinent to system design. Further, this paper provides practical examples of how to use the model. The examples show the effects of air supply wells and air injection wells on the zone of influence.

Theoretical Background

The foundation of the numerical model is an equation which describes steady-state air flow in a 3-D cartesian coordinate system. A few basic assumptions were used in deriving the equation: 1) the ideal gas law applies; 2) the flow is under isothermal conditions; 3) Darcy's law for gas velocity in porous media applies; and 4) the gravity effect on the air flow is negligible. Similar assumptions have been used by other researchers in derivation of their 2-D models (3,10,11).

The air flow equation is basically a combination of three equations: the continuity equation, Darcy's equation, and the equation of state. The continuity equation for steady-state flow in a 3-D cartesian coordinate system can be expressed as (16):

$$\frac{\partial}{\partial x}(\rho u_x) + \frac{\partial}{\partial y}(\rho u_y) + \frac{\partial}{\partial z}(\rho u_z) + \frac{\rho q}{(dV)} = 0 \tag{1}$$

where ρ is the density of the air, (dV) is the control volume, q is the volumetric flow rate of a sink or source (a positive value for a sink and negative for a source), and u_x, u_y and u_z are the superficial velocities in the x, y and z directions, respectively.

Assuming Darcy's law applies and neglecting the gravity effect on the air flow, the air velocity in the x-direction can be expressed as:

$$u_x = - \frac{k_x}{\mu}\left(\frac{\partial P}{\partial x}\right) \tag{2}$$

where μ is the viscosity of the air, P is the pressure, and k_x is the permeability of the soil matrix in the x-direction.

With equation 2 and similar equations for the air velocities in the y- and z-directions, equation 1 becomes

$$\frac{\partial}{\partial x}\left(\rho \frac{k_x}{\mu} \frac{\partial P}{\partial x}\right) + \frac{\partial}{\partial y}\left(\rho \frac{k_y}{\mu} \frac{\partial P}{\partial y}\right) + \frac{\partial}{\partial z}\left(\rho \frac{k_z}{\mu} \frac{\partial P}{\partial z}\right) - \frac{\rho q}{(dV)} = 0 \tag{3}$$

The equation of state for an ideal gas can be described as:

$$\rho = \frac{PM}{ZRT} \tag{4}$$

where M is the molecular weight, Z is the compressibility factor, R is the universal gas constant, and T is the absolute temperature. From kinetic theory, the viscosity of an ideal gas depends solely on temperature. Therefore, viscosity is constant for an isothermal flow. With the assumptions of constant viscosity and compressibility, equations 3 and 4 can be combined into the following form:

$$\frac{\partial}{\partial x}\left[k_x \frac{\partial P^2}{\partial x}\right] + \frac{\partial}{\partial y}\left[k_y \frac{\partial P^2}{\partial y}\right] + \frac{\partial}{\partial z}\left[k_z \frac{\partial P^2}{\partial z}\right] - 2 \frac{\mu P q}{dV} = 0 \tag{5}$$

Description of Numerical Solution

A finite-difference scheme is used to provide a numerical solution for equation 5 in this model. The spatial domain is divided into a block-centered finite-difference grid as shown in Figure 1. The grid points, or block centers, are at locations x_i, $i = 1,....,NX$ in the x-direction; y_j, $j = 1,....,NY$ in the y-direction; and z_k, $k = 1,......,NZ$ in the z-direction.

The first term of equation 5, which represents the flow flux in the x direction across the grid boundary, is approximated as:

$$k_x \left(\frac{\partial X}{\partial x}\right)_{x + dx} = k_{x, i+1/2} \left[\frac{X_{i+1} - X_i}{(\Delta x_{i+1} + \Delta x_i)/2}\right] \tag{6}$$

where X_i is equivalent to P^2 at the center of grid i, Δx_i is the size of the grid i in the x-direction, and $k_{x,i+1/2}$ denotes k_x at the boundary between blocks i and i+1. To correctly relate steady flow rate and pressure drop in the interval for a heterogeneous formation, the permeability in the x-direction at the grid block boundary is chosen as:

$$k_{x,i+1/2} = (\Delta x_i + \Delta x_{i+1})\left[\frac{\Delta x_i}{k_{x,i}} + \frac{\Delta x_{i+1}}{k_{x,i+1}}\right]^{-1} \tag{7}$$

Similar equations can also be derived for fluxes in the y and z directions. Hence, at grid block (i,j,k), equation 5 can be written in a finite difference form as:

$$(T_x)_{i+1/2} [X_{i+1} - X_i] - (T_x)_{i-1/2}[X_i - X_{i-1}]$$
$$+ (T_y)_{j+1/2} [X_{j+1} - X_j] - (T_y)_{j-1/2}[X_j - X_{j-1}]$$
$$+ (T_z)_{k+1/2} [X_{k+1} - X_k] - (T_z)_{k-1/2}[X_k - X_{k-1}]$$
$$- \frac{2 qP}{\Delta x_i \Delta y_j \Delta z_k} = 0 \tag{8}$$

where
$$(T_x)_{i+1/2} = (2/\mu\Delta x_i)[k_{x,i+1/2} / (\Delta x_i + \Delta x_{i+1})]$$
$$(T_y)_{j+1/2} = (2/\mu\Delta y_j)[k_{y,j+1/2} / (\Delta y_j + \Delta y_{j+1})]$$
$$(T_z)_{k+1/2} = (2/\mu\Delta z_k)[k_{z,k+1/2} / (\Delta z_k + \Delta z_{k+1})]$$

The last term on the left side of equation 8 is the source (or sink) term, and represents flows to or from a well within the grid block (i,j,k). Because cartesian coordinates are not as suitable as cylindrical coordinates in representing cylindrical wells, a "hybrid" model is used to represent grid blocks which contain a well. Hereafter, this type of block is referred as a well block. Basically, the hybrid model uses a cylindrical coordinate system to describe specifically the well, and the flow into or from the well. Further discussion of this hybrid model is provided below using an example well system.

As illustrated in Figure 2, a well is located in the center of a well block and is surrounded by well packing. Both the well and the packing are completely inside the well block, and the permeability of the well packing section can be assigned a value different from the rest of the block. The flow between the well (and the well packing) and the well block is handled as a radial flow in a cylindrical coordinate. To represent this flow with a finite difference approximation, the source or sink term in equation 8 takes the following form *(10)*:

$$qP = \frac{\pi \Delta z \, k_r}{\mu} \frac{(P^2 - P_w^2)}{\ln(r_e/r_w)} \tag{9}$$

where Δz is the z-directional grid size of the well block, P_w is the well pressure, r_w is the well radius, r_e is the equivalent radius of the well block, and k_r is the equivalent radial permeability. The equivalent diameter of the well block is taken as the arithmetic average of the well packing diameter and the smaller of the x- and y- directional grid sizes of the well block. The equivalent radial permeability is taken as

$$k_r = \ln(r_e / r_w)[\frac{\ln(r_p/r_w)}{k_p} + \frac{\ln(r_e/ r_p)}{k_h}]^{-1} \tag{10}$$

where r_p is the radial distance from the well center to the periphery of the packing, k_p is the permeability of the well packing section, and k_h is the average horizontal permeability of the well block, i.e., $1/2(k_x + k_y)$. The selection of equation 10 as the equivalent radial permeability will ensure correct representation of the steady-state flow rate and pressure drop *(10)*.

For the flow between the well block and the neighboring blocks, a local analytical model is used similar to that described by Williamson and Chappelear *(17)*. In this model, the flows to the four neighboring blocks are first represented with the radial flow equation for a cylindrical coordinate. The model then evaluates the flows to match the pressure at the neighboring blocks. The model gives the following expression for the flows:

$$q_m P = \frac{2\pi k_m \Delta z}{\mu} g_m (P^2 - P_m^2), \quad m = 1,2,3,4 \tag{11}$$

where q_m is the volumetric flow from the well block to the mth neighboring block, k_m is the averaged permeability between the well block and the mth neighboring

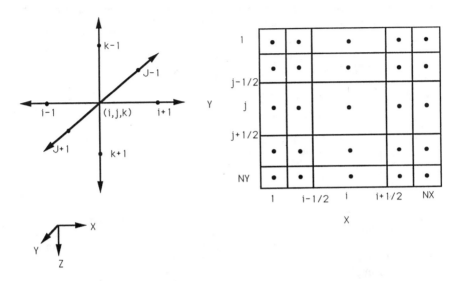

Figure 1. Finite difference Grid of the 3-D Model.

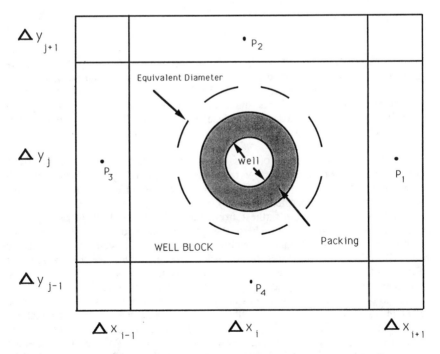

Figure 2. Schematic Diagram of Well Block.

block, P_m is the pressure at the center of the mth neighboring block, and g_m is the mth element of the first row of the inverse of matrix G:

$$
G = \begin{bmatrix}
\ln(r_1 / r_e) & 0 & r_1 & r_1^2 \\
\ln(r_2 / r_e) & r_2 & 0 & -r_2^2 \\
\ln(r_3 / r_e) & 0 & -r_3 & r_3^2 \\
\ln(r_4 / r_e) & -r_4 & 0 & -r_4^2
\end{bmatrix}
\tag{12}
$$

In equation 12, r_m is the distance between the centers of the well block and the mth neighboring block, and r_e is the equivalent radius of the well block as defined by equation 9. The "averaged" permeability, k_m, is chosen to correctly relate steady-state flow rate and pressure drop for a non-homogeneous formation (10). For example,

$$
k_1 = \ln (r_1/ r_e)\left[\frac{\ln(\frac{\Delta x_i/2}{r_e})}{k_{x,i}} + \frac{\ln (\frac{r_1}{\Delta x_i/2})}{k_{x,i+1}}\right]^{-1}
\tag{13}
$$

As a brief summary, flows to or from the well and the well blocks are represented as radial flows in the hybrid model. Therefore, equation 8, which is based on cartesian coordinates can not apply directly as the material balance equation between the well blocks and their adjacent blocks. For blocks adjacent to a well block, equation 11 should replace the corresponding term in equation 8. For the well blocks, flows are represented by equations 9 and 11, instead of equation 8.

The boundary conditions of this simulator can be set at either no flow, constant pressure, or constant flow rate. Typical boundary conditions for soil venting processes would be atmospheric pressure at the surface (or no flow at the portion covered with impermeable liners); no flow at the water table; no flow or constant pressure at the side farthest from the well; and no flow at the wellbore except the perforated section where the pressure is equal to the bottomhole pressure. In this model, either the wellbore pressure or the well flow rate can be specified. When well pressures are specified, the unknowns are the pressures at the center of all grid blocks, and the equations requiring solutions are the material balance equations of the grid blocks. If a well flow rate is specified (the corresponding well pressure becomes an unknown), equation 9 is then included in the set of equations to be solved in the model. Because the number of equations can be quite large for three-dimensional problems, an iterative method is used; specifically, the line successive over-relaxation (LSOR) iterative method implemented in the Black Oil Reservoir Simulation (BOAST) program (21). The velocity profiles can be derived after the pressure profiles are determined.

Model Verification

To verify the accuracy of the model, direct comparisons were made between published field data and model predictions. One set of the field data selected for these comparisons are from the experiment no. 3 of an API-sponsored pilot-scale subsurface venting system study conducted by Crow et al. *(2)*. Only relevant information regarding this pilot test is repeated below.

The site was located at a petroleum fuels marketing terminal. A vapor extraction well was positioned between two air supply wells in a straight line configuration. The distances between the extraction well and two supply wells were 20 and 40 feet, respectively. The wells were constructed of 2-inch ID schedule 40 PVC casing in 4-inch bore holes and perforated from 14 to 20 feet below grade. A plastic membrane liner was used to cover the test cell.

The ground water was at approximately 25 feet below grade. Core samples taken at one foot above the capillary fringe revealed that the average soil porosity was 38 percent and the average permeability was 9.8×10^{-4} cm/sec (which is equivalent to an air permeability of 15.5 darcy at 20 °C). During the experiment no. 3, the test cell was vented for 15 days at a flow rate of 39.8 cfm. Pressure drawdown data at the extraction well and a few other locations were taken at a depth of 20 feet when the air supply wells were capped, and the results are shown in Figure 3. The pressure drawdown at the extraction well was approximately 13-inches of water, as shown in the figure. The paper also reported that the ratio of the combined air inflow rate (through the air supply wells) to the test cell extraction rate ranged from 0.09 to 0.11 during each experiment.

Several model runs were conducted to simulate the field test data. In these runs the flow rate, porosity, well diameter, packing diameter, perforation interval, water table depth, and distance between the extraction and air supply wells were set equal to those of the field tests. For the information not supplied in the Crow's paper, the following values were used in the simulation runs: subsurface temperature was 20 °C; liner dimensions were 80 x 80 feet, centered at the extraction well; permeability of the packing was 50 darcy; and the ratio of horizontal to vertical permeability was arbitrarily chosen as three, a typical value.

The permeabilities in the x-, y-, and z- directions (k_x and k_y were kept equal) were determined by matching the pressure drawdown at the extraction well. The best match was obtained at $k_x = k_y = 37$ darcy ($k_z = 12.3$ darcy). Model predictions of the pressure drawdowns away from the well compare well with the reported data, as shown in Figure 3.

The only reported permeability value was 15.5 darcy from core samples. These cores were taken from vertical borings, and the measured value should represent the vertical permeability. The vertical permeability value predicted by the model is within 20% of the averaged field value. The two values are reasonably close, and it is a surprise to the authors because the ratio of horizontal to vertical permeability was arbitrarily chosen as three. Two major factors contributing the difference are non-homogeneity of the formation and the unknown subsurface temperature.

A model run was also conducted to predict the air inflow rates through the air supply wells to the subsurface. The flow rates of the two wells were determined by the model to be 3.06 and 1.27 cfm, respectively. The total inflow rate, 4.33

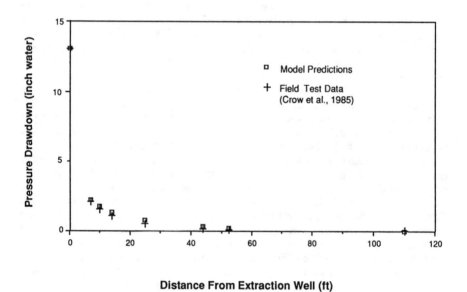

Figure 3. Pressure Drawdown vs. Distance from the Extraction Well.

cfm, is 10.9% of the extraction rate. This value corresponds very well to the field data, which ranged from 0.09 to 0.11.

The pressure drawdown distribution, permeability values, and inflow rate to the air supply wells determined from the model predications fit the published field data well. These results indicate that the model should be able to predict the pressure and air velocity profiles within an acceptable range, provided reasonably accurate input data to the model are provided.

Radius of Influence

The success or failure of a soil venting project is often judged by the amount of time spent to clean up a site. If a project takes longer than expected, it may be deemed a failure. In general, the contaminated zone in the vicinity of a vapor extraction well, in which the air flow rate is high, should be readily remediated. However, the lateral distance limit at which a venting well can induce an air flow is uncertain.

The radius of influence has been defined as the lateral distance from an extraction well where the pressure is equal to the ambient atmospheric pressure *(7)*. Based on this definition, no air beyond this radial distance will flow toward the extraction well. Zone of influence is analogously used for a multiple-well system. The pressure within the zone is less than that of the atmosphere. This definition of radius of influence is often used to space vapor extraction wells. For example, if a radius of influence is measured at 15 feet, wells might be placed 30 feet apart. The strategy has been considered questionable *(9)*, since pressure drawdown and air velocity decrease rapidly as distance from the extraction well increases *(2)*. Applying this definition to the example above, pressure drawdown should still be detectable at 14 feet from the extraction well; however, the air flow rate may not be sufficient to remediate the soils at this location.

Alternatively, radius of influence has been defined as the lateral extent in which sufficient air flow can reduce the contaminant concentrations below an acceptable level within a pre-specified time frame *(10)*. Under this definition, the radius of influence depends not only on the characteristics of the vacuum pump and physical properties of the formation, but also on characteristics of contaminants and desirable remediation time frame. The required air flow rate at the edge of the plume should be determined from bench or pilot scale studies and/or experience from similar projects. The studies described below use this definition of radius of influence.

Example Studies

Four simplified cases are described in this section to illustrate the effect of air supply wells (Case 1), perforation depths of the supply wells (Case 2), locations of the supply wells (Case 3), and air injection wells (Case 4) on the performance of a soil venting system. It should be noted that these examples illustrate simple applications of the model, and other issues such as the effect of perforation interval or the effect of various vapor injection rates can be readily investigated using the model.

In these exemplary studies, unless otherwise specified, a contaminated plume is located at 20 feet below the ground surface with a vertical extent of 20 feet, while the groundwater is at 60 feet. The lateral extent of the plume is 20 feet away from a 4-inch vapor extraction well. The well was installed into a 12-inch diameter borehole, which was packed with sand up to six feet below the ground surface and then cemented. The horizontal permeability of the formation (k_x and k_y) and that of the well packing are chosen to be 1,000 and 5,000 milli-darcy (md), respectively. The vertical permeability of the formation and the packing are 100 and 500 md, respectively. The air flow rate from the extraction well is 20 scfm.

Ideally, a soil venting system will be most effective when the extracted air flows radially toward the perforation. In this ideal case, the contaminated zone at 20 feet away from the well will be swept 3.8 times a day (3.8 air exchanges per day), and the air velocity at that location is 11.5 ft/day. If the bench-scale studies or other similar projects show that 300 air exchanges are necessary to remediate the plume, it is then plausible to say the site will be cleaned in 90 days. In the following examples, zone of influence is defined as the region in which the air flows at a rate greater or equal to 11.5 ft/day toward the extraction well.

Example 1: Effect of Air Supply Wells. This case study illustrates the effect of air supply wells on the performance of a vapor extraction well. Four air supply wells, which are perforated from 20 to 40 feet below the ground surface, are located at four corners of a square with the extraction well in the center. The distances between each supply well and the extraction well are equal to 20 feet. In the simulation runs, the bottomhole pressure of the vapor supply wells was set at atmospheric by neglecting the pressure drop in the well. In addition to generating velocity and pressure profiles, the model calculates the inflow rates of the supply wells. As shown in Table I, the bottomhole pressure of the extraction well is 11.99 psia which is higher than the base case pressure of 11.52 psia (no air supply wells). This difference between the bottomhole pressures implies that the extraction rate from a venting system with air supply wells would be higher than that of the base case by using identical pumps. The results also show that the total inflow rate from supply wells into the subsurface is 11.8 cfm, which accounts for approximately 60% of the total extraction rate. Consequently, only 40% of the extracted flow results from the vertical air fluxes, indicating that "short-circuit" of air from the ground surface is minimized. The total vertical flux from the surface as a function of distance from the extraction is shown in Figure 4.

Table I. Summary of Computer Simulation Results

Case No.	No. of Supply Wells	Perforation of Supply Wells	Distance between Wells	Pressure of Extraction Well	Flow Rate of Supply Wells
Base	0	-	-	11.53 psia	-
1	4	20' - 40'	20'	11.99 psia	11.8 cfm
2	4	40' - 60'	20'	11.67 psia	5.6 cfm
3	4	20' - 40'	28'	11.82 psia	10.0 cfm
4	4	20' - 40'	20'	12.29 psia	20.0 cfm

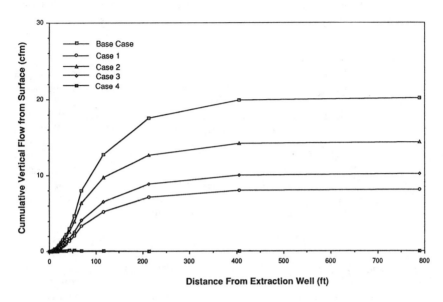

Figure 4A. Cumulative Vertical Flow from Surface.

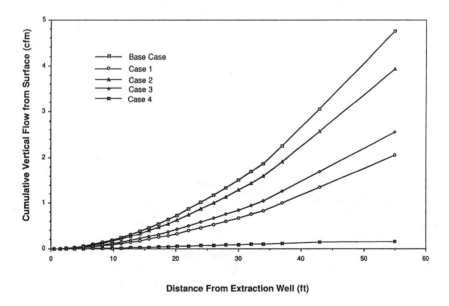

Figure 4B. Cumulative Vertical Flow from Surface (in the Vicinity of the Extraction Well).

Figure 5 shows the first quadrant of the zone of influence (ZI) of this case. The shape of ZI in the other three quadrants would be the same, due to the symmetric nature of the well layout. It should be noted that the ZI in Figure 5 and subsequent figures are located at the center-plane of the well perforation, i.e., 30 feet below the surface. For comparison, the ZI of the base case (without supply wells) is also shown in Figure 5. The ZI of the base case is a circle with a radius of approximately 17 feet, which is identical to that calculated by a 2-D model for the same conditions used in this example *(10)*.

As discussed, presence of air supply wells lessens the short-circuit of air from the ground surface. It extends the ZI directly between the extraction well and the vapor supply wells from 17 to 22 ft, as shown in Figure 5. However, the air supply wells also create a short-circuit of air flow between them and the extraction well. The radius of influence in the direction from the extraction well toward the mid-point of the two neighboring supply wells decreases from 17 feet to 15 feet. The locations beyond 15 feet will not be cleaned as effectively. Therefore, caution should be exercised in using and locating supply wells.

Example 2: Effect of Perforation Depth of Vapor Supply Wells. It is relatively common that contaminants would travel through the unsaturated zone and enter the underlying aquifer. Groundwater monitoring and/or extraction wells would then be installed in those cases. Without well caps, these groundwater wells will also serve as air supply wells if soil venting is operated concurrently. This exemplary study simulates this phenomenon by setting the perforation depth of the supply wells directly above the water table (40 to 60 feet below ground surface).

The computation results show that the air inflow rate of the supply wells is 5.6 cfm, which is equivalent to 28% of the total extraction rate. As discussed in the previous examples, the presence of these air supply wells decreases the vertical air flux from the surface, but to a smaller extent than in example 1. In addition, the ZI of this case decreases substantially to 16 feet, which is even smaller than that of the base case (as shown in Figure 6). This is due to a short-circuit of flow between the vapor extraction well and the supply wells which were perforated at a deeper depth. The majority of the induced air flow does not pass through the whole lateral extent of the plume. The results indicate that presence of near-by groundwater wells does affect the performance of a soil venting system.

Example 3: Effect of Distance Between the Extraction and Supply Wells. This example shows the effect of spacing between the extraction and supply wells. The supply wells are located 28 feet (as compared to 20 feet in example 1) away from the extraction well.

The results indicate that the total air inflow rate of the supply wells is 10 cfm which accounts for 50% of the extraction. Presence of the vapor supply wells again enhances the clean-up along the supply wells and the extraction well as a result of short-circuiting. The ZI of this case is shown in Figure 7. It should be noted that the affected distance from the extraction well toward the mid-point of the two neighboring supply wells is greater than that of example 1.

Example 4: Effect of Air Injection Wells. Four air injection wells are installed at the same locations as those of the vapor supply wells in example 1. The total injection rate is set at 20 cfm, which is equal to the vapor extraction rate.

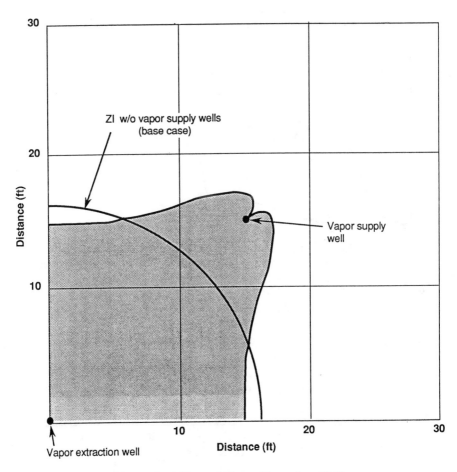

Figure 5. Effect of Vapor Extraction Well.

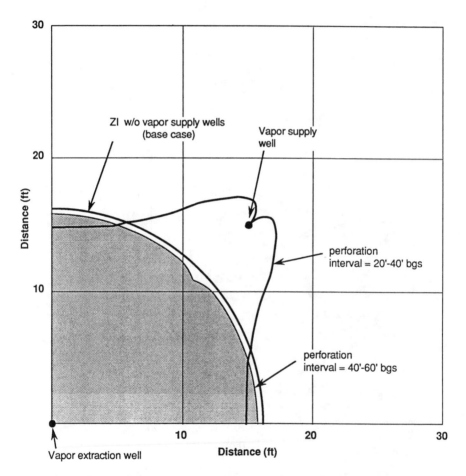

Figure 6.   Effect of Perforation Depth of Vapor Supply Wells.

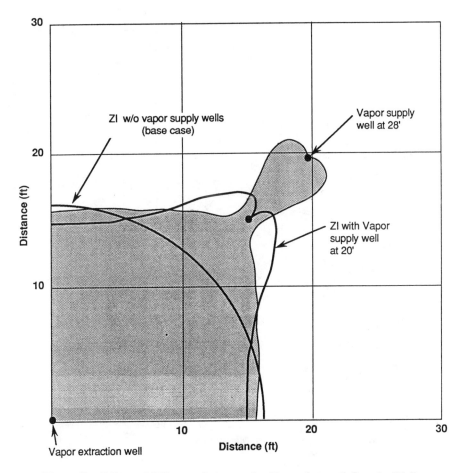

Figure 7. Effect of Distance between the Extraction and Supply Wells.

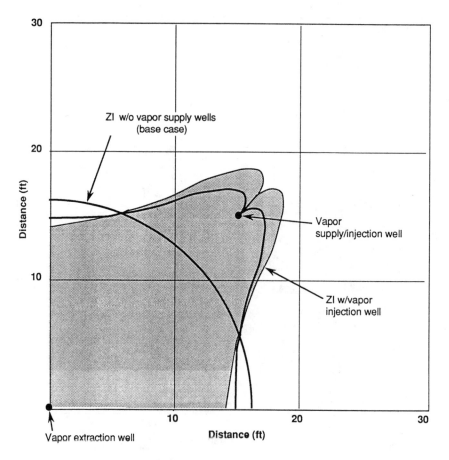

Figure 8. Effect of Vapor Injection Wells.

As shown in Figure 8, the ZI along the extraction well and the injection increases from 22 ft to 24 ft; however, the short-circuit problem worsens. The ambient air fluxes within a radius of 50 feet from the extraction well are insignificant, at a rate of 0.15 cfm (Figure 4). The computation results also indicate that the bottomhole pressure of the injection well is 15.12 psia, implying that a pressure head of 0.42 psi is sufficient to inject the air into the formation in this case.

Concluding Remarks

A three dimensional computer model was developed to estimate the zone of influence, flow rate, and bottomhole pressure, as well as the pressure and air velocity profiles of a soil venting project. The estimated values of these parameters are useful for design of a soil venting system.

Literature Cited

1. Thornton, J. S.; Wootan, Jr., W. L. "Venting for the Removal of Hydrocarbon Vapors from Gasoline Contaminated Soil"; *J. Environ. Sci. Health*, **1982**, *A17(1)*, pp. 31-44.
2. Crow, W. L.; Anderson, E. C.; Minugh, E. M. "Subsurface Venting of Vapors Emanating from Hydrocarbon Product on Ground Water"; *Ground Water Monitoring Review*, **1987**, *VII(1)*, pp. 51-57.
3. Baehr, A. L.; Hoag, G. E.; Marley, M. C. "Removing Volatile Contaminants from the Unsaturated Zone by Inducing Advective Air-Phase Transport"; *J. Contam. Hydrol.* **1989**, *4*, p.1.
4. Stinson, M. K. "EPA SITE Demonstration of the Terra Vac In Situ Vacuum Extraction Process in Groveland, Massachusetts"; *JAPCA*, **1989**, *39*, pp. 1054-1062.
5. Connor, J.R. "Case Study of Soil Venting"; *Pollution Eng.* **1988**, *20(7)*, pp. 74-78.
6. Bailey, R. E.; Gervin, D. "In-Situ Vapor Stripping of Contaminated Soils: A Pilot Study"; Proc. 1st Annual Hazardous Materials Conf./Central, Rosemount, IL, March 15-17, 1988, p.207.
7. Johnson, P. C.; Kemblowski, M. W.; Colthart, J. D. "Quantitative Analysis for the Cleanup of Hydrocarbon-Contaminated Soils by In-Situ Soil Venting"; *Ground Water*, **1990**, *28*, pp. 413-429.
8. Cho, J. S.; DiGuilio, D. "The Simulation of Air Flow in Soil Vacuum Extraction Systems by a 3-D Groundwater Flow Model"; Conf. on Hazardous Waste Research, Kansas State Univ., May 21-22, 1990, Manhattan, Kansas.
9. DiGiulio, D. C.; Cho, J. S.; Dupont, R. R.; Kemblowski, M. W. "Conducting Field Test for Evaluation of Soil Vacuum Extraction Application"; Proc. 4th National Outdoor Action Conf. on Aquifer Restoration, Groundwater Monitoring and Geophysical Methods, May 14-17, 1990, Las Vegas, Nevada.

10. Kuo, J.F.; Aieta, E.M.; Yang, P.H. " A Two Dimensional Model for Estimating Radius of Influence of A Soil Venting Process"; Proc. Hazmacon 90, p. 197-207, Anaheim, CA, April 17-19, 1990.
11. Wilson, J. D.; Clarke, A. N.; Mutch, Jr., R. D. "Mathematical Modeling of In-Situ Vapor Stripping of Contaminated Soils"; *Israel J. Chem.* **1990**, *30*, pp. 281-293.
12. Wilson, D. J.; Clarke, A. N.; Clarke, J. H. "Soil Clean Up by In-Situ Aeration. I. Mathematical Modeling"; *Sep. Sci. & Tech.* **1988**, *23*, pp. 991-1037.
13. Gannon, K.; Wilson, D. J.; Clarke, A. N.; Mutch Jr., R. D.; Clarke, J. H. "Soil Clean Up by In-Situ Aeration. II. Effects of Impermeable Caps, Soil Permeability, and Evaporative Cooling"; *Sep. Sci. & Tech.* **1989**, *24*, pp. 831-862.
14. Wilson, D. J.; Clarke, A. N.; Mutch Jr., R. D. "Soil Clean Up by In-Situ Aeration. III. Passive Venting Wells, Recontamination, and Removal of Underlying Nonaqueous Phase Liquid"; *Sep. Sci. & Tech.* **1989**, *24*, pp. 939-979.
15. Bennedsen, M. B. "Vacuum VOC's from Soil"; *Pollution Eng.* **1987**, *19(2)*, pp. 66-8.
16. Matthews, C.S.; Russell, D.G. *Pressure Buildup and Flow Tests in Wells*; Monograph Volume 1, Henry L. Doherty Series, Society of Petroleum Engineers of AIME, New York, 1967.
17. Williamson, A.S.; Chappelear, J.E. "Representing Wells in Numerical Reservoir Simulation: Part 1 - Theory"; *Soc. Pet. Eng. J.* **1981**, pp. 323-38.
18. Downey, D. C.; Elliott, M. G. "Performance of Selected Soil Decontamination Technologies: An Air Force Perspective"; *Environ. Prog.* **1990**, *9(3)*, pp. 169-173.
19. Ball, R.; Steve, W. "Design Consideration for Soil Cleanup by Soil Vapor Extraction"; *Environ. Prog.* **1990**, *9(3)*, pp. 187-9.
20. Hutzler, N. J.; Murphy, B. E.; Gierke, J. S. "State of Technology review: Soil Vapor Extraction System"; US Environmental Protection Agency, EPA/600/S2-89/024, Jan. 1990.
21. Fanchi, J.R.; Harpole, K. J.; Bujnowski, S.W. "BOAST: A Three-Dimensional, Three-Phase Black oil Applied Simulation Tool"; US Dept. of Energy, DE-AC19-80BC10033, September 1982.

RECEIVED April 5, 1991

Chapter 20

Comparison of Fixation Techniques for Soil Containing Arsenic

Peylina Chu[1], Michael T. Rafferty[1], Thomas A. Delfino[1],
and Richard F. Gitschlag[2]

[1]Geomatrix Consultants, Inc., San Francisco, CA 94105
[2]Rhone-Poulenc, Inc., P.O. Box C95266, Princeton, NJ 08543

Bench-scale fixation studies were performed on soil samples with arsenic concentrations ranging from 1200 to 2100 mg/kg. Fixation reagents tested were Portland cement, fly ash, silicates, and metal (ferric and aluminum) hydroxides. Untreated and treated samples were extracted according to the federal Extraction Procedure-- Toxicity (EPTox) test and the Toxicity Characteristic Leaching Procedure (TCLP), as well as the California Waste Extraction Test (WET). The parameters that appeared to affect leaching included dosage of fixation reagent, the fixation capability of the reagent, and the final pH of the leaching test. Bulking and costs for treatment and disposal were estimated. In general, treatment using silicates reduced arsenic leaching most effectively, resulted in minimal bulking, and had reasonable treatment and disposal costs.

While treatment of soil containing high concentrations of metals has become a pressing issue for uncontrolled hazardous waste site remediation, there are few demonstrated treatment technologies. Since leaching of metals from soil is generally the primary concern with metal-containing soil, fixation is a promising treatment technology. The physical and chemical mechanisms and the chemistry of fixation and leaching is complex and is being studied by several investigators (1-8), and Conner (8) has provided an overview of chemical fixation technologies and associated chemistry. Some of the fixation mechanisms that have been postulated include precipitating chemicals to form relatively insoluble compounds, binding precipitated chemicals into a solidified matrix that decreases the exposed surface area and thereby the leachability, forming a complex with the chemical, or adsorbing the chemical on the surfaces of the fixation materials. Several mathematical models also have been developed for predicting leach rates for various leaching conditions. The models include mathematical descriptions of diffusion, chemical reaction

0097–6156/91/0468–0401$06.00/0

(instantaneous and kinetically controlled), and mass transfer, as well as empirical models (8).

Numerous fixation reagents have been used for treatment of metal-containing soil. Some of the common reagents for treating arsenic-containing soil include cement, fly ash, and silicates (8). Other reagents that are used for removing arsenic and other metals from wastewater also may be appropriate for fixation of metals in soil. For example, ferric hydroxide and aluminum hydroxide are adsorptive precipitates that has been used to remove dissolved arsenic and other metals from wastewater (9, 10). If these hydroxides are mixed into soil or formed in the soil, they may adsorb metals and prevent leaching. In this paper, we present a case study in which cement, fly ash, silicates, and ferric/aluminum hydroxides were tested as fixation reagents on arsenic-containing soil from a site in California. State and federal leaching tests were used to determine the effectiveness of treatment. The results, including bulking considerations and costs, are compared.

Methods and Materials

Source of Soil Samples. Soil samples for the fixation studies were collected from a site adjacent to San Francisco Bay in California. Arsenic compounds, primarily sodium arsenite, had been manufactured at the site from as early as 1929 to the 1960s. The raw material for the arsenic compounds was primarily arsenic trioxide from lead smelter flue dust. Site investigations indicated that the soil is contaminated with inorganic arsenic. The primary source of the contamination is believed to be the sodium arsenite produced at the site and arsenic trioxide from the raw materials. Arsenic has been detected in soil at up to 54,000 mg/kg with average concentrations of approximately 1000 mg/kg. The contaminated soil is predominantly silt and clay.

Five soil samples were used in the fixation studies (Table I).

Table I. Arsenic Concentrations in Soil Samples

Sample Location	Total As Concentration (mg/kg)
A	1200
B	2000
C	1800
D	1600
E	2100

Leaching Tests. Three leaching tests were used to judge the effectiveness of fixation treatment: the federal Extraction Procedure--Toxicity test, the federal Toxicity Characteristic Leaching Procedure, and the California Waste Extraction Test. The primary differences between the three leaching tests are (1) surface area of the waste available for leaching, (2) leaching solution, (3) ratio of leaching solution to waste, and (4) total time of contact between the leaching solution and the waste.

Extraction Procedure--Toxicity Test. Prior to September 1990, the federal Extraction Procedure--Toxicity (EPTox) test was used to determine if a waste was hazardous under the Resource Conservation and Recovery Act (RCRA) (*11*). Prior to extraction, the solid fraction of a waste is crushed to pass a 9.5-mm sieve unless the waste passes a confined compressive strength test, in which case it is tested uncrushed. The waste is then mixed with a leaching solution consisting of 0.5 N acetic acid added to distilled deionized water to a pH of 5.0 and extracted for 24 hours. The weight ratio of waste to leaching solution is 1:16. During extraction, the pH of the solution is maintained at 5.0 by adding 0.5 N acetic acid. After extraction, the solution is filtered, and the resulting EPTox leachate is analyzed for the constituents of interest. Prior to September 1990, a waste that had arsenic concentrations greater than 5.0 mg/l in the EPTox extract was considered to show the characteristic of toxicity.

Toxicity Characteristic Leaching Procedure. After September 1990, the Toxicity Characteristic Leaching Procedure (TCLP) was used to determine if a waste exhibits the characteristic of toxicity under RCRA (*12*). The TCLP requires that solid wastes must be crushed to pass a 9.5-mm sieve, with no exceptions based on the strength of the material. The crushed waste is then mixed with a 0.1 N acetic acid solution at pH 2.9 for moderate to high alkaline wastes or a 0.1 N acetate buffer solution at a pH of 4.9 for other wastes, and extracted for 18 hours. The ratio of waste to leaching solution is 1:20. The solution is filtered, and the resulting TCLP leachate analyzed. If the concentration of arsenic in the TCLP extract is greater than 5.0 mg/l, the waste is considered a RCRA hazardous waste.

California Waste Extraction Test. The Waste Extraction Test (WET) is the leaching test adopted by the state of California (*13*). The WET generally is more stringent than the EPTox test and the TCLP described above. Solid wastes are crushed to pass a 2-mm sieve and then extracted with a 0.2 M buffered sodium citrate solution at pH 5.0. In this test, the ratio of waste to leaching solution is 1:10. The sample is extracted for 48 hours, then the solution filtered to obtain the WET leachate. If the arsenic concentration of the constituent in the WET leachate is greater than 5.0 mg/l, the waste is considered hazardous in California.

Bench-Scale Studies

Ferric and Aluminum Hydroxide Studies. The ferric and aluminum hydroxide fixation studies were performed by Geomatrix Consultants. Soil used in this study was from location C. Before treatment, the soil was crushed and sieved through a 1/4-inch mesh, then mixed to form a homogeneous mixture. Solutions having varying concentrations of ferric chloride or aluminum sulfate were mixed with the soil, and calcium carbonate was used as a neutralizing agent to form the ferric or aluminum hydroxide precipitates. Minimal amounts of water were used in the treatment. The samples were covered and allowed to cure at room temperature for approximately 24 hours. The treated samples were then extracted according to modified WET procedures. The WET procedures were modified in that the sample was not sieved through a 2-mm screen and was extracted for 24 instead of 48 hours. Duplicate samples analyzed by standard WET procedures gave similar concentrations, indicating that the modifications did not significantly change the leaching results. An untreated and a treated sample were also extracted using the TCLP. No EPTox tests were performed on these samples.

Portland Cement and Fly Ash Studies. The portland cement and fly ash studies were performed by a commercial vendor in Pennsylvania. For the study, Portland Type-1 cement and ASTM Class C fly ash were obtained from vendors in the San Francisco Bay Area. Various combinations of cement and fly ash were used to treat soil from location C. The soil was blended with water and cement and/or fly ash and compacted. The amount of water added during treatment was based on visual observation to produce a damp granular product that could be easily handled. The treated samples were then cured in plastic bags for seven days at 100°F. The untreated soil and all treated samples were analyzed by the EPTox test. The treated sample having the lowest arsenic concentration in the EPTox leachate was also analyzed by the WET and TCLP.

Silicate Studies. The silicate studies were performed by a commercial vendor in Arizona. Fixation studies were performed using calcium aluminum silicates and a proprietary cementitious mixture. Soil samples from locations A through E were treated in this study. Samples were mixed with the reagents and allowed to dry and cure for 48 hours at room temperature. All untreated and treated samples were analyzed by the WET. The treated sample from location C was also analyzed by the TCLP.

Results

Ferric and Aluminum Hydroxide Studies. Since very little water was used in treatment, the treated material was a friable, soil-like material. Treatment doses, the final pH of the WET extract, and results of the WET and TCLP analyses for the untreated and treated samples are presented in Table II.

Untreated soil had a WET leachate arsenic concentration of 85 mg/l and a TCLP leachate concentration of 22 mg/l. In general, arsenic leachate concentrations decreased with increasing dosage of ferric hydroxide or aluminum hydroxide. The lowest WET leachate concentration was 5.2 mg/l arsenic, representing a decrease in leaching of approximately 94%. This sample also had the highest value for the final pH of the WET leachate. The TCLP leachate of this sample had an arsenic concentration of 0.02 mg/l, representing a decrease in leaching of more than 99%.

Bulking due to treatment by ferric or aluminum hydroxides was low, approximately 20%. The cost of reagents (ferric chloride and calcium carbonate) was estimated to be $34 to $45/yd^3 of soil treated. Treatment operations were estimated to be $30/yd^3 and would include crushing the soil, mixing reagents and water into the soil, and stockpiling and curing the treated material. Disposal costs were estimated to be $600/yd^3 and included a bulking factor of 1.2, disposal at a Class I facility (in accordance with federal regulations applicable to the site), transportation from the site to the disposal facility, tip fees, and applicable state and local taxes. The total cost for treatment and disposal would range from $664 to $675/yd^3.

Table II. Ferric and Aluminum Hydroxides - Treatment Results
(all samples from location C)

Sample	$Fe(OH)_3$ (g/g soilx10^3)	$Al(OH)_3$ (g/g soilx10^3)	pH of WET Extract	As Concentration in Leachate (mg/l) WET	TCLP
Untreated	-	-	-	85	22
1	-	7.8	5.15	76	-
2	-	15.6	5.11	64	-
3	-	31.2	5.09	50	-
4	17.7	-	5.09	56	-
5	44.4	-	4.92	46	-
6	69.4	-	5.34	17	0.02
7	69.4	-	7.12	5.2	-

Portland Cement and Fly Ash Studies. Fixation treatment using Portland cement and/or fly ash produced a rigid monolith. Treated samples had unconfined compressive strengths ranging from approximately 200 to 1000 psi. Dry density ranged from 80 to 110 lbs/ft^3. Treatment doses, water/cement and/or fly ash ratios, and leaching results are presented in Table III.

Table III. Portland Cement and Fly Ash - Treatment Results
(all samples from location C)

Sample	Portland Cement	Fly Ash	Water/Cement & Fly Ash Ratio	Arsenic Concentration in Leachate (mg/l) EPTox	TCLP	WET
Untreated	-	-	-	0.70	22	85
1	25%	-	0.66	0.35	-	-
2	-	45%	0.48	0.09	0.09	24
3	7%	20%	0.68	0.32	-	-

Water/cement and/or fly ash ratios ranged from 0.48 to 0.68. The best result was obtained for the sample treated with 45% fly ash, which gave an EPTox result of approximately one tenth of the EPTox result for the untreated sample. This sample had the lowest water/cement and/or fly ash ratio of 0.48. This sample was also extracted according to the TCLP and WET. The arsenic concentration in the TCLP extract was 0.09. Compared with the original TCLP leachate concentration of 22 mg/l, this represents more than a 99% decrease in leaching. The arsenic concentration in the WET extract for this sample was 24 mg/l. Compared with the WET leachate concentration of the untreated sample (85 mg/l), leaching was reduced by approximately 72%.

Sample weight increased by approximately 100% after treatment. Reagent costs were estimated to range from $18 to $24/yd³ of untreated soil. Costs for treatment were estimated to be $30/yd³ and would consist of crushing the soil, mixing reagents and water with the soil, and stockpiling and curing prior to disposal. Disposal costs were estimated to be $1000/yd³ and included a bulking factor of 2, disposal at a Class I facility (in accordance with federal regulations applicable to the site), transportation from the site to the disposal facility, tip fees, and applicable state and local taxes. The total cost for treatment and disposal would range from $1048 to $1054/yd³.

Silicate Studies.   Treated samples had a rigid structure with an unconfined compressive strength greater than 50 psi.   Leaching results are presented in Table IV.

Table IV. Silicates - Treatment Results

Sample Location	As Concentration in Leachate (mg/l) WET	TLCP
A - untreated	70	-
A - treated	4.2	-
B - untreated	82	-
B - treated	2.7	-
C - untreated	93	22
C - treated	3.2	0.03
D - untreated	82	-
D - treated	3.8	-
E - untreated	94	-
E - treated	11.9	-

Treated samples A through D had WET leachate concentrations of arsenic less than 5.0 mg/l. Treated sample E had a WET leachate concentration of 11.9 mg/l. The decrease in WET leachate concentrations ranged from 87% to 97%. The

TCLP extract concentration for treated sample C was 0.03 mg/l, a decrease of more than 99% in the leachate concentration.   The final pH of the WET extract was approximately 9.0 for all of the samples.

Bulking due to silicate treatment was low, ranging from 15% to 20%. Reagent costs were estimated to range from \$61 to \$69/yd^3 of untreated soil.   Costs for treatment were estimated to range from \$16 to \$24/yd^3 and would consist of crushing the soil, mixing reagents and water with the soil, and allowing it to cure prior to disposal.   Disposal costs were estimated to range from \$575 to \$600/yd^3 and included a bulking factor of 1.15 to 1.20, disposal at a Class I facility (in accordance with federal regulations applicable to the site), transportation from the site to the disposal facility, tip fees, and applicable state and local taxes.   The total cost for treatment and disposal would range from \$652 to \$693/yd^3.

Comparison of Leaching Results.   The TCLP and WET leachate concentrations before and after treatment of soil from location C are presented in Figures 1 and 2, respectively.   The original TCLP leachate concentration of arsenic was 22 mg/l. The three treatment methods produced samples having TCLP leachate concentrations of arsenic ranging from 0.02 to 0.09 mg/l, corresponding to decreases in leaching of over 99%.   The WET is a much more aggressive leaching test than the TCLP, as illustrated by Figures 1 and 2.   The original WET concentration of arsenic was 85 mg/l for the soil from location C.   Only treatment using silicates produced a sample having a WET leachate concentration less than 5.0 mg/l.

Discussion

Arsenic Chemistry of Leaching Tests.   Since leaching tests are generally used to determine the success of fixation treatment, the aqueous chemistry of the waste is important.   Arsenic chemistry in water and the environment has been characterized by others (14-17); general arsenic chemistry that is pertinent to this study is summarized below.

Arsenic is found primarily in the -3, +3, and +5 oxidation states and forms many inorganic and organic compounds.   The primary arsenicals at the site are believed to be arsenic trioxide (As_2O_3), arsenic pentoxide (As_2O_5), and sodium arsenite ($NaAsO_2$).   Arsenic trioxide is slightly soluble in water (2.1 g/100 g water at 25°F) and requires long periods of time to achieve equilibrium.   As it dissolves, it forms arsenious acid ($HAsO_2$) (16). $HAsO_2$ has a very low disassociation constant ($pK_a = 9.21$); $HAsO_2$ is the predominant form at pH values less than 9.2.   Arsenic pentoxide is also soluble in water and forms arsenic acid (H_3AsO_4) when dissolved in water.   Depending on the pH, the predominant species may be H_2AsO4^-, $HAsO4^{-2}$, or $AsO4^{-3}$ (15).   Sodium arsenite is readily soluble in water (17).

As^{+5} can form relatively insoluble compounds with other metals.   Wagemann (14) has theorized that the presence of barium, chromium, iron, and copper may control dissolved arsenic concentrations by forming insoluble compounds with arsenic.   Wagemann calculated that as pH increases, arsenic is precipitated with these metals, and dissolved arsenic concentrations decrease.   Of particular interest for this

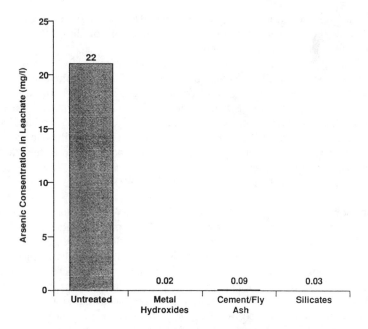

Figure 1. Comparison of Arsenic TCLP Results

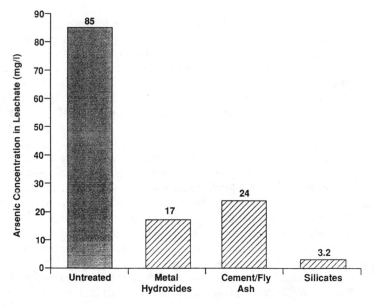

Figure 2. Comparison of Arsenic WET Results

study is the formation of $FeAsO_4$, a relatively insoluble compound ($K_s = 5.7 \times 10^{-21}$). Soil from the site has high concentrations of iron (approximately 25,000 mg/kg), which may control dissolved arsenic concentrations in leaching tests.

In the EPTox test, if the solid waste can be leached as a monolith, rather than be crushed, the potential for leaching is greatly reduced since the surface area available for leaching is much less. The primary characteristic of the EPTox test that affects arsenic leaching is probably pH. As discussed above, arsenic can form insoluble compounds with other metals, with pH controlling the dissolved arsenic concentration. If the solid is leached as a monolith, the pH within the monolith may be very different from the pH in the bulk leaching solution. If the solid is crushed, the formation of insoluble compounds is limited since the pH is controlled in the EPTox test.

The chemistry of the TCLP is very similar to that of the EPTox, since both tests use acetic acid in the leaching solution. Again, pH probably is the primary factor determining arsenic leaching. Since the solid waste must be crushed in the TCLP, the pH in the leaching solution can be very important in determining the dissolved arsenic concentrations. The formation of insoluble arsenical compounds with several metals is favored at higher pH values. If the waste is naturally high in alkalinity or has high buffering capacities, the final pH of the TCLP leachate may be higher than the initial value of 2.9 or 4.9, since pH is not controlled during the TCLP, unlike the EPTox test. With the presence of other metals to form insoluble compounds, arsenic leaching may be controlled.

The WET is generally a more aggressive leaching test than the TCLP for several reasons. In the WET, solid wastes are crushed to pass a smaller sieve (2 mm as compared to 9.5 mm). This provides additional surface area for leaching. The contact time between leaching solution and waste is also longer, which allows more time for arsenic trioxide to dissolve, and the ratio of leaching solution to waste is lower, which concentrates the leachable constituents in the leachate. The WET is particularly aggressive on metal-containing wastes, since citric acid is used in the leaching solution. The citrate ion is a particularly aggressive chelating agent for arsenic, keeping the arsenic dissolved in the leachate. As with the EPTox test and the TCLP, pH is also an important factor. As described above, the formation of insoluble arsenical compounds is favored at higher pH values. If the waste is naturally high in alkalinity or has high buffering capacities, the final pH of the WET leachate may be higher than the initial value of 5.0, and arsenic may remain insoluble in the leaching solution or may precipitate out of solution.

Ferric and Aluminum Hydroxide Studies.   Ferric hydroxide and aluminum hydroxide are adsorptive precipitates that have been used to remove dissolved arsenic and other metals from wastewater (*9, 10*). These hydroxides can be formed by hydrating ferric chloride or aluminum sulfate to form $Fe(OH)_3$ or $Al(OH)_3$, which precipitate and tend to adsorb or trap arsenic ions within the precipitates. In the bench-scale fixation study, these hydroxides were formed in the soil to adsorb or trap the arsenic and reduce its leachability.

The results of this study indicate two important parameters affecting arsenic leaching: dosage of reagent and pH. As shown in Table II, leaching of arsenic decreased with increasing doses of aluminum or ferric hydroxide. Several studies have shown that pH is an important variable when using coprecipitation/adsorption in removing arsenic from wastewater (*17*), although the optimum pH generally varied for the different studies. The importance of pH in this study is noted in comparing results of Samples 6 and 7. These samples were treated with the same dose of ferric hydroxide; however, more calcium carbonate was added to sample 7, producing a higher pH in the final extract. The arsenic concentration in the leachate from sample 7 (5.2 mg/l) was much lower than the concentration in the leachate from sample 6 (17 mg/l). The higher pH may produce a more favorable environment for forming the ferric hydroxide, which may have increased adsorption of the metal, or adsorption of arsenic onto the ferric hydroxide may be enhanced by higher pH values. The higher pH also may have caused arsenic to form $FeAsO_4$ with the iron originally in the soil and precipitate out of solution.

Portland Cement and Fly Ash Studies. Portland cement and fly ash are two of the most common ingredients used in fixation treatments. The behavior and chemistry of metals and the various hydrolysis reactions in portland cement/fly ash fixation systems is complex and has been studied by several investigators (*8, 19*). Fixation of metals using Portland cement and fly ash is believed to be a combination of producing a impermeable monolith, which reduces the surface area available for leaching; creating a high pH environment that generally limits the solubility of most metals and limits their leachability (*4*); or by formation of metal complexes with the cement/fly ash matrix (*7*). The water to cement ratio has been found to be an important factor in fixation treatment. The addition of too much water may reduce the strength as well as increase the permeability of the final product (*20*), which may reduce the effectiveness of the fixation treatment (*21*). Conner (*8*) has noted that the minimum ratio of water to Portland cement is approximately 0.4 for the cement to be workable, although more water may be used depending on the waste characteristics.

There are several variables that may have affected the results of cement/fly ash fixation in this study, including the fixation capabilities of cement as compared with fly ash, dosage of cement and/or fly ash, and the ratio of water to cement and/or fly ash. Since the EPTox test was the primary leaching test in this study and pH is controlled in this leaching test, the effects of pH can not be determined. The effects and interaction of the other variables are complex, and the separate effect of each variable can not be determined based on the results of this study. However, several observations can be made with regards to the results.

The amount of cement used to treat Sample 1 was approximately the same as the amount of fly ash used to treat Sample 3, and the arsenic concentrations in the EPTox leachate of both these samples were approximately the same. This suggests that the fixation capabilities of the fly ash and cement used in this study may not differ significantly. Also, the amount of reagent used to treat Sample 2 was almost twice the amounts used for Samples 1 and 3, and leaching from Sample 2 decreased significantly more than for Samples 1 and 3.

The water/cement and/or fly ash ratios in the experiment (ranging from 0.48 to 0.68) are higher than the minimum ratio of 0.4 noted by Conner. Since the soil consisted primarily of silt and clay and samples were mixed with water to a visually determined optimum moisture content, more water may have been added for easier handling characteristics. Sample 2 had the lowest water/cement and/or fly ash ratio of 0.48 and the lowest leachate concentration of arsenic; Samples 1 and 3 had ratios of 0.66 and 0.68, respectively. Since Samples 1 and 3 had higher ratios, the strength and permeability of Samples 1 and 3 after treatment may have been less than that of Sample 2. This also may be a reason why leaching from Samples 1 and 3 was significantly greater than leaching from Sample 2.

Silicate Studies. Silicate solutions have been used in combination with cement systems for fixation of metals. The chemistry between the silicates, cement, and metals is complex. In addition to the chemistry of cement and characteristics of cement systems that reduce leachability of metals, fixation of metals in silicate/cement systems is believed to occur by formation of low-solubility metal oxide/silicates, encapsulation of metal ions in a silicate or metal silicate matrix, and adsorption of metal ions on silica surfaces. Silicate solutions also reduce the permeability of the cement matrix, which reduces the available leaching surface and assists in reducing leachability of waste (Conner, 1990).

The results of the studies showed that silicate treatment reduced arsenic leaching more than the other treatments. The two parameters that may affect leaching from silicate-treated samples are pH and the fixation capabilities of silicates. The final pH of the WET leachate was approximately 9.0 for all of the samples, which is much higher than the final WET leachate pH of the ferric/aluminum hydroxide studies (ranging from 4.92 to 7.12). This supports the theory that a higher pH reduces the amount of arsenic leaching. However, the results of this study are not sufficient to determine if pH or the fixation capabilities of silicates was the primary parameter governing arsenic leaching.

Summary and Conclusions

Treatment using silicates appears to most effectively reduce arsenic leaching from soil. For soil containing approximately 1800 mg/kg arsenic, leaching by the TCLP is reduced from 22 to 0.03 mg/l (greater than 99% decrease), and leaching by the WET is reduced from 85 to 3.2 mg/l (96% decrease). Treatment by cement/fly ash or metal hydroxides results in TCLP leachate concentrations less than 5 mg/l, with decreases in leaching greater than 99%; WET leachate concentrations of the treated soil were reduced 72%.

Bulking due to treatment by silicates and metal hydroxides was low, approximately 20% or less, while bulking after treatment by cement/fly ash was high, approximately 100%. Treatment and disposal costs are summarized in Table V and are approximately \$664-\$675/yd^3 for metal hydroxide fixation, \$1048-\$1054/yd^3 for cement/fly ash fixation, and \$652-\$693/yd^3 for silicate fixation.

Based on the results of this study, the primary parameters governing arsenic leaching appeared to be the fixation capabilities of the reagents, dosage of reagent, and pH. Although the effects of these parameters could not be separated using the results of this study, several observations can be made. In general, as dosage of reagent was increased, arsenic leaching decreased. The fixation capabilities of cement and fly ash appear to be approximately the same, and the fixation capabilities of silicates may be greater than the other reagents used in this study.

The final pH of the leachate appeared to be an important factor. This is consistent with other studies on leaching and fixation and general arsenic chemistry. The final pH of the WET leachate for silicate treatment was higher than the final WET leachate pH for metal hydroxide treatment and corresponded to less leaching. A higher pH environment is more favorable for the formation of $FeAsO_4$, which would precipitate out of solution. Since high concentrations of iron were present in the soil, arsenic leaching may have been controlled by the formation of $FeAsO_4$.

Table V. Cost Comparison

	Metal Hydroxides	Cement/ Fly Ash	Silicates
Reagent Cost/yd^3:	$34 to $45	$18 to $24	$61 to $69
Treatment Cost/yd^3 [1]:	$30	$30	$16 to $24
Disposal Cost[2] (bulking factor x Class I disposal cost):	1.20 x 500 $ 600	2.00 x 500 $1,000	1.15 to 1.20 x 500 $575 to $600
TOTAL COST/yd^3:	$664 to $675	$1,048 to $1,054	$652 to $693

[1] Treatment costs include crushing the soil, mixing reagents and water with the soil, and stockpiling and curing the treated material.

[2] Disposal costs include transportation, tip fees and taxes for disposal at Kettleman City, California. Quoted cost is $300 per ton; a soil density of 124 lb/ft^3 assumed.

Literature Cited

1. Shively, W.; Bishop, P.; Gress, D.; Brown, T.; "Leaching Tests of Heavy Metals Stabilized with Portland Cement"; *J. WPCF* **1986**, *vol. 58*, pp. 234-241.
2. Ortego, J.D.; Jackson, S.; Yu, G.; Cocke, D.; McWhinney, H.; "Solidification of Hazardous Substances - A TGA and FTIR Study of Portland Cement Containing Metal Nitrates"; *J. Env. Sci. & Eng.* **1989**, vol 24.
3. Akher, H.; Butler, L.G.; Branz, S.; Carledge, F.K.; Tittlebaum, M.E.; "Immobilization of As, Cd, Cr, and Pb-Containing Soils using Cement or

Pozzolanic Fixing Agents"; *2nd Annual Symposium, Gulf Coast Hazardous Substance Research Center, Solidification/Stabilization Mechanisms & Applications, Proceedings,* February 15-16, 1990.

4. Cheng, K.Y., and Bishop, P.L.; "Developing a Kinetic Leaching Model for Solidified/Stabilized Hazardous Waste"; *2nd Annual Symposium, Gulf Coast Hazardous Substance Research Center, Solidification/Stabilization Mechanisms & Applications, Proceedings,* February 15-16, 1990.

5. Batchelor, B.; "Leaching Models: Theory and Application"; *2nd Annual Symposium, Gulf Coast Hazardous Substance Research Center, Solidification/Stabilization Mechanisms & Applications, Proceedings,* February 15-16, 1990.

6. Bridle, T.R.; Côté, P.L.; Constable, T.W.; Fraser, J.L.; "Evaluation of Heavy Metal Leachability from Solid Wastes"; *Wat. Sci. Tech.,* **1987**, *vol. 19,* pp. 1029-1036.

7. Bishop, P.L.; "Leaching of Inorganic Hazardous Constituents from Stabilized/Solidified Hazardous Wastes"; *Haz. Waste & Haz Mat.* **1988**, *vol. 5,* pp. 129-143.

8. Conner, J.R. *Chemical Fixation and Solidification of Hazardous Wastes*; Van Nostrand Reinhold: New York, NY, 1990.

9. Merrill, D.T.; Manzione, M.A.; Parker, D.S.; Petersen, J.J.; Chow, W.; Hobbs, A.O.; "Field Evaluation of Arsenic and Selenium Removal by Iron Coprecipitation"; *Env. Prog.,* **1987**, *vol. 6,* pp. 82-90.

10. McIntyre, G.; Rodriguez, J.J.; Edward L.; Thackston, E.L.; Wilson, D.J.; "Inexpensive Heavy Metal Removal by Foam Flotation"; *J. WPCF,* **1983**, *vol. 55,* pp. 1144-1149.

11. U.S. EPA, 1986, Test Methods for Evaluating Solid Wastes, Physical/Chemical Methods, SW-946, 3rd edition, Office of Solid Waste and Emergency Response, November.

12. U.S. EPA, 1990, Federal Register, Vol. 55, No. 61, pp. 11798-11877, March 29.

13. California Code of Regulations (CCR), 1985, Title 21, Social Security Division 4, Environmental Health Criteria, Chapter 30, Sections 66680 and 6670, 12 January.

14. *Wagemann, R.; "Some Theoretical Aspects of Stability and Solubility of Inorganic Arsenic in the Freshwater Environment"; Water Res.,* **1978***, vol. 12,* pp. 139-145.

15. Lemmo, N.V.; Faust, S.D.; Belton, T.; Tucker, R.; "Assessment of the Chemical and Biological Significance of Arsenical Compounds in a Heavily Contaminated Watershed. Part 1. The Fate and Speciation of Arsenical Compounds in Aquatic Environments - A Literature Review"; *J. Environ, Sci. Health,* **1983**, *vol. A18,* pp. 335-387.

16. Arsenic, Committee on Medical and Biologic Effects of Environmental Pollutants, National Academy of Sciences: Washington, D.C., 1977.

17. Latimer, W.M.; Hildebrand, J.H. *Reference Book of Inorganic Chemistry*, The Macmillan Company: New York, NY, 1940.
18. Trace Element Removal by Coprecipitation with Amorphous Iron Oxyhydroxide: *Engineering Evaluation*; Electric Power Research Institute, EPRI CS-4087, Project 910-2, 1985.
19. Cocke, D.L.; "The Binding Chemistry and Leaching Mechanisms of Hazardous Stubstances in Cementitious Solidification/Stabilization Systems"; *2nd Annual Symposium, Gulf Coast Hazardous Substance Research Center, Solidification/Stabilization Mechanisms & Applications, Proceedings*, February 15-16, 1990.
20. Mehta, P.K. *Concrete: Structure, Properties, and Materials*; Prentice-Hall, Inc.: Englewood Cliffs, NJ, 1986.
21. Daniali, S.; "Solidification/Stabilization of Hazardous Waste Substances in Latex Modified Portland Cement Matrices"; *2nd Annual Symposium, Gulf Coast Hazardous Substance Research Center, Solidification/Stabilization Mechanisms & Applications, Proceedings*, February 15-16, 1990.

RECEIVED April 5, 1991

Chapter 21

Abovegrade Earth-Mounded Concrete Vault
Structural and Radiological Performance

G. R. Darnell[1], R. Shuman[2], N. Chau[2], and E. A. Jennrich[2]

[1]EG&G Idaho, Inc., Idaho Falls, ID 83415
[2]Rogers and Associates Engineering Corporation, Salt Lake City, UT 84110–0330

The analysis of the long-term structural and radiological performance of a hypothetical abovegrade earth-mounded concrete vault used for the disposal of low-level radioactive waste is presented. The vault structure is designed based on the application of accepted standard engineering codes. The degradation of the concrete vault and the grouted waste forms over time and the resultant changes in the leaching and migration of radionuclides through the environment are modeled using a combination of the HELP, BARRIER, PATHRAE and PRESTO-CPG computer codes. The resultant radiological doses to an adjacent farmer and to several types of inadvertent intruders are calculated.

Increasingly stringent requirements for the treatment and disposal of low-level radioactive waste (LLW) have prompted interest in alternatives to disposal by shallow land burial. The majority of these disposal alternatives employ engineered barriers, typically concrete, in an attempt to further isolate waste contaminants from the surrounding environment. Among the predominant disposal technologies under consideration by state and federal entities, three are most prevalent. These include modular concrete canisters, and belowgrade and abovegrade concrete vaults. Until now, abovegrade disposal options were generally perceived as being more susceptible to exposure of the waste through erosion, flooding and increased degradation rates of concrete barriers such as vaults. This has been due, in part, to past abovegrade disposal concepts not incorporating the use of an earthen cover system designed to protect the disposal technology from continuous exposure to the natural elements (1). The Department of Energy Defense Low-Level Waste Management Program (DOE-DLLWMP) is evaluating the feasibility of a LLW treatment and disposal complex, which includes an Abovegrade Earth-Mounded Concrete Vault (AGEMCV) as the disposal system (2).

0097–6156/91/0468–0415$06.00/0

This paper addresses the performance of an AGEMCV and its ability to comply with performance objectives set forth by DOE Order 5820.2A (3). It provides an optimum implementation of the design basis of the disposal vault system and models the deterioration of the system over time.

Disposal Facility Description

For the purpose of performing a comparative radiological assessment of an AGEMCV disposal facility, it is considered to be located in the same humid northeastern United States hypothetical site as the one developed for the DOE's comparative report on alternative disposal technologies (2).

The hypothetical disposal site receives approximately 1 m of precipitation annually. Annual infiltration is 0.73 m for non-cap portions of the site. Infiltrating water intercepts the aquifer at a depth of 28 m below natural grade, at which point it may flow horizontally to regional wells. Average wind speed at the disposal site is 5 m/s under moderately stable conditions, Pasquill stability Class D. Complete site characteristic data, as used in the performance assessment, are included in Reference 1.

An estimated 22,370 m^3 of waste are assumed to be disposed of in the AGEMCV during the 30-year operational period. The source term used in the assessment, listed in Table I, is derived from disposal records for 1988 for the Radioactive Waste Management Complex at the Idaho National Engineering Laboratory. Decay daughters were included at the appropriate activities. Some short-lived nuclides have been excluded from the original data, several others have been considered in the assessment. Approximately 95 percent of the total activity listed in Table I results from radionuclides with half-lives of less than 10 years.

All dry waste is wrapped/packaged to preclude the inadvertent release of contaminants. This waste is delivered to the waste treatment facility (WTF) by truck in reusable cargo containers or bins. Dirt, rock, sand, concrete, asphalt, and other inorganic materials from decommissioning and decontamination are also transported in reusable bins. Waste items too large for these containers are received separately after prior approval.

The WTF treats the waste for disposal in the AGEMCV facility. Volume reduction of the waste is achieved through sorting, shredding, plasma-arc sizing, shearing, incinerating, and grouting. All waste processed in the WTF is fully treated, regardless of DOE or Nuclear Regulation Commission waste classification, and is in the form of grouted inorganics when processing is complete. Approximately 98% of the treated waste is inorganic. Additional details on waste treatment may be found in Reference 4.

Packages of waste exceeding 500 mrem/h (at contact) are grouted in WTF overpacks or in high-integrity containers by the generator prior to shipment to the WTF. Resin, sludge, and zirconium chips and other pyrophoric materials are grouted in the WTF overpack and shipped directly to the disposal vault after being approved by WTF personnel.

The waste disposal facility consists of three vaults, arranged end to end. Each vault is approximately 72 m long, 16.5 m wide, and 9.6 m tall. The vault is

Table I. Solid Low-Level Waste Disposal Source Term

Radionuclide	Inventory[a] (Ci)	Radionuclide	Inventory[a] (Ci)
H-3	4.7E+00	Po-212	1.4E+00
C-14	1.2E-02	Bi-214	1.0E-01
Mn-54	5.0E+04	Pb-214	1.0E-01
Fe-55	1.1E+03	Po-214	1.0E-01
Ni-59	5.4E+00	Po-216	2.2E+00
Co-60	1.6E+04	Rn-220	2.2E+00
Ni-63	3.3E+03	Rn-222	1.0E-01
Sr-90	3.0E+01	Ra-224	2.2E+00
Y-90	1.8E+01	Ra-226	1.0E-01
Nb-95	5.5E+00	Ac-228	1.8E-04
Zr-95	5.5E+00	Ra-228	1.8E-04
Tc-99	1.0E-06	Th-228	2.2E+00
Rh-106	1.8E+01	Th-232	1.8E-04
Ru-106	1.8E+01	Th-234	3.6E-01
Sb-125	7.8E+00	U-232	2.2E+00
Te-125m	7.8E+00	U-234	7.6E-08
Cs-134	1.2E+00	U-235	1.8E-02
Cs-137	2.0E+01	Np-237	4.9E-05
Ba-137m	2.0E+01	U-238	3.6E-01
Eu-152	1.0E-05	Pu-238	2.1E-04
Tl-208	7.9E-01	Pu-239	1.8E-02
Bi-210	1.0E-01	Pu-240	1.6E-03
Pb-210	1.0E-01	Am-241	3.5E-03
Po-210	1.0E-01	Pu-241	2.1E-02
Bi-212	2.2E+00	Pu-242	4.1E-05
Pb-212	2.2E+00	Am-243	2.0E-04

[a]Inventory reflects omission of several short-lived radionuclides and the ingrowth of daughter products.

constructed using high-grade Type V Portland cement with reinforcing steel and are situated above natural grade and above the probable maximum flood plain.

The roof is sloped to the sides of the facility to promote drainage. Roof thickness ranges from 91 cm at the outer edges to 122 cm at centerline. Number 8 steel reinforcement is placed on 10-cm centers in the upper and lower faces of the roof. A 15-cm internal wall runs down the center of each vault, for a roof span length of 7.9 m. The roof is fully supported by the vault contents.

The vault floors and exterior walls are 107 cm thick. Number 8 steel reinforcement is placed on 10-cm centers in the floor and exterior face of the walls. Steel reinforcement is placed on 15-cm centers in the interior face of the vaults.

The treated waste, packaged in carbon steel overpacks, is stacked inside the covered vaults using a bridge crane. The lack of appendages on the overpacks permits the formation of a tightly stacked monolith. As shown in Figure 1, stacking of the waste begins at the side walls of the vaults and proceeds inward. A vertical gap of approximately 15 cm is left in the center, and is filled with concrete prior to vault closure.

As sections of each vault are filled, impervious membranes are placed on the waste stack and a concrete roof slab is poured. When the entire vault is filled, the end wall is poured against the waste and the roof is completed. For final closure, the entire vault is covered with an impervious membrane and a 4-m earthen cover system, shown in Figure 2.

Analysis of Performance

The performance assessment of the AGEMCV considered exposures to members of the general public and individuals who inadvertently intrude into the facility. While occupational exposures were not included in the analysis, final selection of a disposal facility would be required to consider the doses to these individuals as well.

The pathways by which radionuclides will be released and transported from the disposal facility will be strongly influenced by the disposal site location. The humid northeastern United States has relatively abundant rainfall, shallow aquifers, and lush vegetation. Therefore, radionuclide release and transport via groundwater and biotic pathways can be expected to predominate over atmospheric pathways in most situations.

The release of waste contaminants will be dominated by the dissolution of radionuclides in groundwater percolating through the disposal facility. The rate of release will be a function of the strength of adsorption of waste nuclides by the waste form and the solubility of the contaminants in the groundwater. Those portions of the inventory released from the waste form will migrate vertically to the aquifer with the infiltrating water.

Upon reaching the aquifer, dissolved nuclides will travel within the saturated zone until intercepted by groundwater wells or bodies of surface water. From either of these groundwater sources, contaminants may enter the foodchain through the irrigation of food crops, ingestion by livestock, and direct consumption by humans.

The AGEMCV must satisfy the performance objectives of DOE Order 5820.2A (*see Reference 3*). These objectives are:

1. The annual effective dose equivalent to any member of the public shall not exceed 25 mrem.

2. Radioactive materials in liquid effluents released by DOE facilities shall not cause private or public drinking water systems to result in any member of the general public receiving an annual dose equivalent exceeding 4 mrem to the whole body or to any organ.

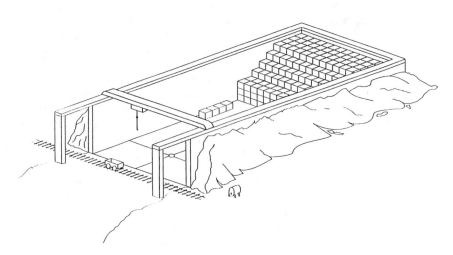

Figure 1. AGEMCV facility and filling procedure (not drawn to scale). (Reprinted from ref. 2.)

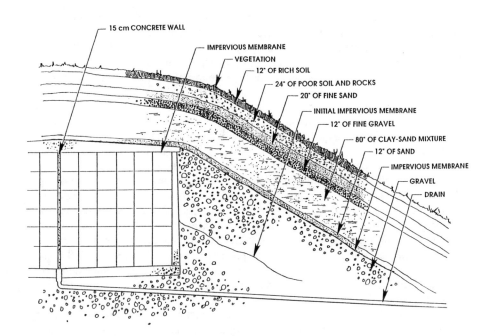

Figure 2. AGEMCV earthen cover system (not drawn to scale). (Reprinted from ref. 2.)

3. The annual effective dose equivalent received by individuals
    who may inadvertently intrude into the facility after the loss of
    institutional control shall not exceed 100 mrem for continuous
    exposure or 500 mrem for a single acute exposure.

Four exposure scenarios were formulated to ensure that these dose criteria were satisfied. The adjacent farmer exposure scenario was used to assess doses to members of the general public. For this scenario, an individual is assumed to reside at a location 100 m from the disposal vaults at the end of the 30-year operational period of the facility. Food crops, which supply the resident farmer with 50 percent of his food requirements, are grown. Contaminants are released due to leaching of the waste form. Nuclide releases are transported to the off-site farm via groundwater. Surface water transport is not considered to be a likely path of contaminant migration from the site as the extensive cover system and low potential for significant erosion at the site will prevent exposure of the waste.

The exposures for the adjacent farmer scenario include both internal and external pathways. Internal exposures include inhalation of contaminated dust suspended from the soil surface and ingestion of contaminated water, vegetables grown in contaminated soil, and animal products derived from animals grazing on contaminated feed. External exposures are due to direct radiation from the soil surface.

Three inadvertent intrusion exposure scenarios were analyzed to demonstrate compliance with the intruder dose criteria. Two of these, the intruder-explorer and intruder-construction exposure scenarios, address the acute dose limit of 500 mrem, while the intruder-agriculture scenario represents a continuous or chronic exposure.

For the intruder-explorer exposure scenario it is assumed that a person arrives at the waste disposal facility following the 100-year institutional control period and spends 1,000 hours over the course of a year exploring or wandering about the site surface. No attempt is made to dig into the disposal vaults during this time, thus limiting exposures to direct radiation from (1) buried waste while the facility is intact, and (2) waste exposed through erosion or vault failure. Following vault failure, the possibility for inhalation of contaminated particulates also exists.

In modeling the intruder-construction scenario, an intruder is assumed to construct a house over a disposal vault following both the end of institutional control and the failure of the concrete roof. The intruder excavates a basement 100 m^2 in area to a depth of 3 m below the ground surface (over or into the waste) and drills a well through the waste. Excavated material brought to the surface is mixed with the surface soil. The construction process is assumed to require 500 hours over a three month period. Exposure pathways during this time include direct radiation from the contaminated surface soil and inhalation of contaminated soil suspended during the construction process.

The intruder-agriculture scenario assumes that an individual moves onto the waste disposal site and resides in the house built in the intruder-construction scenario. This scenario is assumed to occur following the 100-year institutional control period and following failure of the vault roofs. The individual plants food crops which supply 50 percent of his food requirements. Waste radionuclides are taken up by these crops from contaminated soil unearthed during home construction.

Exposure pathways include inhalation of contaminated dust suspended from the soil surface, ingestion of contaminated water, vegetables grown in contaminated soil, and animal products derived from animals grazing on contaminated feed, and direct radiation from the soil surface.

The performance of the AGEMCV facility was assessed using the BARRIER (5), HELP (6), PRESTO-CPG (7), and PATHRAE (8) computer codes. The portions of the assessment for which each code was used are illustrated in Figure 3.

BARRIER was used to analyze the long-term hydrologic and structural performance of the disposal vaults. The code models the degradation of concrete structures over time due to chemical and physical forces, and the cracking and ultimate failure that accompany this deterioration. Further, the code models the degradation of solidified waste forms and the consequent contaminant release rates.

The HELP code was used to project rates of water infiltration through the intact concrete disposal vaults and through the earthen cover system. Infiltration through intact concrete was calculated to be approximately 0.01 cm/yr due to the very low porosity and hydraulic conductivity of the concrete. The minimum cover percolation rate calculated by the HELP code was 2.8 cm/yr). This percolation rate was modeled in addition to the percolation rate of 0.1 cm/yr prescribed by the French (*see Reference 4*), to examine the impact of a reduced level of effectiveness in excluding water from the disposed waste. HELP was developed to facilitate rapid, economical estimation of the amounts of surface runoff, subsurface drainage, and leachate that may be expected to result from the operation of a wide variety of landfill designs. The program models the effects of hydrologic processes including precipitation, surface storage, runoff, infiltration, percolation, evapotranspiration, soil moisture storage, and lateral drainage using a quasi-two-dimensional approach.

Contaminant migration from the disposal vaults, and the resulting doses, were modeled using the PRESTO-CPG and PATHRAE computer codes. These codes, developed for use by the U.S. Environmental Protection Agency (EPA) in establishing LLW disposal regulatory standards, evaluate maximum annual doses to a critical population group. The codes account for radionuclide migration via groundwater, surface water, atmospheric, and biotic pathways, and calculate doses for specified times as well as for the years of maximum exposure.

Results of Analysis

Each aspect of concrete and waste package deterioration weakens the roof, floor, and walls of the disposal vaults. Concrete cover over the reinforcing steel is considered effectively lost when cracking due to corrosion occurs. Concrete is also lost as sulfate attack proceeds and the strength of the remaining concrete is reduced through $Ca(OH)_2$ leaching.

Steel Corrosion. The rates of corrosion of steel reinforcement vary for the upper and lower faces of the roof and floor as exposure and design conditions differ within and outside the vaults. For the upper face of the roof and the lower face of the floor, de-passivation of the steel reinforcement is complete 650 years after site closure. Upon initiation, corrosion of the steel in each of these faces proceeds rapidly and cracking due to corrosion is seen 700 years after closure of the site.

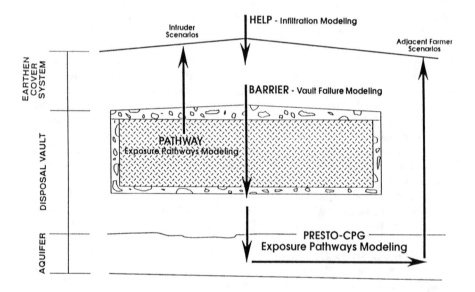

Figure 3. Computer codes used in the AGEMCV performance assessment (not drawn to scale). (Reprinted from ref. 2.)

The cracks extend from the steel to the surface of the concrete and run parallel with the steel.

Corrosion of steel reinforcement begins at a later time inside the vault, reflecting the lower chloride ion concentrations. These less severe exposure conditions are offset partially by a thinner concrete cover over the steel reinforcement, minimizing the distance chloride ions must penetrate. Corrosion begins 710 years after site closure for the inner faces of the roof and floor, with cracking due to corrosion occurring within an additional 40 years. Once again, these cracks run along the steel and extend outward from the reinforcement.

Sulfate Attack. The loss of concrete due to sulfate attack is limited to the upper and lower faces of the roof and floor, respectively, until the steel waste overpacks have failed. Corrosion of the 12 gauge carbon steel overpacks begins at the end of the 30-year operational period and is complete within 230 years. At that time deterioration of the grouted waste form begins. Complete degradation of the grouted waste occurs within 4,000 years of site closure. After year 230 of the simulation, chemical conditions inside the vault are altered to reflect the failure of the waste packages. Internal conditions are now assumed to reflect the characteristics of the waste, and deterioration of the inner faces of the vault now begins. Approximately 5.6E-3 cm of concrete on each face of the structure are lost annually to sulfate attack.

As the solidified waste degrades, the rate of release of waste contaminants due to leaching is enhanced. This is due to the fact that the degraded waste undergoes advective leaching, while contaminants are released from the solidified waste form through much slower diffusive processes.

Ca(OH)$_2$ Leaching. The loss of concrete yield strength due to leaching of Ca(OH)$_2$ results in a 10% loss within 6,500 years after site closure. This rate of loss is constant throughout the simulation because pertinent exposure conditions do not change.

Vault Failure. Vault failure is considered to have occurred when the roof has either lost sufficient strength to no longer withstand the forces acting on it or when 75 percent of the original thickness of the roof has been consumed by chemical and physical attack. In the first case, cracking due to bending occurs. The uniform support of the roof by the contents of the vault, and the lack of subsidence of those contents with time, minimize bending of the roof. Under these conditions, much of the roof's strength must be lost before cracking due to bending is possible. Under conditions where 75 percent of the roof thickness has deteriorated, the roof/s bearing strength is seriously compromised and the structural properties of the roof are considered lost. This loss in strength results from the gradual decay of the concrete via sulfate attack.

The effects of deterioration upon the roof and floor of each disposal vault differ significantly from each other. Whereas the floor of each disposal vault fails due to loading within 1,200 years of site closure, the roof does not fail until just after 5,000 years.

The additional time required for roof failure reflects the relatively mild loading conditions on this portion of the structure. As the waste overpacks corrode, limited expansion of the vault contents occurs, which results in an upward bending of the vault roof. The expansion projected by the BARRIER code, however, is quite small, insufficient to cause cracking due to bending.

The time to failure of the vaults is not affected by the rate of infiltration through the earthen cover. While the concrete is intact it represents the limiting factor in terms of flow through the structure. Water in excess of the 0.01 cm/yr concrete infiltration rate drains laterally along the peaked roofs. Consequently, groundwater chemical concentrations and rates of $CaOH_2$ leaching are unaffected.

The structural integrity of the disposal vaults has two important roles in the performance of the disposal facility. First, the intact concrete vaults provide a barrier which discourages excavation into the waste and subsequent dispersal of contaminants. Secondly, the low permeability of the concrete itself minimizes the amount of water percolating through the waste. As the predominant mechanism of contaminant release is leaching of the waste, this, in turn, minimizes potential exposures.

While the floor fails structurally, and hydraulically, rather early in the assessment period, the fact that the roof remains intact for several thousand years minimizes the importance of this failure. That is, the intact roof still represents an intruder barrier and effectively isolates the waste from increased infiltration rates. These advantages are lost, however, when the roof fails, just after 5,000 years. At this time, the amount of water percolating through the waste increases to post-failure values, either 0.1 or 2.8 cm/yr.

Protection of the General Public. The maximum annual doses (effective dose equivalent) for the adjacent farmer scenario for the two post-failure percolation rates used in the analysis are given in Table II. Under conditions in which the earthen cover system permits 0.1 cm of annual percolation through the waste following roof failure, no doses are projected for this pathway. This is due to the fact that the contaminant travel times to the well for even the most mobile radionuclides exceed 10,000 years. In fact, while C-14 and Tc-99 are projected to arrive at the well within 11,500 years, long-lived nuclides do not arrive at the well until 200,000 or more years after the start of the simulation. A number of radionuclides undergo radioactive decay and never reach the well.

In contrast, doses are projected to occur within the 10,000-year simulation period for this scenario when a post-failure percolation rate of 2.8 cm/yr is assumed. In this case, contaminant travel times for C-14 and Tc-99 are less than 5,250 years (vault failure plus a travel time of about 210 years), while the remaining radionuclides arrive at the well in 13,000 or more years. All projected doses, however, fall well within the proposed EPA groundwater standard of 4 mrem/yr.

The doses for the 2.8-cm/yr infiltration rate exposure scenario as a function of time are illustrated in Figure 4. As discussed above, contaminants do not reach the offsite well until about 5,250 years after site closure. A peak dose is seen shortly after arrival, at which point exposures decline for the duration of the simulation due to decay of C-14 and Tc-99.

Table II. Radionuclide Dose Summary for the Adjacent Farmer Exposure Scenario

	Dose Summary Post-Failure Percolation Rate	
Exposure Scenario	*0.1 cm/yr*	*2.8 cm/yr*
Adjacent Farmer		
Year of maximum dose	---	5249
Maximum effective dose equivalent (mrem/yr)	0.0E+00	7.4E-04
Dominant pathway	---	Ingestion
Dominant nuclide	---	C-14

It is evident from these results that the performance of the disposal facility depends on the functional relationship between the cover system and the concrete vaults themselves. That is, while the presence of the concrete vaults helps in minimizing the percolation of water through the waste and, hence, the release of contaminants, the importance of these effects is ultimately dependent upon facility hydraulic properties after failure. These properties are determined by the performance of the earthen cover.

The interrelationship of the cover and the vaults can, perhaps, be better appreciated if the components are considered separately. In the absence of concretevaults, the most mobile contaminants leached from the waste would arrive at the offsite well within 5,500 and 210 years for the 0.1-cm/yr and 2.8-cm/yr percolation rates, respectively. Conversely, waste disposed of in concrete vaults would not reach the well until after 5,000 years, regardless of the performance of the cover system.

From these data, it is evident that the presence of the concrete vaults is relatively more important as the performance of the cover system degrades. For highly impermeable cover systems, as typified by the 0.1-cm/yr percolation rate, the very long contaminant travel times reduce the significance of the benefits gained from the structures.

These comparisons pertain specifically to the hypothetical AGEMCV disposal facility analyzed and consider only the most mobile contaminants. In terms of long-lived, immobile radionuclides, the importance of the benefits gained through the use of concrete vaults diminishes dramatically. While contaminant travel times increase significantly, for either post-failure percolation rate, the use of concrete vaults still provides a maximum of 5,000 years delay time.

Thus, the performance of the earthen cover system is an important part of overall facility performance. The performance of the cover system will depend on the materials and methods used in its construction, and the underlying support conditions. A major strength of the AGEMCV facility lies in the use of the tightly

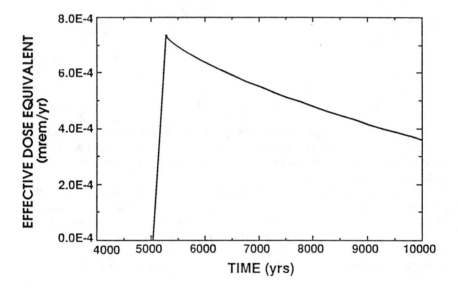

Figure 4. Doses for the adjacent farmer scenario as a function of time (2.8 cm/yr infiltration rate). (Reprinted from ref. 2.)

packed, grouted waste form to provide uniform support for the vault roofs and overlying cover system.

Protection of the Inadvertent Intruder. The maximum doses (effective dose equivalent) for the intruder exposure scenarios for each post-failure percolation rate are given in Table III. Doses for the intruder-explorer and intruder-construction scenarios fall well below the acute exposure limit of 500 mrem/yr. Similarly, continuous exposures for the intruder-agriculture scenario are much less than the applicable 100-mrem/yr standard.

Table III.   Radionuclide Dose Summary for the Intruder Exposure Scenarios

	Dose Summary Post-Failure Percolation Rate	
Exposure Scenario	*0.1 cm/yr*	*2.8 cm/yr*
Intruder-Explorer		
Year of maximum dose	100	100
Maximum effective dose equivalent (mrem/yr)	1.3E-13	1.3E-13
Dominant pathway	Gamma Radiation	Gamma Radiation
Dominant nuclide	Tl-208	Tl-208
Intruder-Construction		
Year of maximum dose	5,008	5,008
Maximum effective dose equivalent (mrem/yr)	1.7E-04	1.7E-04
Dominant pathway	Dust Inhalation	Dust Inhalation
Dominant nuclide	U-238	U-238
Intruder-Agriculture		
Year of maximum dose	5,008	5,204
Maximum effective dose equivalent (mrem/yr)	5.8E-04	1.3E-03
Dominant pathway	Dust Inhalation	Ingestion
Dominant nuclide	U-238	C-14

The maximum doses for the intruder-construction and intruder-explorer scenarios occur at that time when each scenario is first considered to be feasible. Following these times, doses for these scenarios decline due to decay of the waste radionuclides.

For the intruder-explorer scenario, the time of maximum exposure corresponds with the end of the institutional control period. While the intruder-construction scenario can also occur at the end of the 100-year institutional period, no exposures will be realized from this scenario until the roof of each vault has lost its structural integrity. Prior to that time, the intruder would not be able to drill a well into the waste. While the foundation for the house could be excavated, the depth of the earthen cover (about 4.0 m) would preclude contacting the waste.

Exposures for the intruder-explorer and intruder-construction scenarios are unaffected by the post-failure percolation rate. This is due to the fact that they occur before or at the time of vault failure.

Similar to the intruder-construction scenario, maximum exposures for the intruder-agriculture scenario do not occur until the vault roofs have failed. In contrast, projected exposures for the intruder-agriculture scenario are affected by changes in post-failure percolation rates. At the lower rate of post-failure percolation (0.1 cm/yr), all exposures to the agricultural intruder arise through dust inhalation, direct radiation, and ingestion of contaminated food. No doses are received through the consumption of contaminated water because contaminant travel times preclude the arrival of radionuclides at the well within 10,000 years. As discussed earlier, these travel times are significantly reduced at the higher percolation rate (2.8 cm/yr). Consequently, projected doses increase, reflecting the ingestion of contaminated water.

Conclusions

The contributions of the waste form, vault, and cover system to the performance of the AGEMCV are:

- The use of tightly packed, grouted waste forms provides uniform support for the vault roof and the overlying cover system, thereby eliminating major subsidence and subsidence-induced degradation of the roof and cover system.

- The AGEMCV provides just over 5,000 years delay in the time at which radionuclides begin to move through the environment.

- The AGEMCV delays inadvertent intrusion for just over 5,000 years barring a catastrophic disruption of the structure.

- The more permeable cover system (2.8 cm/yr) provides a transit time of approximately 210 years (from the vault to the adjacent farmer well) for the most mobile radionuclides, C-14 and Tc-99, in the waste inventory.

- The less permeable cover system (0.1 cm/yr) provides a 5,500-year transit time (vault to well) for the C-14 and Tc-99.

Based on the performance assessment calculations, the AGEMCV more than satisfies the DOE performance objectives for protection of the public and inadvertent intruders.

The maximum calculated dose (effective dose equivalent) to a member of the public occurs 5250 years after site closure and is 7.4E-04 mrem/yr, more than

30,000 times less than the DOE performance objectives. This dose results from C-14 that is ingested and occurs for a facility using a cover system that allows 2.8 cm/yr infiltration. No dose to the public occurs within the 10,000 year analysis period when the less permeable cover system (0.1 cm/yr) is used.

The maximum dose to an inadvertent intruder never approaches the limits of 100 or 500 mrem. The maximum dose occurs 5,200 years after site closure at a magnitude of 1.3E-03 mrem/yr. This dose, which pertains to the intruder-agriculture scenario and results from ingestion of C-14, is more than 70,000 times less than the 100 mrem/yr DOE performance objective. Doses for the intruder-construction and intruder-explorer scenarios are smaller still, falling below regulatory requirements by a factor of more than 100,000.

The hypothetical AGEMCV performance reported above represents a significant improvement over the performance of the abovegrade vault (AGV) concept that was modeled previously (*1*). While the performance of AGEMCV and the AGV cannot be quantitatively compared due to differences in waste source term and leaching characteristics, and due to the lack of a detailed analysis of concrete performance for the AGV (vault failure was assumed to occur 500 years after site closure), a qualitative comparison of facility performance can be made. The AGV facility design did not provide uniform support conditions for the vault roof and did not employ an earthen cover over the disposal vaults. Consequently, when the roof of each vault failed, disposed waste was exposed to the environment. The waste was readily suspended into the atmosphere and transported away from the facility with surface runoff.

The greatly improved performance of the AGEMCV over the AGV is based on a combination of several design features which were not included in the AGV. The design features most critical to the improved performance concern the support conditions for the AGEMCV's roof and the use of an earth cover system. The engineered earthen cover and the unique support conditions used in the design of the AGEMCV facility act together to negate the major exposure pathways calculated for the AGV. The grade of the cover and the establishment of vegetation minimizes erosive processes, while the support conditions for the concrete roof minimize the potential for localized failure of the cover system. Further, upon failure of the vault, the waste within remains covered by several meters of cover. These features act to ensure the isolation of the waste even after many thousands of years.

In light of the high level of performance achieved by the AGEMCV, a cost-benefit analysis of the disposal facility is warranted. Such an analysis would maintain the same successful design approach while optimizing construction practices employed in facility implementation. This approach would alleviate unnecessary, and costly, construction procedures without compromising the health and safety of the humans and the environment.

Literature Cited

1. Department of Energy, "Conceptual Design Report: Alternative Concepts for Low-Level Radioactive Waste Disposal," National Low-Level Waste Management Program, DOE/LLW-60T, June 1987.

2.   Shuman, R., N. Chau, and E.A. Jennrich, "Long-Term Structural and Radiological Performance Assessment for an Enhanced Abovegrade Earth-Mounded Concrete Vault," Department of Energy Defense Low-Level Waste Management Program, DOE/LLW-78T, October 1989.

3.   U.S. Department of Energy Order 5820.2A, "Radioactive Waste Management," U.S. Department of Energy, September 26, 1988.

4.   Darnell, G.R., M.M. Larsen, and J.D. Dalton, "Application of Existing Low-Level Waste Technology Offers 17-to-1 Volume Reduction and Enhanced Disposal at Low Cost," Idaho National Engineering Laboratory, EGG-LLW-8054, October 1988.

5.   Electric Power Research Institute, "BARRIER: A User's Guide," Research Report RP 2691-1, November 1988.

6.   Schroeder, P.R., et al., "The Hydrologic Evaluation of Landfill Performance (HELP) Model," U.S. Environmental Protection Agency, Vol. 1, PB85-100840, 1983.

7.   Environmental Protection Agency, "PRESTO-EPA-CPG: A Low-Level Radioactive Waste Environmental Transport and Risk Assessment Code, Documentation and Users Manual," Rogers and Associates Engineering Corporation, RAE-8706/1-4, November 1987.

8.   Merrell, G.M., et al., "The PATHRAE-RAD Performance Assessment Code for the Land Disposal of Radioactive Wastes," Rogers and Associates Engineering Corporation, RAE-8511-28, August 1986.

RECEIVED April 5, 1991

Author Index

Affiliation Index

Subject Index

Production: Kurt Schaub
Indexing: Deborah Steiner
Acquisition: Cheryl Shanks
Cover design: Neal Clodfelter

Printed and bound by Maple Press, York, PA

*Paper meets minimum requirements of American National Standard
for Information Sciences—Permanence of Paper for Printed Library
Materials, ANSI Z39.48–1984* ∞

Bestsellers from ACS Books

The ACS Style Guide: A Manual for Authors and Editors
Edited by Janet S. Dodd
264 pp; clothbound, ISBN 0–8412–0917–0; paperback, ISBN 0–8412–0943–X

Chemical Activities and Chemical Activities: Teacher Edition
By Christie L. Borgford and Lee R. Summerlin
330 pp; spiralbound, ISBN 0–8412–1417–4; teacher ed. ISBN 0–8412–1416–6

Chemical Demonstrations: A Sourcebook for Teachers,
Volumes 1 and 2, Second Edition
Volume 1 by Lee R. Summerlin and James L. Ealy, Jr.;
Vol. 1, 198 pp; spiralbound, ISBN 0–8412–1481–6;
Volume 2 by Lee R. Summerlin, Christie L. Borgford, and Julie B. Ealy
Vol. 2, 234 pp; spiralbound, ISBN 0–8412–1535–9

Writing the Laboratory Notebook
By Howard M. Kanare
145 pp; clothbound, ISBN 0–8412–0906–5; paperback, ISBN 0–8412–0933–2

Developing a Chemical Hygiene Plan
By Jay A. Young, Warren K. Kingsley, and George H. Wahl, Jr.
paperback, ISBN 0–8412–1876–5

Introduction to Microwave Sample Preparation: Theory and Practice
Edited by H. M. Kingston and Lois B. Jassie
263 pp; clothbound, ISBN 0–8412–1450–6

Principles of Environmental Sampling
Edited by Lawrence H. Keith
ACS Professional Reference Book; 458 pp;
clothbound; ISBN 0–8412–1173–6; paperback, ISBN 0–8412–1437–9

Biotechnology and Materials Science: Chemistry for the Future
Edited by Mary L. Good (Jacqueline K. Barton, Associate Editor)
135 pp; clothbound, ISBN 0–8412–1472–7; paperback, ISBN 0–8412–1473–5

Personal Computers for Scientists: A Byte at a Time
By Glenn I. Ouchi
276 pp; clothbound, ISBN 0–8412–1000–4; paperback, ISBN 0–8412–1001–2

Polymers in Aqueous Media: Performance Through Association
Edited by J. Edward Glass
Advances in Chemistry Series 223; 575 pp;
clothbound, ISBN 0–8412–1548–0

For further information and a free catalog of ACS books, contact:
American Chemical Society
Distribution Office, Department 225
1155 16th Street, NW, Washington, DC 20036
Telephone 800–227–5558